HUMBOLDT GLACIER, GREENLAND.

ELEMENTARY

GEOLOGY.

BY EDWARD HITCHCOCK, D.D., LL.D.,
PROFESSOR IN AMHERST COLLEGE,

AND

CHARLES H. HITCHCOCK, A.M.,
LECTURER ON ZOÖLOGY AND CURATOR OF THE CABINETS IN AMHERST COLLEGE.

A NEW EDITION,

REMODELED, MOSTLY REWRITTEN, WITH SEVERAL NEW CHAPTERS, AND BROUGHT UP TO THE PRESENT STATE OF THE SCIENCE.

FOR USE IN

SCHOOLS, FAMILIES, AND BY INDIVIDUALS.

NEW YORK:
IVISON, PHINNEY & CO., 48 & 50 WALKER ST.
CHICAGO: S. C. GRIGGS & CO., 39 & 41 LAKE ST.
BOSTON: BROWN & TAGGARD. PHILADELPHIA: SOWER, BARNES & CO., AND J. B. LIPPINCOTT & CO. CINCINNATI: MOORE, WILSTACH, KEYS & CO. SAVANNAH: J. M. COOPER & CO. ST. LOUIS: KEITH & WOODS. NEW ORLEANS: E. R. STEVENS & CO. DETROIT: RAYMOND & LAPHAM.
1862.

OTHER WORKS BY EDWARD HITCHCOCK, LL.D.

1. Religion of Geology and its Connected Sciences. A new and enlarged edition. 1859. $1 25. Crosby, Nichols, Lee & Co., Boston.
2. The Same Work: the Author's Copyright Edition. 1859. Two Shillings. James Blackwood, London.
3. Religious Truth Illustrated from Science. 1857. $1 25. Crosby, Nichols, Lee & Co., Boston.
4. Religious Lectures on Peculiar Phenomena in the Four Seasons. 1859. One Shilling and Sixpence. James Blackwood, London. *The American edition is out of print.*
5. History of a Zoological Temperance Convention in Central Africa. Pp. 160. 18mo. 1854. $0 25. Bridgman & Childs, Northampton.
6. Outlines of the Geology of the Globe. 1 vol. 8vo. Pp. 136, and Six Plates. With a Geological Map of the Globe, and another of North America. 1853. $1 25. Crosby, Nichols, Lee & Co., Boston.
7. Ichnology of New England. Quarto. Pp. 220; with 60 Plates. 1858. $5 00. J. S. & C. Adams, Amherst, Mass.
8. Illustrations of Surface Geology. Quarto. Pp. 155; with 12 Plates. Second edition. 1860. $3 00. J. S. & C. Adams, Amherst, Mass.
9. Final Report on the Geology of Massachusetts. Quarto. Pp. 831; with 55 plates. 1841. $6 00. Bridgman & Childs, Northampton.
10. Elementary Geology. 31st Edition. Rewritten, and over 200 Illustrations added; making 420 in all. 1 vol. 12mo. Pp. 430. $1 25. 1860. Ivison & Phinney, New York.

ELECTROTYPED BY
SMITH & McDOUGAL,
82 & 84 Beekman-st.

PREFACE

TO THE THIRTY-FIRST EDITION.

THE steady demand for thirty editions of this work, during the last twenty years, has called for frequent revisions, lest it should fall behind the state of the science. In making the present revision—so rapidly has geology advanced of late—we found it desirable to rewrite nearly the whole, as we have done. We have also modified the plan not a little, to give it greater unity. Some of the old Sections have been left out, but more new ones added. The wood-cut illustrations have been increased from 208 to 418.

We have tried to reduce the size of the work, by leaving out every thing not indispensable. But the great amount of new matter and of new illustrations, as well as the expansion of the typography to improve the appearance, have well nigh thwarted our efforts. We are aware that teachers sometimes can not find time to carry their pupils through the whole of a science which so abounds in facts and important applications as does geology. Two methods have been adopted by authors to meet this difficulty. One is, to leave out a large part of the science, and retain only the most striking and brilliant parts. But this is too much like attempting to perform the play of Hamlet with Hamlet left out. And teachers and pupils ought to know, that the study of such abridgments does not make them acquainted with geology, which still remains to be learned. We prefer to give a condensed view of the whole subject, and to designate the parts most important to be thoroughly studied, by a larger type. We have often, too, placed the most difficult reasoning in small type, that the teacher may allow those pupils not mature enough to master it to pass over it. He can also, when necessary, do the same with some whole sections ; say Part I., Section 6, on Metamorphism ; Part II., Section 3, the Laws of Palæontology ; and Section 4, the Inferences from the facts ; although in truth these parts, to those who mean to master the whole subject, are indispensable. Part IV., on Economical Geology, can also be omitted, as well as Part V., on American Geology. And it may be that some might profitably study the descriptive and phenomenal parts of the subject, who are too young to understand its religious applications in Part III. Thus while we present the whole subject, both for the sake of teachers and private individuals who wish to study the science, we put it into such a shape that it can be accommodated to the age and time of the pupil, and also lead him to see that until he has mastered the whole, he does not understand geology.

In this revision I have associated with me my youngest son, who has borne a large share in the work. As assistant in the geological survey of Vermont, a large part of the last three years has been devoted by him to the study of the rocks in the mountains, and in writing out their descriptions. Should it happen, as most likely it will, that this is the last revision of the work I shall ever make, and he should survive me, I trust he will be found fully competent to keep the work up to the advancing state of the science, should the public call for its continued publication.

EDWARD HITCHCOCK.

AMHERST, *June* 1, 1860.

CONTENTS.

PART I.

DESCRIPTION AND DYNAMICAL GEOLOGY.

PART II.

PALÆONTOLOGY.

PART III.

PART IV.

PART V.

INTRODUCTORY NOTICE

TO A FORMER EDITION

By DR. J. PYE SMITH, OF LONDON.

In a manner unexpected and remarkable, the opportunity has been presented to me of bearing a public testimony to the value of Dr. Hitchcock's volume, Elementary Geology. This is gratifying, not only because I feel it an honor to myself, but much more as it excites the hope that, by this recommendation, theological students, many of my younger brethren in the evangelical ministry, and serious christians in general, who feel the duty of seeking the cultivation of their own minds, may be induced to study this book. For them it is peculiarly adapted, as it presents a comprehensive digest of geological facts and the theoretical truths deduced from them, disposed in a method admirably perspicuous, so that inquiring persons may, without any discouraging labor, and by employing the diligence which will bring its own reward, acquire such a knowledge of this science as cannot fail of being eminently beneficial. It is no exaggeration to affirm that Geology has close relations to every branch of Natural History and to all the physical sciences, so that no district of that vast domain can be cultivated without awakening trains of thought leading to geological questions; and, conversely, the prosecution of knowledge in this department, cannot fail to excite the desire and to disclose the methods of making valuable acquisitions to the benefit of human life. In our day, through every degree of extensiveness, from the perambulation of a parish to the exploring of an empire, TRAVELING has become a "universal passion," and action too. Within a very few years, the interior of every continent of the earth has been surveyed with an intelligence and accuracy beyond all example. Who can reflect, for instance, upon the activity now so vigorously put forth, for introducing European civilization, the arts of peace, the enjoyment of security, and the influence of the most benign religion, into the long sealed territories of Central Asia, and not be filled with astonishment and delightful anticipation? Similar labors are in progress upon points and in directions innumerable, reaching to the heart of all the other vast regions of the globe; and the men to whom we owe so much, and from whom so much more is justly expected, are geologists, as well as transcendent naturalists in the other departments. Whoever would run the same career must possess the same qualifications. Even upon the smallest scale of provincial traveling for health, business or beneficence, acquaintance with natural objects opens a thousand means of enjoyment and usefulness.

The spirit of these reflections bears a peculiar application to the ministers of the gospel. To the pastors of rural congregations, no means of recreating and preserving health are comparable to these and their allied pursuits; and thus, also, in many temporal respects, they may become benefactors to their neighbors. In large towns the establishment of libraries, lyceums, botanic gardens, and scientific associations, is rapidly diffusing a taste for these kinds of knowledge. It would be a perilous state for the interests of religion, that "precious jewel" whose essential characters are "wisdom, knowledge, and joy," if its professional teachers should be, in this respect, inferior to the young and inquiring members of their congregations. For those excellent men who give their lives to the noblest of labors, a work which would honor angels, "preaching among the heathen the unsearchable riches of Christ;" a competent acquaintance with natural objects, is of signal importance, for both safety and usefulness. They should be able to distinguish mineral and vegetable products, so as to guard against the pernicious and determine the salubrious; and very often geological knowledge will be found of the first utility in fixing upon the best localities for missionary stations; nor can they be insensible to the benefits of which they may be the agents, by communicating discoveries to Europe or the United States of America.

To answer these purposes, and especially in the hands of the intelligent and studious ministers of CHRIST, this work of Professor Hitchcock appears to me especially suited. Though I flatter myself that I have studied with advantage the best English treatises on Geology, and find ever new improvement and pleasure from them; and have also paid some attention to French and German books of this class; I think it no disparagement to them to profess my conviction that, with the views just mentioned, this is the book which I long to see brought into extensive use. The plan on which it is composed, is different from that of any other, so far as I know, in such a manner and to such a degree, that it is not an opponent or rival to any of them. Yet, in this arrangement of the matter, there is no affectation: all is plain, consecutive, and luminous. It is more comprehensive with regard to the various relations and aspects of the science, than any one book with which I am acquainted; and yet, though within so moderate limits, it does not disappoint by unsatisfactory brevity or evasive generalities. Such is the impression made upon me by the first edition of the "Elementary Geology," and I cannot entertain a doubt but that the ample knowledge and untiring industry of the author will confer every practicable improvement upon his proposed new edition.

I received a deep conviction of the Professor's extraordinary merits, from his "Report" upon the Geology, Botany, and Natural History generally, of the Province of Massachusetts, made by the command of the State government; a large volume, published in 1833, and the second edition in 1835; and from his papers in the "American Biblical Repository," which were of great service to me in composing a book on "The Relation between the Holy Scriptures and some parts of Geological Science." But I did not till recently know that he was a "faithful brother and fellow-laborer in the gospel of Christ." An edifying manifestation of this, it has been my privilege to receive in Dr. Hitchcock's "Essay and Sermon on the Lessons taught by Sickness," prefixed to "A Wreath for the Tomb, or Extracts from Eminent Writers on Death and Eternity." It is my earnest prayer that great blessings from the God of all grace may attend the labors of my honored friend.

J. PYE SMITH.

HOMERTON COLLEGE, NEAR LONDON,
March 16, 1841.

ELEMENTARY GEOLOGY.

PART I.

DESCRIPTIVE AND DYNAMICAL GEOLOGY.

SECTION I.

GENERAL STRUCTURE OF THE EARTH, AND PRINCIPLES OF CLASSIFICATION.

GEOLOGY, from the Greek γῆ, *earth, and* λόγος, *discourse*, is the history of the mineral masses that compose the earth, and of the organic remains which they contain.

The two primary divisions of the science relate to the mineral masses and the organic remains: hence Part I. will embrace exclusively the description of the structure and composition of the rocks, and the forces concerned in their production. This is *Descriptive and Dynamical Geology.* Part II. will treat of the character and distribution of organic remains. As Geology has an important bearing upon other subjects, we shall consider in Parts III. and IV. the *Relations of Geology to Religion and to the Economical interests of society.* A brief account of the *Geological structure of North America*, in Part V., will conclude the treatise.

From these statements it will appear that an acquaintance with *Chemistry* and *Mineralogy* is necessary for a thorough knowledge of the mineral masses of the earth, and an acquaintance with the structure of animals and plants, or *Zoology* and *Botany*, for the study of the organic remains. We shall presume upon some knowledge of these branches in the student.

Geology is a history, not merely in the sense of description, but as a record of events. It narrates the condition of things from the period previous to the existence of organic life, through successive dynasties of more perfect races, to the dominion of man. Physical catastrophes, and the birth and extinction of races, are indelibly written upon the stony leaves of natures' volume. But this record is much less perfect than the written history of man. It is what the history of empires would have been, had our means of knowledge been confined to the works of man's art, like the sculptures of Nineveh and Egypt, obscurely fashioned by successive nations.

Every part of the globe, which is not animal or vegetable, including water and air, is regarded as *mineral.*

The term *rock*, in its popular acceptation, embraces only the solid parts of the globe; but in geological language it includes also the loose materials, the soils, clays, and gravels, that cover the solid parts.

The Earth as a Whole.—The form of the earth is that of a sphere, flattened at the poles: technically, an oblate spheroid. The polar diameter is about 26 miles shorter than the equatorial. Not only does astronomy prove this theoretically, but the measurement of the degrees of the meridian in different latitudes shows it to be true.

Hence it is inferred that the earth must have been once in a fluid state; since it has precisely the form which a fluid globe, revolving on its axis with the same velocity as the earth, would assume.

Taken as a whole, the earth is from five to six times heavier than water; or 2.5 times heavier than common rocks.

Proof 1. Careful observations upon the relative attracting power of particular mountains and the whole globe, with a zenith sector. 2. The disturbing effect of the earth upon the heavenly bodies.

We hence learn that the density of the earth increases from the surface to the centre; but it does not follow that the nature of the internal parts is different from its crust. For in consequence of condensation by pressure, water at the depth of 362 miles, would be as heavy as quicksilver; and air as heavy as water at 34 miles in depth; while at the centre, steel would be compressed into one fourth, and stone into one eighth of its bulk at the surface.

Configuration of the surface.—The surface of the earth, as well beneath the ocean as on the dry land, is elevated into ridges and insulated peaks, with intervening valleys and plains.

The highest mountains are about 29,000 feet above the ocean level, and the mean height of the dry land is about 1,000 feet.

The highest mountain in Asia is Mt. Everest, one of the Himalayahs, 29,002 feet; the highest in Europe is Mt. Blanc, 15,700 feet; the highest in North America is Mt. Elias, 17,850 feet; the highest in South America is Aconcagua, in Chile, 23,910 feet. The mean height of land in Asia is 1,100 feet; in Europe, 600 feet; in North America, 710 feet, and in South America, 1,000 feet.

Fig. 1.

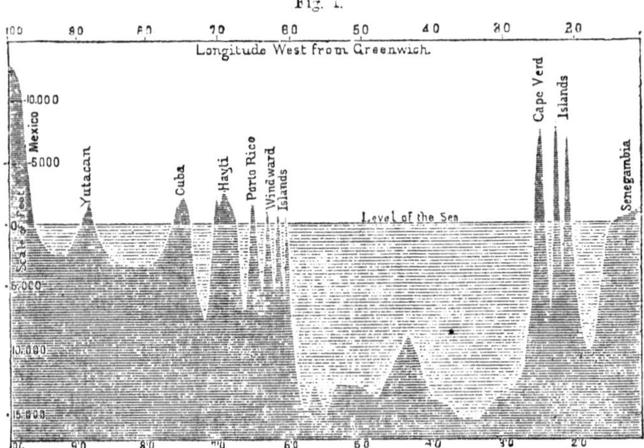

Bottom of the Atlantic Ocean.

The mean depth of the ocean is probably between two and three miles. Fig. 1 represents the configuration of the bottom of the Atlantic Ocean between Southern Mexico and Northern Africa.

Occasionally parts of the interior of a continent are below the ocean level. The Caspian Sea is 84 feet below the Black Sea, and the Dead Sea is 1,350 feet below the Mediterranean.

Hence it appears that the present dry land might be spread over the bottom of the ocean, so that the globe would be entirely covered with water. For nearly three-fourths of the surface is at present submerged.

The dry land is mostly situated in one hemisphere. For if we place the north pole at London, as may be illustrated upon an artificial globe, the northern hemisphere will be seen to embrace most of the land, while there will be little but water in the southern hemisphere.

In carrying out the order already indicated, we shall treat of the general structure and arrangement of the materials composing the exterior crust of the earth in Section I.; their chemical and mineralogical characters in Sections II. and III. The remainder of Part I. will relate to the forces which have modified these mineral masses.

STRATIFIED ROCKS.

The rocks that compose the globe are divided into two great classes, the STRATIFIED and UNSTRATIFIED, or AQUEOUS and IGNEOUS.

Stratification consists of the division of a rock into regular masses, by nearly parellel planes, occasioned by a peculiar mode of deposition. Strata vary in thickness from that of paper to many yards.

The term *stratum* is sometimes employed to designate the whole mass of a rock, while its parallel subdivisions are called *beds* or layers. The term bed is also employed to designate a layer, whose shape may be more or less lenticular, or wedge-shaped, included between the layers of a more extended rock; as a bed of gypsum, a bed of coal, a bed of iron, etc. In this case the bed is sometimes said to be subordinate.

When beds of different rocks alternate, they are said to be interstratified.

A *seam* is a thin layer of rock that separates the beds or strata of another rock; ex. gr., a seam of coal, of limestone, etc.

A bed or stratum is often divided into thin laminæ, which bear the same relation to a single bed as that does to the whole series of beds. This division is called the *lamination* of the bed; and always results from a mechanical mode of deposition.

The lamination is sometimes parallel to the planes of stratification; sometimes the layers are much inclined to each other; and often they are undulating and tortuous.

Fig. 2, shows the different kinds of lamination.

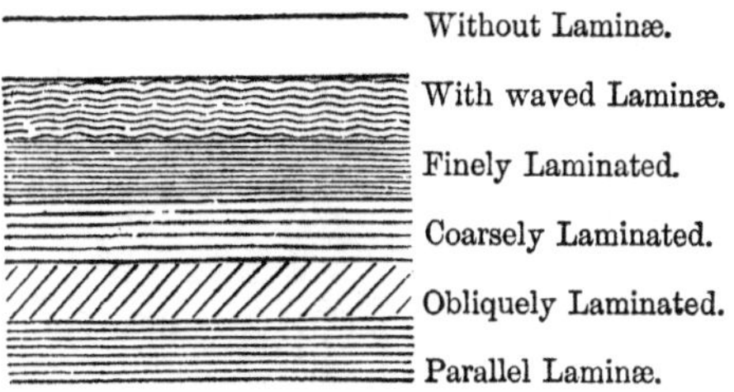

Fig. 3, is a sketch of a block of sandstone, six feet long, from Mount Tom, in East Hampton. Its face is a fine example of the oblique lamination above described, resulting from counter currents and depositions of coarse sand on surfaces sloping in different directions. Such examples are common in that locality.

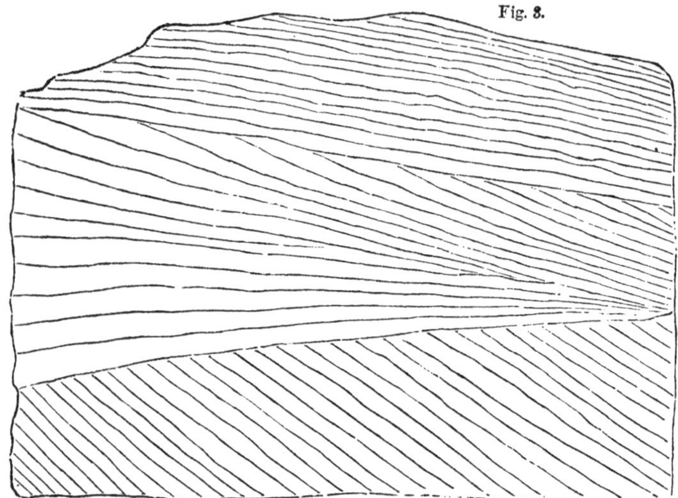

Fig. 3.

Origin of lamination.—All the lamination of stratified rocks was undoubtedly produced originally by deposition in water, and the varieties have resulted from modifying circumstances. 1. The parallel laminæ are the result of quiet deposition upon a level surface. 2. The waved lamination, in many instances, is nothing but *ripple marks;* such as are seen constantly upon the sand and mud at the bottom of rivers, lakes, and the ocean. In the secondary rocks this is too manifest to be mistaken. 3. The oblique lamination has generally been the result of deposition upon a steep shore, where the materials are driven over the edge of an inclined plane. 4. Highly contorted lamination has often resulted from lateral and vertical pressure, as illustrated by Fig. 4. This is sometimes seen in deposits of clay.

Fig. 4.

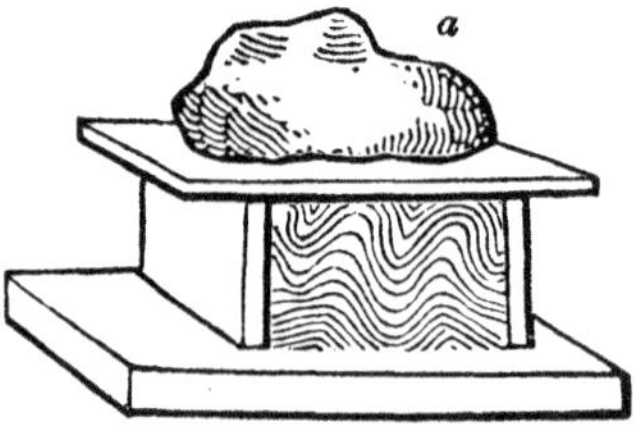

Illustration If pieces of cloth of different colors be placed upon a table *c*, and covered by a weight, *a*, and then lateral forces, *b*, *b*, be applied; while the weight will be somewhat raised, the cloth will be folded and contorted precisely like the laminæ of many rocks; as is shown in the figure.

How to distinguish between strata and laminæ.—This cannot be done by the relative thickness, since strata are sometimes as thin as laminæ. But strata can, and laminæ can not, be easily split apart. A stratum marks some pause or change in the deposition; but the laminæ were formed rapidly between the pauses. Hence the latter are more closely compacted together, and generally the rock will break more easily in any other direction than in that of its laminæ.

Inclination of strata.—The angle which the surface of a stratum makes with the plane of the horizon is called its *inclination* or *dip;* and the direction of its upturned edge is called its *strike* or *bearing.*

Of course horizontal strata have neither strike nor dip. The exposure of a stratum at the surface is called in the language of miners its *outcrop* or *basseting.* An *outlier* is a detached ledge or mass of strata.

As a general fact, the newer or higher rocks are less inclined than those below. The highest are usually horizontal; while the oldest are often perpendicular. But this is not an universal rule.

The instrument employed for ascertaining the dip of a stratum, is called a *clinometer.* The inclination may be determined by the eye either by itself, or with the help of the hands situated as in Fig. 5. The person must stand opposite the strata, and placing the hands in the range between the eye and the rock, notice the position of the planes when compared with the lines of reference. Each dotted line incloses with either hand an angle of 45°. The strike may be determined with a compass.

Fig. 5.

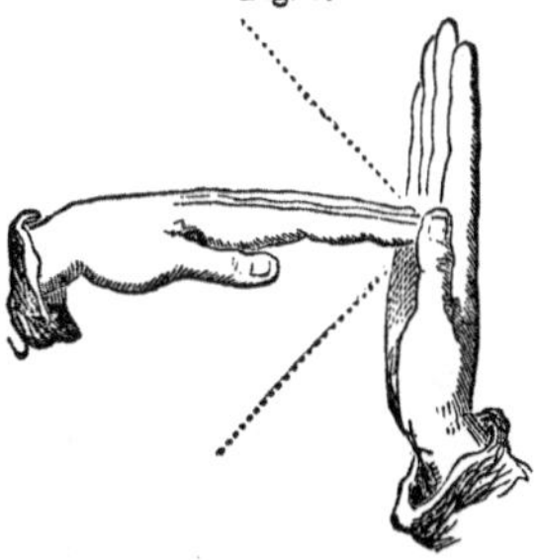

Axes.—The line along which the strata dip in opposite directions is called an *anticlinal line*, or *anticlinal axis.* In Fig. 6, *a* represents a simple anticlinal; *b* and *c* show the contour of the surface when denudation has removed the ridge, and *d* represents a *complex* anticlinal. In some instances the strata have been folded together on a vast scale, and in such a manner as to bring some of the newer rocks beneath the older. Fig. 7 is a section of this character. Originally the strata were probably folded, as is shown by the curved lines passing from 1 to 1, 2 to 2, and so on. But their upper parts

Fig. 6.

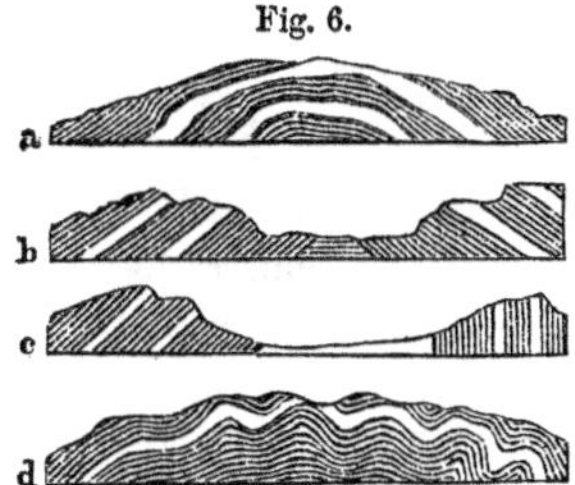

have been denuded, so that the present surface is *a*, *a*. The oldest strata are now found to be 6, 6; and they correspond outward on each side of these; as, 5, 5; 4, 4; etc. Such an example as this has been called a *folded axis*, or an *inverted anticlinal.*

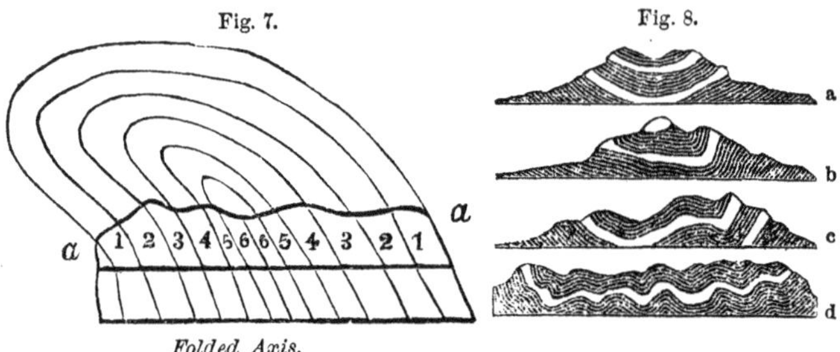

Folded Axis.

When the strata dip toward each other they constitute a *synclinal axis.* In Fig. 8, *a* is a *shallow* synclinal, *b* a *sharp* synclinal, *c* and *d complex* synclinals.

When the strata dip from any point in all directions outward, (a) around the crater of a volcano,) the dip is said to be *quaquaversal.*

Metamorphic Stratified Rocks.—According to the views of the ablest geologists at the present time, we ought perhaps to limit both the terms stratification and lamination to rocks whose mechanical texture proves them to have been deposited from water. But there is a large class of rocks that have been powerfully metamorphosed, so as to become crystalline, yet are divided by parallel planes very analogous to stratification and lamination; and it is usual to regard the former structure, that is, stratification, as extending through them all, and to have resulted from original deposition in water. But the subdivisions of the strata, viz.: *cleavage, foliation, and joints,* which often cross the strata, appear to have been for the most part *superinduced:* that is, they were produced after the original deposition of the strata by other agencies than water alone; although some of them, as foliation and cleavage, in some instances seem to be mere modifications of original lamination.

Joints.—Both the stratified and unstratified rocks are traversed by divisional planes, called *joints;* which divide the mass into determinate shapes, which are different from beds and their subdivisions.

The most important of these joints, called *master-joints*, are more or less parallel, and so extended as to imply some general cause of production.

When these joints cross the beds obliquely, as they usually do, and there are two sets of them, they divide the rock into rhomboidal masses of considerable regularity; though wanting in that perfect equality in the corresponding angles of the prisms which is found in crystals of a simple mineral. They do the same in the unstratified rocks, producing a pseudo-stratification, and are of great help in quarrying.

Figs. 9 and 10 are examples of joints in unconsolidated clay, in West

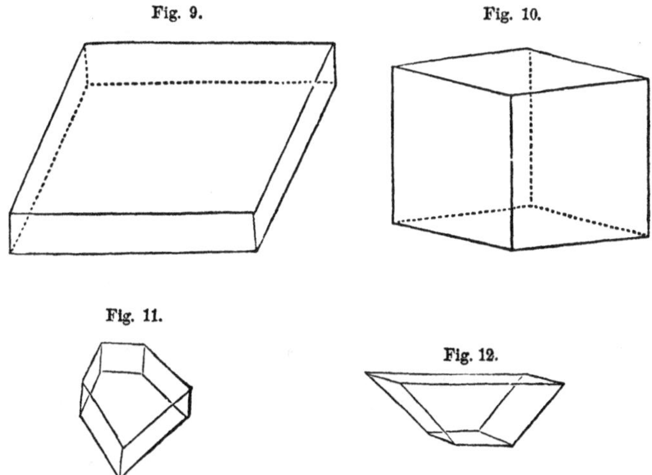

Springfield, Massachusetts. Figs. 11 and 12 are more complicated forms from the quartz rock of Bernardston, in Massachusetts.

Sometimes fissures are quite irregular in direction; but they assist in breaking the rock into fragments. The fissures are sometimes occupied by a foreign mineral, such as calcite; but these are properly veins.

Cleavage.—Rocks of homogeneous composition, especially clay slate, are often divided by parallel planes, sometimes conforming in dip and direction to the bedding or stratification, and sometimes not. They differ from joints in causing the rocks to split into plates indefinitely thin, and also by being far more extensive, and but rarely crossed by other planes as joints are. Roofing slate is a good example.

We venture to doubt, however, whether the indefinitely thin plates, generally regarded as an essential property of cleavage, are always present. For we have not unfrequently met in quartz rock and in some siliceous slates with parallel divisions, which could not properly be referred to joints or stratification, where the plates could not be split thinner than half an inch, and often not so thin; and if not cleavage, we can give them no name. May we not omit thinness of the plates in our definition of cleavage, and still not confound cleavage with joints?

The cleavage planes may be inclined to the planes of stratification at any angle from 0° to 90°, and sometimes the two planes dip in opposite directions. The cleavage planes are remarkable for their almost perfect parallelism, while strata, laminæ and folia are often contorted.

Fig. 13.

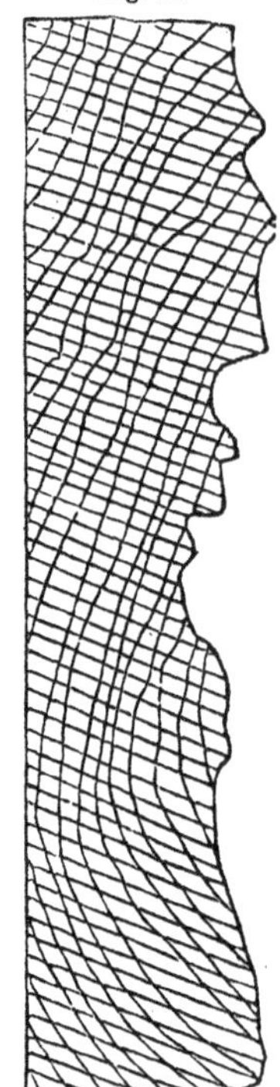

Fig. 13 represents cleavage planes, *bb*, crossing irregular strata, *aa*.

In Fig. 14 are represented the planes of stratification, B B, B B; the joints A A, A A; and the slaty cleavage, *dd*.

Fig. 14.

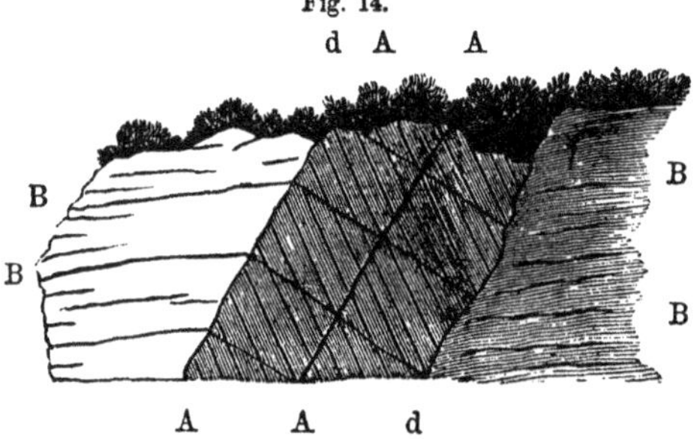

Foliation.—A change in metamorphic rocks analogous to cleavage is called *foliation.* It is *a crystalline lamination,* or a separation of the different mineralogical compounds into distinct layers, much resembling strata. In districts where these crystalline rocks have not been much disturbed, the foliation coincides with the stratification. In regions much corrugated or disturbed the foliation often intersects the strata at a considerable angle, like cleavage planes. In fact, foliation appears to be the result of the same forces as cleavage, except that in the former the process was carried so far that crystallization resulted.

Fig. 15.

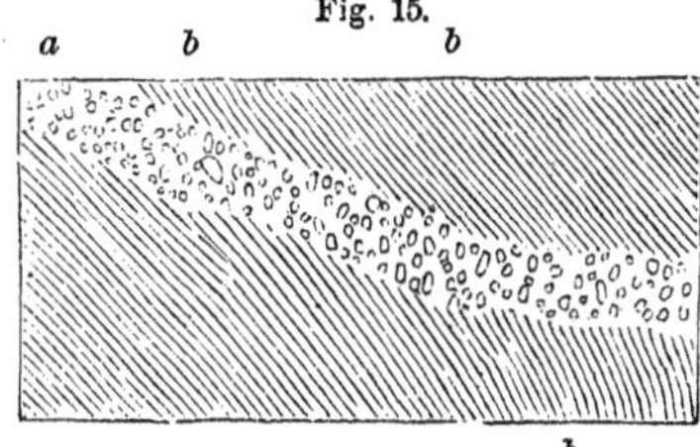

Fig. 15 represents foliation as it is seen in talcose conglomerate in Richmond, Vermont: *aa* shows the position of the strata, and *b b* the inclination.

The rocks in which foliation exists are called *schists*, as mica schist, talcose schist. Gneiss, however, is foliated, and some contend that foliation is sometimes produced in unstratified or igneous rocks. The term *slate* ought to be limited to those fissile rocks that are homogeneous, and schist to those where the materials are heterogeneous, and are arranged in alternate layers. Few geologists, however, have as yet carried out these new views rigidly, so that their works still speak of mica slate, hornblende slate, &c. The theory of the origin of the various superinduced structures will be deferred to the chapter on Metamorphism.

Fig. 16.

Contorted Laminæ of Gneiss: Colbrook, Ct.

Plication and Contortion.—The laminated rocks sometimes, but the foliated and metamorphic much oftener, present examples of folding, plication and contortion most remarkable, and in general the more thorough the metamorphism the greater the curvatures and tortuosities. Fig. 16 was sketched from a block of gneiss lying by the roadside in Colebrook, Connecticut, and is no unusual example of plication in the folia of that rock.

Fig. 18, for a sketch of which we are indebted to Mr. Eben A. Knowlton, shows a remarkable specimen belonging to the cabinet of Amherst College, from Shelburne Falls, in Massachusetts. It is six feet long, weighs a ton, and was worn smooth by the water and ice of Deerfield river. It consists of beautifully contorted or plicated strata; or more properly, perhaps, folia, of white gneiss and black hornblende schist alternating. The minute flexures, which frequently become saw-like, can not be exhibited, and actual inspection can alone give a correct idea of its beauty. We shall refer to it again under Metamorphism.

These delicate curves in foliation are a miniature representation of what occurs in the strata of most of the great mountain ranges of the globe. Fig. 17, is an actual section in the Alps, extending southeasterly from the top of the well-known Righi. Here we have mountains thousands of feet high, looking as if crumpled together by some Mighty Hand. Doubtless it was done by lateral forces in the hand of Nature.

In this country we have the same phenomena on a magnificent scale. From Canada to Alabama, a distance of at least 1200 miles along the Appalachian Mountains, the strata have been folded into numerous anticlinal and synclinal axes by a force crowding them from southeast to northwest, making the southeasterly slopes quite gentle, and the northwest ones steep and abrupt. A section across the Appalachian chain, say through New Jersey and Pennsylvania, is given in Fig. 19; and though it be an ideal section, it will convey a good idea of the structure of this chain of mountains almost any

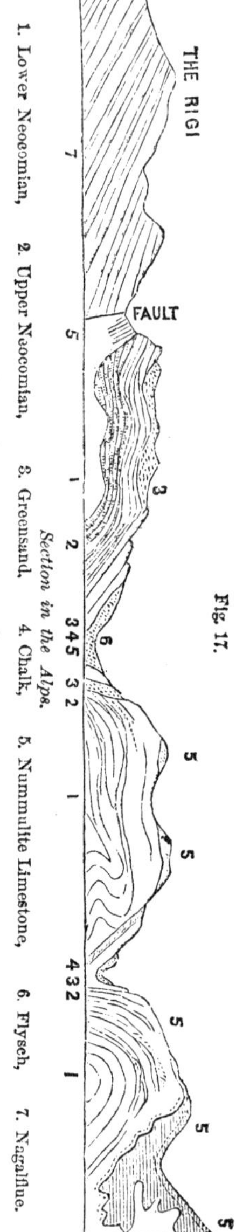

Fig. 17.

Section in the Alps.

1. Lower Neocomian, 2. Upper Neocomian, 3. Greensand, 4. Chalk, 5. Nummulite Limestone, 6. Flysch, 7. Nagalflue.

Fig. 18.

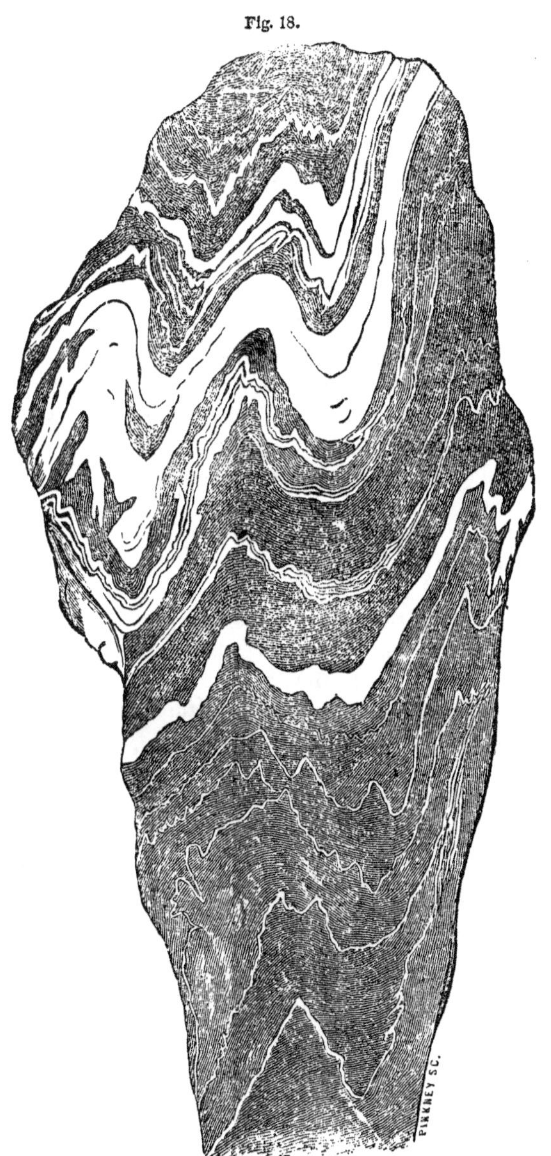

where between Canada and Alabama. How stupendous must have been the force, thus to fold up the vast strata of the mountains, as if they were merely the leaves of a book! Yet how easy for Him who directs and energizes the forces of nature! The manner in which these forces have operated will be better understood after we have developed the doctrine of internal heat.

Fig. 19.

Section across the Appalachian Mountains.

1. Tertiary, 2. Cretaceous, 3. Gneiss and Schist, 4. Jurassic or Triassic.
5. Anthracite Coal Measures. 6. Appalachian Chain, 7. Silurian, 8. Appalachian Coal Fields.

Atlantic.

Concretionary Structures.

In clay beds containing disseminated carbonate of lime, we frequently find nodules of argillo-calcareous matter, sometimes spherical, but more usually flattened. These are generally called *claystones*, and the common impression is, that that they were rounded by water. But they are the result of a tendency of particles to gather about a common center, called molecular attraction. The slaty divisions of the clay extend through the concretions; and on spliting them open, a leaf, a fish, a shell, or some other organic relic is frequently, but not invariably found. In New England, however, the slaty structure, and the organic nucleus are generally wanting.

Fig. 20.

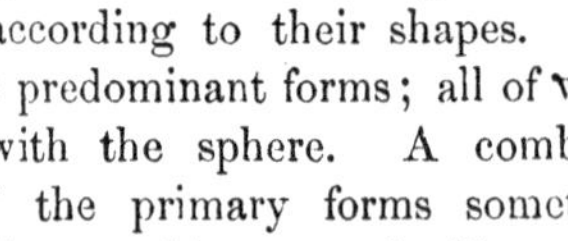

Fig. 20 will convey an idea of the manner in which these concretions are situated in the clay.

The claystones of New England have been classified according to their shapes. There are at least six predominant forms; all of which seem to start with the sphere. A combination of several of the primary forms sometimes produces mimic resemblances to familiar objects.

Fig 21 shows one from Walpole, New Hampshire, which mimics

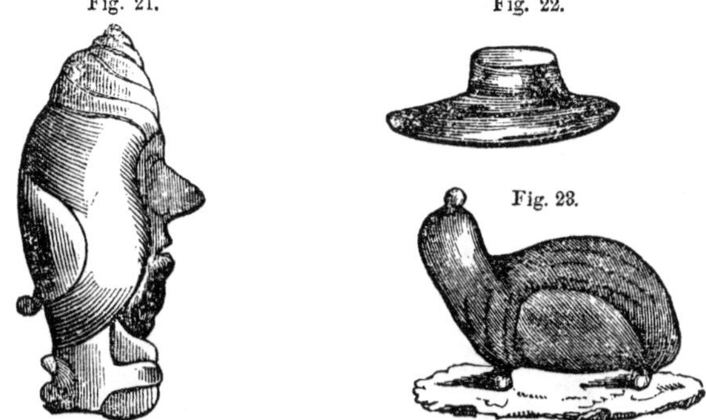

Fig. 21. Fig. 22. Fig. 23.

a human head in relief very closely, with the head-dress and cue behind. Fig. 22 resembles a hat or bonnet, and Fig. 23 a cat.

Fig. 24.

Fig. 24 shows a perfect ring from Rutland, Vermont. The original is 11 inches in diameter.

Similar concretions abound in argillaceous iron ore, which is often disseminated in clay beds or shale. These nodules are usually made up of concentric coats of ore; but sometimes the slaty structure of the rock containing them extends through them, and organic relics are found to form their nucleus.

Fig. 25.

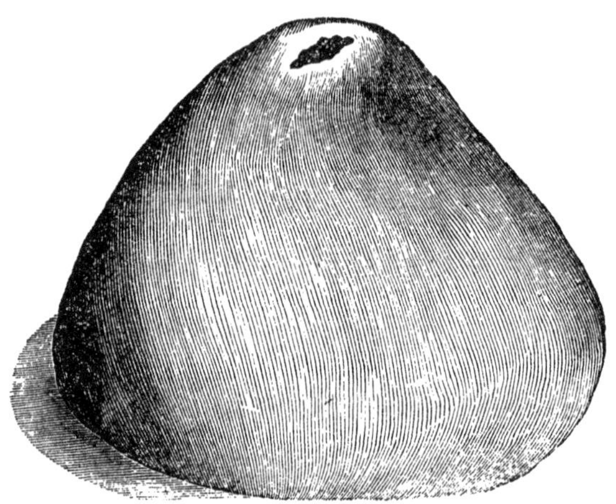

Fig. 25 is a concretion of iron ore with a nucleus of lignite, from Gay Head, in Massachusetts, 7 inches across.

The internal parts of these concretions of limestone and hydrate of iron often exhibit numerous cracks, which sometimes divide the matter into columnar masses, but more frequently into irregular shapes. When these cracks are filled with calcareous spar, as is often the case in calcareous concretions, they take the name of *ludus helmontii*, *turtle stones*, or more frequently of *septaria*. From these is prepared in England the famous Roman cement. Fig. 26 shows a section of one of these.

Fig. 26

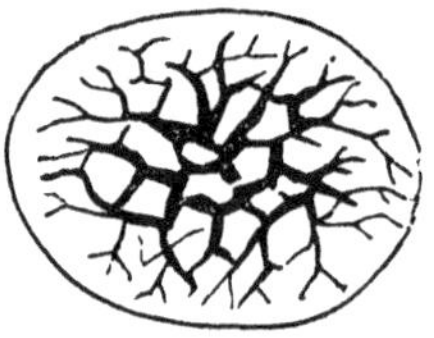

Certain limestones called oolites, are often almost entirely composed of concretions made up of concentric layers; but the spheres are rarely so large as a pea.

The concretionary structure, however, often exists in limestone on a very large scale, forming spheroidal masses not only many feet, but many yards in diameter. Fig. 27 represents some

Fig. 27.

Concretions in Sandstone, Iowa.

large concretions of carboniferous limestone, at Muscatine, in Iowa, as described by Professor Owen.

Unstratified Rocks.

The unstratified rocks occur in four modes. 1. As irregular masses beneath the stratified rocks. 2. As veins crossing both the stratified and unstratified rocks. 3. As beds of irregular masses thrust in between the strata. 4. As overlying masses. Fig. 36 illustrates these modes.

The phenomena of veins, being very important, require a more detailed explanation.

Veins are of two kinds. 1. Those of segregation. 2. Those of injection. The former appear to have been separated from the general mass of the rock by elective affinity, when it was in a fluid state; and consequently they are of the same age as the rock. Hence they are often called contemporaneous veins

Fig. 28. represents a bowlder of granitic gneiss, in Lowell, Massachusetts, about five feet long, traversed by several veins of segregation, whose composition differs not greatly from that of the rock, except from being harder and more distinctly granitic. Where veins of this description cross one another, they coalesce so that one does not cut off the other.

Fig. 28.

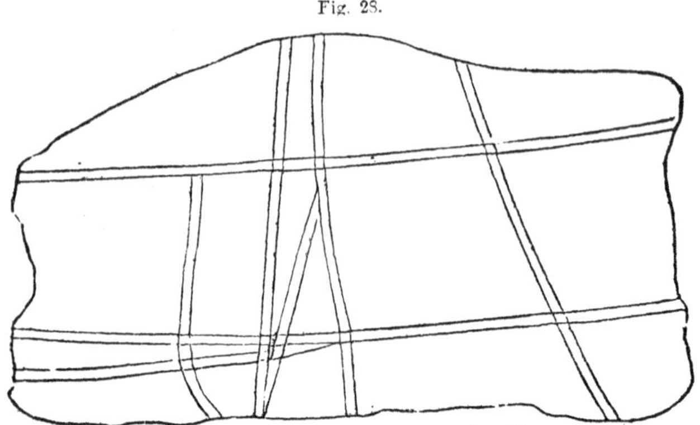

Veins of Segregation in Gneiss, Lowell.

The second class were once open fissures, which at a subsequent period were filled by injected matter.

Veins of segregation are frequently insulated in the containing rock; they pass at their edges by insensible gradations into that rock, and are sometimes tubercular or even nodular.

Injected veins can often be traced to a large mass of similar rock, from which, as they proceed, they ramify and become exceedingly fine, until they are lost. Usually, especially in the oldest rocks, they are chemically united to the walls of the containing rock; but large trap veins have often very little adhesion to the sides.

Fig. 29 exhibits granite veins protruding from a large mass of granite into hornblende schist, in Cornwall.

Fig. 29.

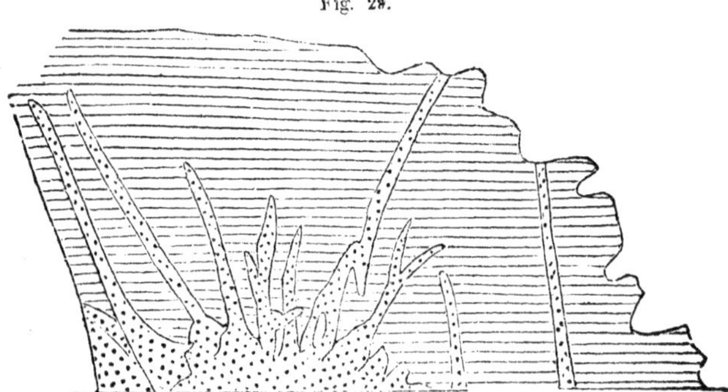

Granite Veins in Hornblende Schist, Cornwall, England.

The large veins that are filled with trap rock or recent lava are usually called *dykes*. These differ from true veins, also, by rarely sending off branches. Dykes of trap are sometimes several yards wide, and nearly a hundred miles long; as in England and Ireland.

Dykes and veins frequently cross one another; and in such a case the one that is cut off is regarded as the oldest. By this rule it may be shown that granite has been injected at no less than four different epochs.

Fig. 30 represents a bowlder of granite in Westhampton, Massachusetts, whose base was the product of the earliest epoch of eruption. This is traversed by the granite vein, *a*, *a*, *a*, which was injected at a second epoch; *b*, is a granite vein cutting *a*, and was therefore produced at a third epoch; while *b*, as well as *a*, are cut off by the granite veins *c*, and *d*, of a fourth epoch.

Fig. 30.

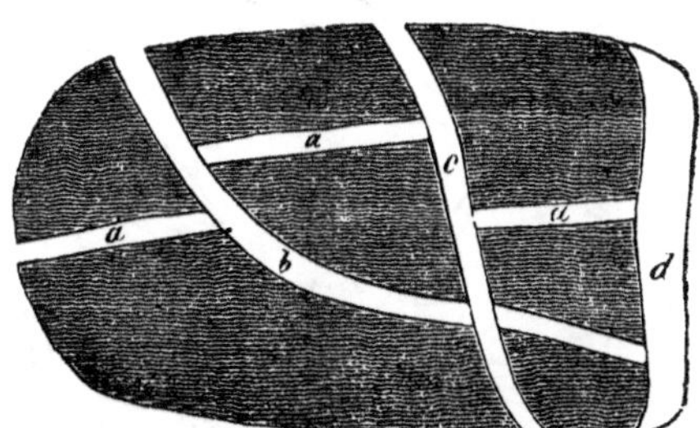

Granite Veins in Granite, Westhampton.

By the same rule can be proved successive eruptions of the trap rocks, as well as other igneous veins. In one remarkable example of veins of different kinds, *eleven epochs* of the injection of unstratified rocks can be traced. This case is in the city of Salem, Massachusetts, near the entrance of the bridge leading to Beverly, on the west side. It is shown upon Fig. 31. The age of the veins is indicated by the figures (1, 2, 3, etc.) attached. No. 1, the basis rock, is syenitic greenstone. The others are mostly granite and greenstone.

Veins and dykes usually cross the strata at various angles. But not unfrequently for a part of their course they have been intruded between the strata; and hence have been mistaken for beds, and have given rise to the inquiry whether granite is not stratified.

Fig. 31.

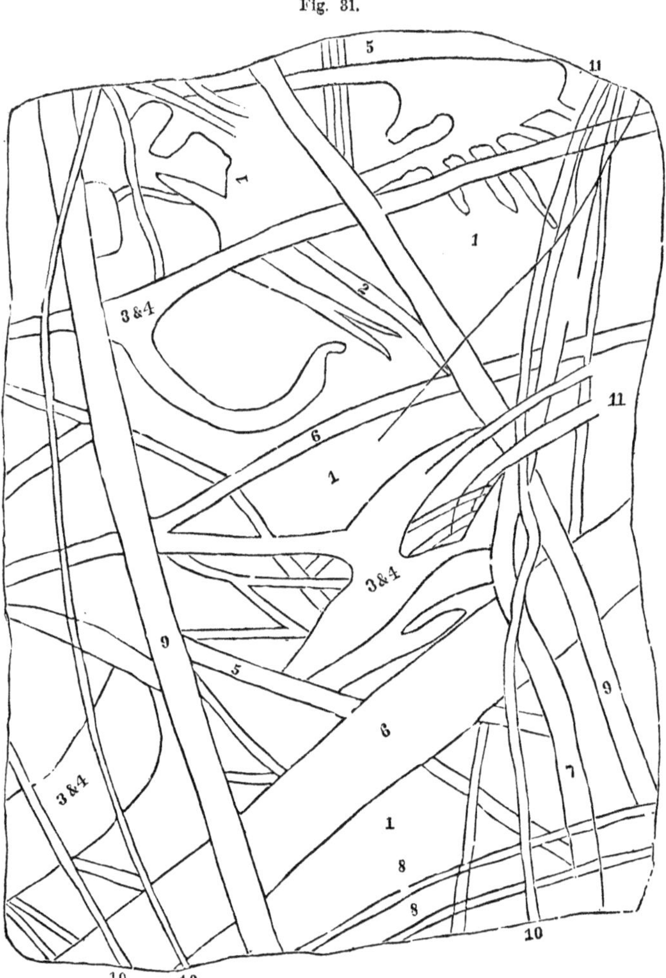

Dykes are usually nearly straight; but granite veins are sometimes very tortuous.

Fig. 32.

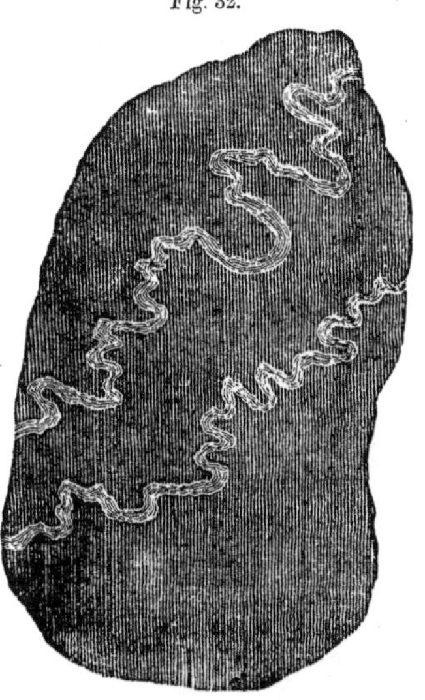

Granite Veins in Micaceous Limestone; Colrain, Massachusetts.

Fig. 33.

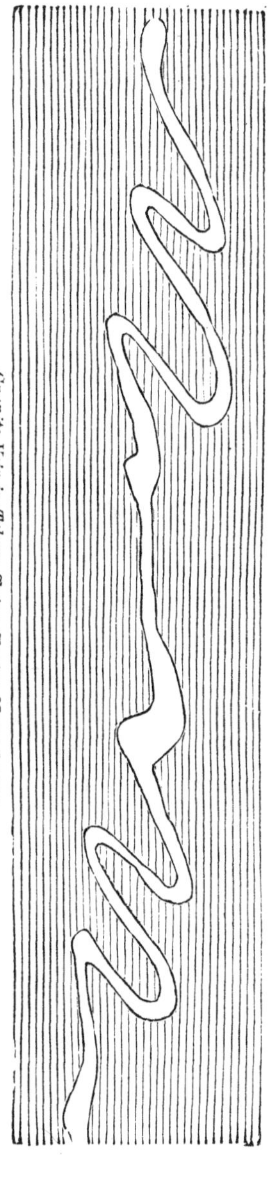

Granite Vein in Talcose Slate; Chester, Massachusetts.

Fig. 32, shows two small but very distinct granite veins in homogeneous micaceous limestone in Colrain, Massachusetts.

Fig. 33 is a tortuous vein of granite in talcose schist, in Chester, Massachusetts, crossing the strata irregularly.

The unstratified rocks, especially when exposed to the weather, are usually divided into irregular fragments by fissures in various directions.

Sometimes, however, these rocks have a concretionary structure on a large scale; that is, they are composed of concreted layers whose curvature is sometimes so slight that they are mistaken for strata.

Cases of this sort can be distinguished from stratification, first, by the concreted divisions not extending through the whole rock; secondly, by the want of a foliated structure in the parallel masses.

A fine example of this concreted structure occurs at one of the quarries in syenite near Sandy Bay, on Cape Ann. Another is at the Lower Falls, upon the Lower Ammonoosuc River, in New Hampshire, among the White Mountains. It is in granite.

An interesting variety of jointed structure in some of the unstratified rocks, is the prismatic, or columnar, by which large masses of rocks are divided into regular forms, from a few inches to several feet in diameter; but with no spaces between them. This curious phenomenon will be more particularly described in a subsequent section.

Fig. 34 is copied from a pebble of black slate, traversed by almost innumerable veins of calcite, from the shores of Lake Champlain, in Vermont. Some of them are cut off and slightly removed laterally, so that they must be veins of injection—doubtless filled by aqueous infiltrations. Many rods square of jet black slate may be seen thus traversed and checkered by these snow white veins.

Fig. 34.

Fig. 35 shows a feldspathic vein conforming to the tortuosities of mica schist, in Conway, Massachusetts. It ought probably to be regarded simply as a layer of the rock, rather than a vein, and a result of metamorphism. But it was probably formed just as some veins are, and is, moreover, a fine example of the plications of mica schist.

Fig. 35.

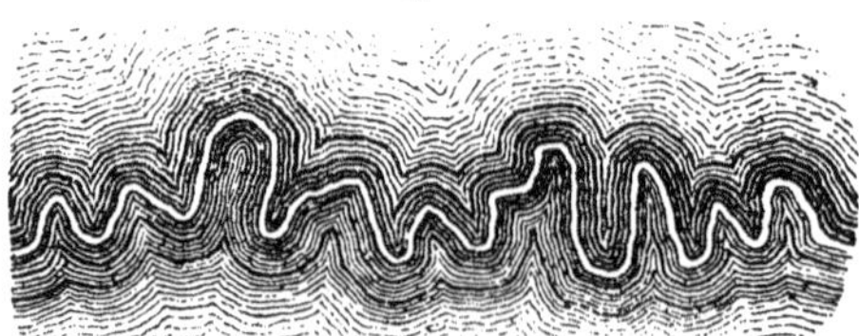

The unstratified rocks, both the masses and the veins and dykes, undoubtedly had an igneous origin, either from dry heat or more usually from aqueo-igneous fusion. But the theory will be more fully stated in the Section on Metamorphism.

Amount of Unstratified Rocks.—Unstratified rocks do not probably occupy one-twentieth part of the earth's surface. In Great Britain they do not cover a thousandth part of the superficies of the island. In Massachusetts, they occupy less than a quarter of the surface.

But there is reason to suppose that these rocks occupy the internal parts of the earth to a great depth, if not to the centre; over which the stratified rocks are spread with very unequal thickness, and sometimes are entirely wanting.

Fig. 36 will convey a better idea than language, of the relative situation of the two classes of rocks. The different groups of stratified rocks are seen resting upon each other in successive order, and the whole upon the unstratified series. Granite is represented as the foundation, but intrusive masses of syenite and porphyry, of granite, of trap, and lastly of lava, are shown to have successively pushed up from beneath the granite, and spread themselves over the surface. A variety of granite is seen rising to the top of the Mesozoic, trap to the top of the Mesozoic, slightly lapping over upon the Tertiary; and finally the lava comes up from the very bottom of the whole, and spreads itself over the Alluvium. Although this is not a section of any particular portion of the earth's crust, it will give a correct idea of the relative situation of the two great classes of rocks, and the reason why the unstratified rocks occupy the whole of the interior of the earth, while they barely reach the surface. We shall refer to this section again after stating the names of the successive formations.

Fig. 37.

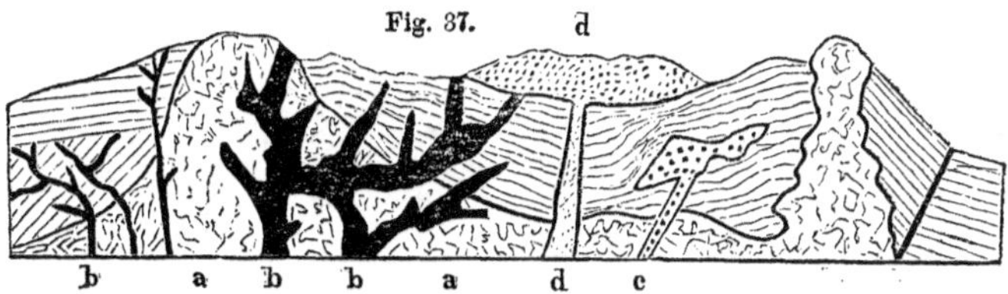

In addition to the last more general figure, we add Fig. 37, specially devoted to the unstratified rocks.

a, a, Irregular masses beneath the stratified rocks.

b, b, Veins (the black irregular lines) crossing both kinds of rocks.

c, Irregular beds between strata.

d, Overlying mass. *e,* A mass injected forcibly, thereby uplifting the strata upon both sides, and causing them to break at *f, f.*

ALLUVIAL
TERTIARY
CRETACEOUS
OOLITIC
TRIASSIC
PERMIAN
CARBONIFEROUS
DEVONIAN
UPPER SILURIAN
LOWER SILURIAN
CAMBRIAN
AZOIC
GRANITIC
QUARTZ ROCK
LIMESTONE, CLAY SLATE
MICA & TALCOSE SCHIST
GNEISS, SERPENTINE
PORPHYRY
GRANITE
SYENITE
PROTOGINE
TRAP
GRANITE
KAINO ZOIC
MESO ZOIC
PALAEO ZOIC
AZOIC
UNSTRATIFIED
LAVA

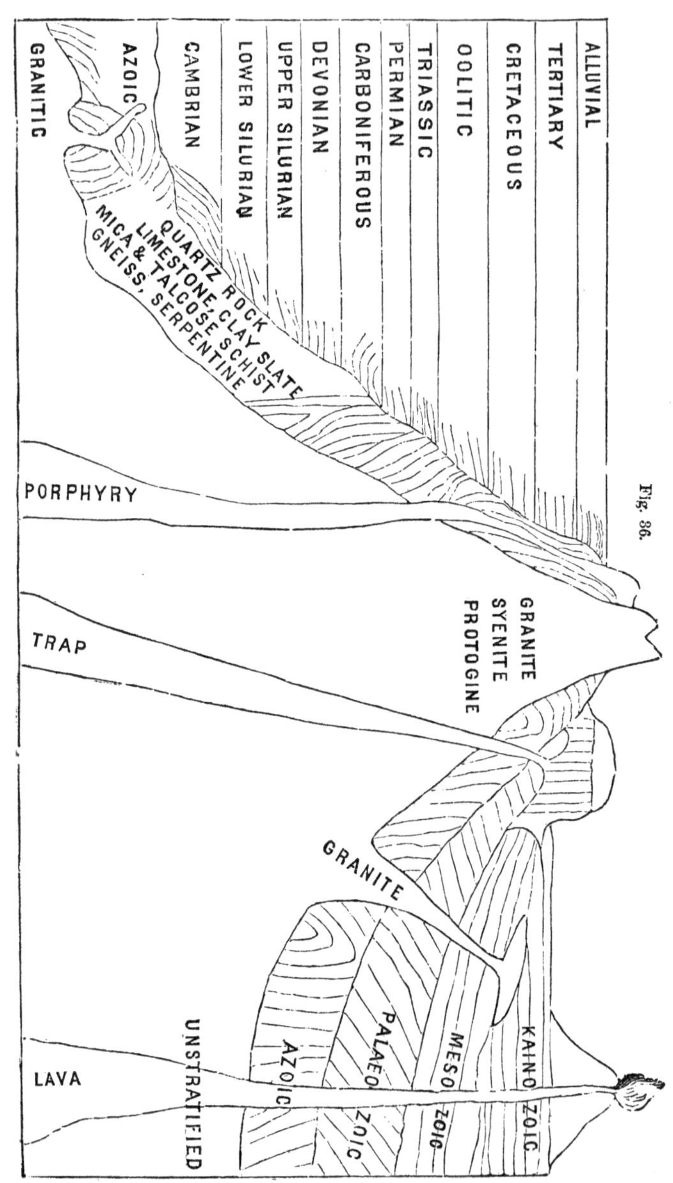

Fig. 36.

FORMATIONS.

Each rock, in its most extended sense, consists of several varieties, agreeing together in certain general characters, and occupying such a relative situation with respect to one another as to show that all of them were formed under similar circumstances, and during the same geological period. Such a group constitutes a *formation*. Ex. gr. Cretaceous formation.

This term often embraces several distinct rocks, when there is reason to suppose that they were produced during the same geological period.

Fig. 38 will give an idea of a formation. It represents the *Lias formation* of England, lying below the oolite, and above the triassic formations.

Fig. 38.

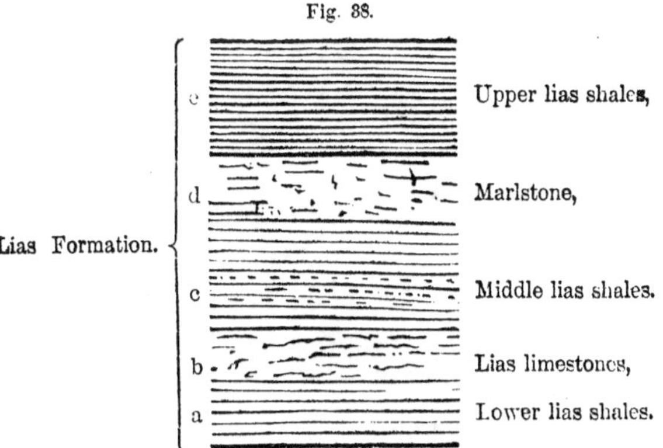

The French word *terrain*, and the English word *group*, are nearly synonymous with formation.

A *series* is a natural group of formations distinguished from all others by characteristic organic remains. It is less comprehensive than *system*, which applies to the greater divisions. Thus in Fig. 36 the beds of stratified rocks upon the left hand side, as Alluvium, etc., are *series* of lesser formations, which are not enumerated: but the four great divisions of the same rocks upon the right hand side, as Azoic, etc., are *systems*. Through inadvertence the terms series, formation, and system, are often used as synonymous by geologists.

When the planes of stratification are parallel to one another in different formations, the stratification is said to be *conformable:* when not parallel, it is *unconformable.*

The stratification in different formations is usually unconformable, as is shown in the position of the azoic and fossiliferous formations, in Fig. 36.

It is hence inferred that the stratified rocks were elevated at different epochs: in other words, those formations which are the most highly inclined, must have been partially elevated before the others were deposited upon them.

These numerous elevations of the strata have produced in them a great variety of cracks, fissures, and slides.

When the continuity of the strata is interrupted by a fissure, so that the same stratum is higher on one side than on the other, or has been slidden laterally, that fissure is called a *fault*, or a *trouble,—a slip,—a dyke,*—etc.: as *a*, *a*, in Fig. 39 and 40.

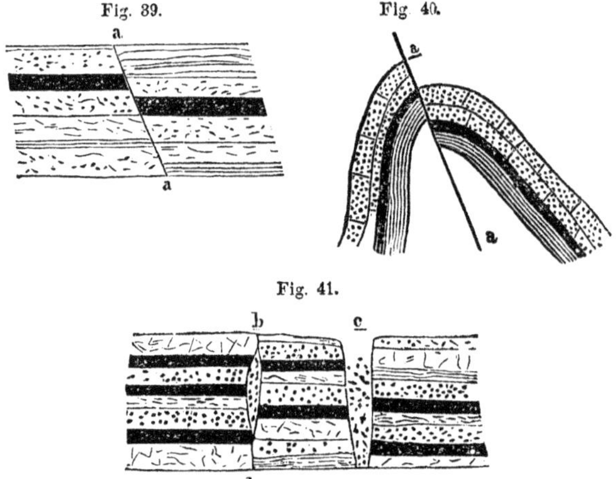

Fig. 39. Fig. 40. Fig. 41.

A fault is sometimes filled with fragments of rocks, clay, etc., as *b*, in Fig. 41; in which case it occasions great trouble in the working of mines, because, when the fragments are reached, it is impossible to decide whether the continuation of mineral sought is above or below the level, or to the right or left.

There are two kinds of faults, the *transverse*, or those that cross the strata at right angles to the strike, or transversely; and the *longitudinal*, or those that are parallel to the strike. The former are usually local, and quite short, the latter are sometimes of great length, and produce much confusion over large districts.

If the fissure is open and of considerable width, and is succeeded at each extremity by a wider valley, it is called a *gorge*, as *c*, in Fig. 41.

If it be still wider, with the sides sloping or rounded at the bottom, a *valley* is produced.

In a similar way most of the valleys of primitive countries were formed.

Classification of Rocks.

One of the great objects sought by geologists is a complete knowledge of the classification, or the order of the succession of the different formations. There are two difficulties in their way. First, there is no district in the world where all the formations are found placed directly over another; and, secondly, the rocks in one country have sometimes little or no external resemblance to those of the same age in another country; or if developed at all their thickness varies greatly. It is even the case that large formations are developed upon one continent which are entirely wanting in another. Could the successive formations be placed upon one another in regular order in any one place, and an excavation be made through them which a geologist might examine, the task of classification would be comparatively easy.

Among all the fossiliferous formations there exists an invariable order of superposition. Rocks of different age may be brought together by the absence of an intermediate group; but the newer never underlies the older, except in a few cases of folded axes, or inverted anticlinals.

A formation in America is identified with the corresponding strata in Europe chiefly by means of organic remains *characteristic* of that group. Previously, the European strata had been thoroughly examined, both as to their mineralogical and organic characters; and had received a local name. Thus one of the lowest series of the Palæozoic system was first explored in Wales. Hence it was called *Silurian* (an ancient name for the district). Now when *synchronous** strata are found in America they are

* *Synchronous* strata are those which are formed simultaneously in different terrains.

termed Silurian; that is, strata of the same age with those in Siluria. Nearly every formation has thus received a local appellation. American geologists have carried this principle so far that nearly fifty local names have been proposed for the different divisions of our series.

The first division of all the rocks is into the *Stratified* and *Unstratified*, corresponding to the Aqueous, including Metamorphic, and Igneous of some geologists. The stratified class is divided into the *Fossiliferous* and *Unfossiliferous*, or those which contain, and those devoid of organic remains. The latter all belong to one series which is termed the *Azoic*, because without life; or *Hypozoic*, *beneath* all evidences of life. The fossiliferous division is divided into three great systems, according to the times in which the organism flourished: the *Palæozoic*, or the *ancient* type of organic life; *Mesozoic*, or the life that flourished during the *middle* periods of geological time; and *Cainozoic*, or the *recent* economies of life.

The following tabular view of the rocks embraces all the important formations and groups described in the more recent works on geology. Where we have made any changes it is simply with the hope of escaping difficulties which embarrass all systems of classification. The arrangement which we give we shall follow in this work.

Class 1.—STRATIFIED OR AQUEOUS ROCKS.

1. Fossiliferous.

Cainozoic.

1. Alluvium, including Drift.
2. Tertiary.

Mesozoic.

1. Cretaceous, with Green Sand.
2. Oolitic or Jurassic, with Wealden and Lias.
3. Triassic or New Red Sandstone.

Paleozoic.

1. Permian.
2. Carboniferous.
3. Devonian.
4. Upper Silurian.
5. Lower Silurian.
6. Cambrian or Huronian.

2. Azoic or Unfossiliferous.

Interzoic and Hypozoic or Laurentian.	1. Quartz Rock, } also fossiliferous. 2. Clay Slate, } 3. Mica Schist. 4. Talcose Schist, including Steatite. 5. Serpentine. 6. Hornblende Schist. 7. Gneiss. 8. Saccharoid Azoic Limestone.

Class 2.—UNSTRATIFIED OR IGNEOUS ROCKS.

1. Granitic Group.

1. Granite. 2. Syenite. 3. Protogine.

2. Trappean Group.

1. Porphyry. 2. Greenstone. 3. Amygdaloid, etc.

3. Volcanic Rocks

1. Basalt. 2. Trachyte. 3. Pumice. 4. Tufa. 5. Peperino. 6. Volcanic Ashes. 7. Vesicular Lava, etc.

Of the Azoic series, quartz rock and clay slate are sometimes found in the Palæozoic system containing fossils, and rocks possessing the same characters are occasionally found interstratified with fossiliferous rocks; that is, gneiss, mica schist, etc., are not confined to the Azoic group, but wherever found they are always devoid of organic remains. Below the Azoic series are the unstratified rocks, which extend to unknown depths.

In this country the Silurian and Devonian series have been subdivided into twenty-three formations, by the State Geologists of New York, who have given them names from the localities where they are best developed. In other States names more or less local have been given to the divisions of other series. In the following table we present the most important subdivisions of all the systems both in Europe and in this country, with the thickness, so far as it is reliable.

	EUROPEAN FORMATIONS.	THICKNESS IN FEET.	NORTH AMERICAN FORMATIONS.	THICKNESS IN FEET.
ALLUVIUM,	Recent, Pleistocene,	500	Alluvium proper, Drift,	20[illegible]—500
TERTIARY,	Pliocene, Miocene, Eocene,	2,547	Yorktown group, Vicksburg group, Claiborne group,	1,200
CRETACEOUS.	Chalk, Gault, Greensand,	2,430	Clays and Greensand,	1,000—3,000
JURASSIC,	Wealden, Upper oolite, Middle oolite, Lower oolite, Lias,	1,300 2,270 1,100	Connecticut River	5,000
TRIASSIC,	Triassic,	[illegible],100	Sandstone,	
PERMIAN	Permian,	1,010	Permian of Kansas, &c.	861
CARBONIFEROUS.	Coal Measures, Millstone grit, Mountain limestone,	915 to 15,000	Coal Measures, Conglomerate, Carboniferous limestone, Conglomerate,	7,000 to 13,000
DEVONIAN,	Upper, Middle, Lower,	10,000	Catskill red sandstone, Chemung group, Portage group, Genesee slate, Hamilton group, Marcellus shales, Upper Helderberg limestone, Scoharie grit, Cock-tail grit, Oriskany sandstone,	5,000 [illegible],200 1,700 [illegible]00 600 [illegible]00 350 [illegible]00 2[illegible]
UPPER SILURIAN.	Upper Ludlow rock, Aymestry limestone, Lower Ludlow rock, Wenlock limestone, Wenlock shale, Woolhope limestone, Denbighshire sandstone, Tarannen shales, May Hill Beds,	650 100 1,000 [illegible]00 1,500 50 2,000 1,[illegible]00 1,000	Lower Helderberg limestone, Water lime group, Onondaga salt group, Niagara group, Clinton group, Upper Hudson river group, Medina sandstone, Oneida conglomerate,	[illegible]00 1,000 2 400 1,000 1,450
LOWER SILURIAN.	Lower Llandovery beds, Caradoc sandstone, Llandeilo flags, Lingula flags,	1,000 9,[illegible]0 5,000? 5,000?	Lower Hudson River group, Utica slate, Trenton limestone, Chazy limestone, Calciferous sand rock, Potsdam sandstone,	2,000 500 2,500 100 300
CAMBRIAN, AZOIC,	Cambrian, Hypozoic	26,000	Huronian, Laurentian,	12,000 20,000
		[illegible]8,8[illegible]-9[illegible],917		[illegible]0,[illegible]61-[illegible]9,061

Professors H. D. and W. B. Rogers have adopted a different classification for the Palæozoic system, as it occurs in the States of Pennsylvania and Virginia. The system is called the *Appalachian Palæozoic Day*, and is divided like the different parts of a day. We present it in a Table, placing along with it the names of the corresponding formations elsewhere, according to the nomenclature of the New York State Geologists. It is copied from the magnificent "Geology of Pennsylvania," by Professor H. D. Rogers.

APPALACHIAN PALÆOZOIC DAY.

NEW YORK SYSTEM.	ROGERS' CLASSIFICATION.	THICKNESS IN PENN.
	Primal Series.	
	Primal conglomerate,	150
	Primal older slate,	1200
Potsdam sandstone,	Primal white sandstone, Primal upper slate.	300 700
		2350

New York System.	Rogers' Classification.	Thickness in Pen:
	Auroral Series.	
Calciferous sandrock,	Auroral calciferous sandrock,	60
Chazy limestone, etc.	Auroral magnesian limestone,	2500
		2560
	Matinal Series.	
Trenton limestone,	Matinal argillaceous limestone,	300—550
Utica slate,	Matinal black slate,	300—400
Lower Hudson River group,	Matinal shales.	1200?
		1800—2150
	Levant Series.	
Oneida conglomerate,	Levant gray limestone,	250—400
Medina sandstone,	{Levant red sandstone.	500—700
	{Levant white sandstone.	450
		1200—1550
	Surgent Series.	
Clinton group,	Surgent lower slate,	200
	Surgent iron sandstone,	80
	Surgent upper slate,	250
	Surgent lower ore shale,	760
	Surgent ore sandstone,	10—30
	Surgent upper ore shale,	300
	Surgent red marl.	350
		1950—1970
	Scalent Series.	
Niagara group,	Scalent variegated marls,	400
Onondaga salt group	Scalent gray marls,	800
Water lime group,	Scalent limestone.	50
		1250
	Pre-meridian Series.	
Lower Helderberg limestone,	Pre-meridian limestone.	80—100
	Meridian Series.	
	Meridian slate,	170
Oriskany sandstone,	Meridian sandstones,	150
		320
	Post Meridian Series.	
Scoharie grit, } Cock-tail grit, }	Post meridian grits,	300
Upper Helderberg limestone,	Post meridian limestones.	80
		380
	Cadent Series.	
Marcellus slate,	Cadent lower black slate,	250
Hamilton group,	Cadent shales,	600
Genesee slates,	Cadent upper black slates.	300
		1150
	Vergent Series.	
Portage group,	Vergent flags,	1700
Chemung group,	Vergent shales.	3200
		4900
	Ponent Series.	
Catskill red sandstone,	Ponent red sandstone.	5000

New York System.	Rogers' Classification.	Thickness in Penn.
	Vespertine Series.	
Conglomerate,	Lower conglomerate of the carboniferous.	2660
	Umbral Series.	
Carboniferous limestone in other States.	Red shales and limestone.	3000
	Seral Series.	
	Conglomerate,	1100
	Coal measures,	2500
	Permian or upper coal measures.	

Several attempts have been made to make out a classification founded upon Palæontology; that is, organic remains. This subject will be better understood after that of Palæontology has been described. But we will present it in brief in this place, premising only, that in the different formations we find different groups of animals and plants, often characterized by the great predominance of some particular races. These life periods correspond in general to the other characters by which different formations are distinguished; so that a palæontological division will correspond essentially to one that is lithological, and this to one founded on the conformity or unconformity of stratification. The system below is that adopted by Prof. Pictet. His large Groups he calls Periods, and the subdivisions, Epochs. Properly speaking, however, an Epoch is the point of time when an event takes place, and a Period the interval between successive Epochs.

1. Palæozoic Period.

First Epoch, Silurian.
Second Epoch, Devonian.
Third Epoch, Carboniferous.
Fourth Epoch, Permian.

2. Secondary Period.

First Epoch, Triassic.
Second Epoch, Jurassic.
Third Epoch, Cretaceous.

3. Tertiary Period.

Tertiary Epoch.

4. Quaternary and Modern Period.

Diluvian and Modern Epoch.

The Palæozoic Period was distinguished, 1. By the entire absence of mammiferous animals and birds. 2. By the presence of many genera of shells called Cephalopods, like the Nautilus, of a peculiar structure, not found afterwards; also by a great number of another family called Brachiopods, which subsequently mostly disappeared; 3, by the existence of large numbers of crustaceans, called Trilobites, of which we find no trace afterwards; 4, by the presence of a great number of singular animals, called Crinoids, which nearly all disappeared in the subsequent formations; and 5, by Polypi or corals of peculiar types and characters.

The Secondary Period was characterized, 1, by the fewness and small size of the Marsupial Quadrupeds, which then existed; 2, by an enormous development, both as to numbers and size, of reptiles of peculiar character; 3, by beautiful groups of the shells called Ammonites, (like the Nautilus), of a peculiar structure; 4, by tribes of Echinoderms (like Sea Stars), altogether different from those of the first Period; 5, by Polypi, with peculiar characters.

The Tertiary Period was characterized, 1, by the appearance of great numbers of mammiferous animals; 2, by an approach to living forms in the reptiles and fishes; 3, by the total disappearance of Ammonites and Belemnites, so abundant in the Secondary Period.

The Quaternary Period is specially distinguished by the appearance of Man, the most remarkable of all terrestrial animals.

We might point out characters almost equally striking and peculiar in each of the nine epochs. Indeed, most of these might be again divided, and still the Faunas and Floras would be quite distinct and peculiar, showing that the earth has been the seat of not a few life periods since organic beings first appeared upon it.

It is sometimes customary to characterize a Period or an Epoch by the name of the predominant race that then lived. Thus the Secondary Period has been called the *Palæosaurian*, or the reign of ancient Saurian Reptiles; the Tertiary Period as *Mammiferous*, or the reign of Mammals, etc. We might carry this nomenclature through all the nine epochs above mentioned, as follows: To begin still lower, we might call the Azoic rocks, *Crystaliferous*, or crystal bearing; the Silurian rocks as *Brachiopodiferous*, *Cephalopodiferous*, and *Trilobiferous*, from the predominence of those three tribes of invertebrates; the Devonian, as *Thaumichthiferous*, from the prevalence of strange fishes; the Carboniferous Epoch, as COAL-BEARING, or *Acrogeniferous*, from the abundance of flowerless trees; the Permian, as *Lacertiferous*, or lizard-bearing; the Triassic Epoch, as *Enaliosauriferous* and *Labyrinthodontiferous;* the Jurassic Epoch, as *Ichniferous*, (track-bearing), and *Palæosauriferous*; the Cretaceous, as *Echiniferous* (bearing Sea Stars,) and *Foraminiferous;* the Tertiary, as *Mammaliferous;* and the Modern Epoch, as *Homoniferous*, or Man-bearing. These designations, however, are more poetical and popular than scientific.

Instruments Convenient for the Practical Geologist.—For determining the position of strata, the Clinometer and Pocket Compass are needed. Still more indispensable are hammers. There should be two or three of these of different sizes, with rounded faces on one side, and wedge shaped or pointed at the other. The largest should be a somewhat heavy sledge, and the smallest of only a few ounces weight for trimming specimens.

In some departments of geological research, a knowledge of heights is requisite. As the heights of but comparatively few elevations in our country are known, a levelling apparatus or barometer is essential. A new kind of barometer, called the *Aneroid*, we have found by long experience to be ad-

Fig. 41.

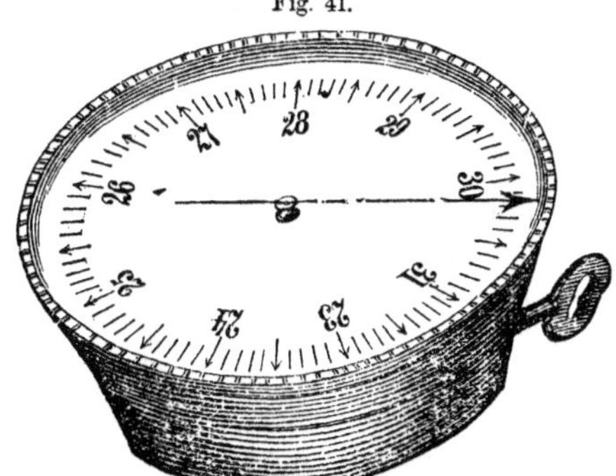

Aneroid Barometer.

mirably adapted to the work. It is less accurate than the mercurial, but much less liable to injury. It is generally of little value above an elevation of 5,000 feet. As is seen in the representation, (Fig. 41), the inches and subdivisions of the common mercurial barometer are marked upon it, and the index points to the different marks, according to its change of elevation. Every new Aneroid should be tested before much reliance can be placed upon it. To ascertain the height of a mountain above a valley by this instrument, multiply the space passed over by the index, (expressed in thousandths of an inch), by the number of feet of elevation requisite to move the index one-tenth of an inch, and cut off four right hand figures for decimals.

A Pedometer and an apparatus for sketching are also desirable.

SECTION II.

THE CHEMISTRY AND MINERALOGY OF GEOLOGY.

Of the sixty-two simple substances hitherto discovered, sixteen constitute, by their various combinations, nearly the whole of the matter yet known to enter into the composition of the globe. They are as follows, arranged in three classes, according to their amount; the first in each class being the most abundant.

1. *Non-Metallic Substances.*—Oxygen, Hydrogen, Nitrogen, Carbon, Sulphur, Chlorine, Fluorine, and Phosphorus.

2. *Metalloids, or the bases of the earths and alkalies.*—Silicium, Aluminium, Potassium, Sodium, Magnesium, and Calcium.

3. *Metals Proper.*—Iron, Manganese.

The metalloidal substances mentioned above, united with oxygen, constitute the great mass of the rocks, consolidated and unconsolidated, accessible to man. Oxygen also forms twenty per cent. of the atmosphere, and one-third part by measure of water. Hydrogen forms the other two-thirds of this latter substance; and it is evolved also from volcanos, and is known to exist in coal. Nitrogen forms four-fifths of the atmosphere, and enters into the composition of animals, living and fossil. It is found also in coal. Carbon, however, forms the principal part of coal; and it exists likewise in the form of carbonic acid in the atmosphere, though constituting only one thousandth part, and it forms an important part of all the carbonates, and is produced wherever vegetable and animal matters are undergoing decomposition. Sulphur is found chiefly in the sulphurets and sulphates that are so widely disseminated. Chlorine is found chiefly in the ocean, and in the rock salt dug out of the earth. Fluorine occurs in most of the rocks, though in small proportion. Phosphorus is widely diffused in the rocks and soils, and is abundant in organic remains, in the form of phosphates.

Nearly all the simple substances above mentioned have entered into their present combinations as binary compounds; that is, they were united two and two before forming the present compounds in which they are found. The following constitute nearly all the binary compounds of the accessible parts of the globe: Silica, Alumina, Lime, Magnesia, Potassa, Soda, Oxide of Iron, Oxide of Manganese, Water, and Carbonic Acid.

It is meant only that these binary compounds, and the sixteen simple substances that have been enumerated, constitute the largest part of the known mass of the globe: for many other binary compounds, and probably all the known simple substances, are found in small quantity in the rocks; but not enough to be of importance in a geological point of view.

It has been calculated that oxygen constitutes 50 per cent. of the ponderable matter of the globe, and that its crust contains 45 per cent. of silica, and at least 10 per cent. of alumina. Potassa constitutes nearly 7 per cent. of the unstratified rocks, and enters largely into the composition of some of the stratified class. Soda forms nearly 6 per cent. of some basalts and other less extensive unstratified rocks; and it enters largely into the composition of the ocean. Lime and magnesia are diffused almost universally among the rocks in the form of silicates and carbonates —the carbonate of lime having been estimated to form one-seventh of the crust of the globe; at least three per cent. of all known rocks are some binary combination of iron, such as an oxide, a sulphuret, a carburet, etc.; manganese is widely diffused, but forms less than one per cent. of the mass of rocks.

A few simple minerals constitute the great mass of all known rocks. These are Quartz, Feldspar, Mica, Hornblende, Pyroxene, Calcite, embracing all carbonates of lime, Talc, embracing Chlorite, Steatite, and Serpentine. Oxide of iron is very common as an impurity; but it does not usually show itself till the decomposition of the rock commences.

Quartz, or silica in the pure state, is transparent, and is known as rock crystal. It is the hardest of all the minerals enumerated, easily scratching all of them. Quartz is also known, when mixed with other substances, under other names; as *amethyst*, when it is colored purple: *Rose*, *Smoky*, *and Ferruginous*, when pink, blackish, and yellowish red; *Chalcedony*, *and Agate*, when there are several colors exhibited in the same specimen, generally arranged fantastically; *Jasper*, when it is bright red. Figs. 42 and 43 represent the most common form of quartz crystals. Quartz is the most abundant of all minerals; there are but few rocks in which it is not the predominant ingredient. All the forms of quartz are absolutely insoluble in water, acids and most liquids, except hydrofluoric acid, a substance that does not appear ever to have been concerned in the formation and alteration of mineral substances.

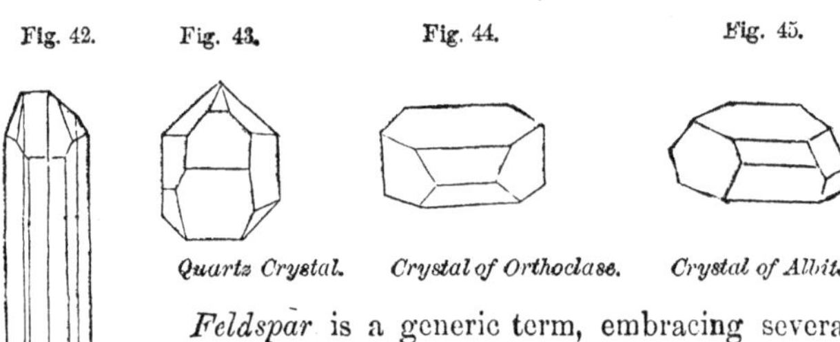

Fig. 42. Quartz Crystal. Fig. 43. Quartz Crystal. Fig. 44. Crystal of Orthoclase. Fig. 45. Crystal of Albite.

Feldspar is a generic term, embracing several silicates of alumina and an alkali. The most common variety is the *potash-feldspar*, or *Orthoclase*, Fig. 44, which is a double silicate of alumina and potassa. The *soda-feldspar*, or *Albite*, Fig. 45, a double silicate of alumina and soda, differs from orthoclase but little in appearance, except in its crystalline form. Both species have a beautiful pearly lustre. The *lime feldspar* or *Labradorite*, a double silicate of alumina and lime, has a still more brilliant luster. Other species of feldspar are given in the table upon page 51. It is important to be able to distinguish these species, since particular rocks are characterized by the kind of feldspar most common in them.

Mica is also a generic term, including many species. They are divided into two classes according to the inclination of their axes of polarization to each other—in the common mica, *Muscovite*, inclining at a large angle, and in the others at a small angle. Muscovite occurs in plates, which scale off in very thin laminæ. It is commonly called *isinglass;* and is well known from its use instead of glass in the doors of stoves. Fig. 46 represents a crystal of Muscovite. Chemically it is a double silicate of alumina and potassa, in which a part of the alumina is usually replaced by iron. *Phlogopite* is a double silicate of alumina magnesia and potassa; and *Biotite* is a double silicate of alumina, iron, magnesia and potassa.

Fig. 46.

Crystal of Muscovite.

Fig. 47.

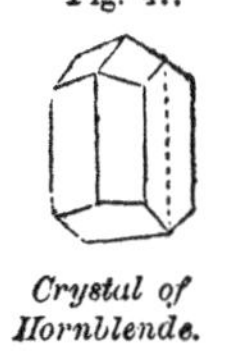

Crystal of Hornblende.

Hornblende is usually a tough, black or dark colored mineral, crystalizing as in Fig. 47, and being a double silicate of alumina or iron and lime.

There are many varieties of hornblende, the most common being the *Tremolite, Asbestus, and Actynolite;* the second of which is often of a soft texture, and can be woven like cotton.

Pyroxene, including *Augite*, *Sahlite* and *Diopside*, is a simple silicate of either lime, magnesia, protoxide of iron or manganese, or soda, and differing externally from the hornblendes, principally in the form of its crystal, (Fig. 48). *Hypersthene* is an important variety of pyroxene, occurring chiefly in the Lawrentian Series.

Calcite, or the simple carbonate of lime, is very widely diffused as crystalline or sedimentary limestone. Its primary crystalline form is rhombohedral, Fig. 49, but it is often modified into the

Fig. 48.

Crystal of Pyroxene.

Fig. 49.

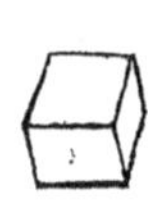

Crystal of Calcite.

Fig. 50.

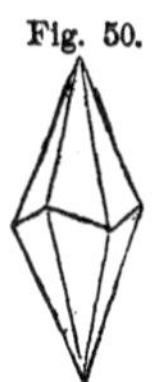

Crystal of Calcite.

shape of Fig. 50. The species dolomite, a double carbonate of lime and magnesia, is also rhombohedral, but it more nearly approaches a square prism in its form. Carbonate of lime may always be known by its effervescence with acids.

Talc is a soft, green or whitish hydrous silicate of magnesia. It has a very greasy or soapy *feel*. An impure form called *steatite*, or soapstone, is well known from its power of retaining heat, and as a non-conductor. *Chlorite* is of a dull emerald-green color, and is a double hydrated silicate of alumina and magnesia.

Serpentine or Ophiolite.—Serpentine is a mottled rock, the predominant color green, and is a hydrous silicate of magnesia. It is distinctly stratified in some localities, and though formerly regarded as a purely igneous rock, is now generally admitted to be a metamorphic rock; perhaps an altered dolomite. It is elegant as an ornamental rock, though not much used in this country, where it exists in immense quantities.

Serpentines generally contain so many foreign mineral matters as to form with them distinct varieties; as, *garnetiferous*, *diallagic*, *hornblendic*, and *chromiferous serpentines*. *Ophicalce* is a mixture of calcite or dolomite, with serpentine, talc, and chlorite, often brecciated. In the latter form are included the beautiful *verde antique marbles*, such as occurs at Roxbury and Proctorsville, Vermont; Newbury and Middlefield, Massachusetts; and New Haven and Milford, Connecticut. When chromic iron is disseminated through serpentine, giving it a peculiar mottled appearance, it is called *ophyte*, from its resemblance to the skin of a snake.

The following table shows the composition of the most common or important minerals:

	Silica.	Alumina.	Oxide of Iron.	Oxide of Manganese.	Lime.	Magnesia.	Potassa.	Soda.	Total.
Orthoclase	65.72	18.57	trace.	trace.	0.34	0.10	14.02	1.25	100.
Albite	69.36	19.26	0.43		0.46			10.50	100.
Oligoclase	62.61	24.11	0.30		2.74	0.55	0.75	8.89	99.95
Andesine	59.60	24.28	1.58		5.77	1.00	1.08	6.53	99.92
Labradorite	53.48	26.46	1.60	0.89	9.49	1.74	0.22	4.10	98.40
Muscovite	47.50	37.30	8.20	0.90	Water. 2.63		9.6		100.93
Biotite	40.00	16.16	7.05			21.54	10.83	Water. 3.00	99.03
Phlogopite	41.30	15.35	1.77	Fluorine. 8.30	Water. 0.28	28.79	9.70	0.65	101.14
Hornblende (calc.)	41.00	16.00	15.00		14.00	14.00			
Pyroxene	60.65	4.20	4.18	0.79	4.97	25.20			99.99
Calcite					56.13	Carbonic acid. 43.87			100.00
Talc	62.35		1.34			31.32		Water. 4.48	99.79
Chlorite	31.47	17.14	4.55	0.53		34.40		Water. 12.12	98.88
Serpentine (calc.)	44.02					43.11		Water. 12.87	100.00

Other minerals forming rocks of small extent, or entering so largely into their composition as to modify their character, are the following: gypsum, the hydrous sulphate of lime (of which a crystal is represented in Fig. 51), diallage, common salt, coal, bitumen, garnet, tourmaline, staurotide, epidote, olivine, and pyrites.

A few of these minerals exist in so large masses as to be denominated rocks; *e. g.*, quartz, carbonate of lime, etc., but in general, from two to four of them are united to form a rock; *e. g.*, quartz, feldspar and mica, to form granite. In some instances the simple minerals are so much ground down, previously to their consolidation, as to make the rock appear homogenous; *e. g.*, shale and clay slate.

Fig. 51.

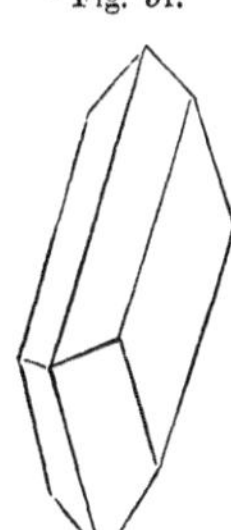

Water constitutes a part of nearly all rocks, either chemically combined with the component minerals, or as a mechanical constituent of the rock itself. The latter is the more usual case. The more common hydrated minerals are talc, chlorite, gypsum, serpentine, diallage, and the zeolites. It is remarkable that the latter, occurring in volcanic or igneous rocks almost exclusively, should contain so much water, while many that are formed in sedimentary rocks have none.

GEOLOGICAL SITUATION OF USEFUL ROCKS AND MINERALS.

The rocks and minerals useful in an economical point of view are in a few instances found in almost every part of the rock series: but in a majority of cases they are confined to one or more places in that series.

EXAMPLES.—*Granite, Syenite, and Porphyry :* found intruded among all the stratified rocks as high in the series as the Tertiary strata; but they are almost entirely confined to the Hypozoic rocks.

Greenstone and Basalt are found among and overlying all the Hypozoic and fossiliferous rocks; but they are mostly connected with the latter.

Lava, some varieties of which, as *Peperino*, are employed in the arts, being the product of modern volcanos, is found occasionally overlying every rock in the series.

Clay: the common varieties used for bricks, earthern ware, pipes, etc., occur almost exclusively in the Tertiary and Alluvial strata. Porcelain clay results from the decomposition of granite, and is found in connection with that rock.

Marl, or a mixture of carbonate of lime and clay, is chiefly confined to the Alluvial and Tertiary strata; and differs from many varieties of limestone, only in not being consolidated.

Limestone, from which every variety of marble, one variety of alabaster, and every sort of quicklime are obtained, is found in almost every rock, stratified and unstratified. In the oldest stratified rocks and in the unstratified it is highly crystalline; and in the newer strata (*e. g.*, chalk), it is often not at all crystalline. The most esteemed marbles are obtained from the Palæozoic rocks, either unaltered or metamorphic.

Serpentine is connected with metamorphic rocks, either Hypozoic or Palæozoic. It is not unfrequently associated with trap rocks in later periods.

Gypsum, or *Plaster of Paris*, is found in Europe in all the Mesozoic and

Tertiary strata. In this country a little is found in the Palæozoic rocks, but in greatest abundance in the Mesozoic and Cainozoic formations. Upon the Red River, in Texas, more than a hundred miles square of surface are underlaid by *selenite*, a transparent variety.

Rock Salt (Chloride of Sodium) is frequently found associated with gypsum, in the New Red sandstone. It occurs also in the Supercretaceous or Tertiary strata; as at the celebrated deposit at Wieliczka in Poland, and in Sicily, and Cordona (Spain), in Cretaceous strata; in the Tyrol, in the Oolites; and in Durham, England, salt springs occur in the Coal series. In the United States they issue from the Silurian rocks.

Forms of Vegetable Matter.—If vegetable matter be exposed to a certain degree of moisture and temperature, it is decomposed into the substance called *peat*, which is dug from swamps, and belongs to the alluvial formation.

Lignite or Brown Coal, the most perfect variety of which is jet, is found in most of the series above the coal formations; and appears to be vegetable matters like *peat* which have long been buried in the earth, and have undergone certain chemical changes. It generally exhibits the vegetable structure.

Bituminous Coal appears to be the same substance which has been longer buried in the earth, and has undergone further changes. The proportion of bitumen is indefinite, varying from 10 to 60 per cent., and the coal is said to be *dry* or *fat*, according to the amount of bitumen present.

There are several varieties of the bituminous coal.

Pitch, or *caking coal*, is a velvet black, highly bituminous mass, which cakes or runs together during combustion. *Cherry coal* is like caking coal, but it does not soften and cake. It breaks so readily that much of it is lost in the mining process. *Cannel coal* is nearly black, with a fine compact texture and a conchoidal fracture. It burns readily like a candle, hence its name. *Splint coal* is a coarse variety of cannel coal. The *Albert coal* of Nova Scotia is perhaps to be regarded as a species of bitumen, because the latter so much predominates. It has a bright, shining lustre, and ignites instantly upon contact with flame. *Coke* is bituminous coal artificially deprived of its bitumen. It is light, and approaches charcoal in appearance. Coke is occasionally found in nature; especially in the neighborhood of dikes. All these varieties are found in the coal formation, and even in the Mesozoic and Tertiary series.

Anthracite is bituminous coal that has been deprived of its bitumen, usually by heat, under pressure. It thus forms a compact heavy mass, igniting with some difficulty. The anthracite of Pennsylvania, of enormous extent, is in the true coal measures, and it is a curious fact, that as we pass westward—that is, recede from the metamorphic and unstratified rocks of the Atlantic coast—the quantity of bitumen increases; so that within a few hundred miles the coal is highly charged with it. The fact makes it extremely probable that the heat, which changed the metamorphic rocks, also drove off the bitumen.

The anthracite of Rhode Island and of Massachusetts, is in what may be called a *metamorphic Coal Field;* that is, the strata have been more acted upon and hardened by heat than is usual. In Rhode Island and in Bristol county in Massachusetts, the fossil remains are still found; but in Worcester, where the bed of coal is seven feet thick, no trace of fossil vegetables has been discovered; and the rocks are considerably hardened and crystalline. The coal also is much more stony, and is partially changed into plumbago.

Graphite, Plumbago, or Black Lead, appears to be anthracite which has undergone still further mineralization; at least, in some instances, when coal has been found contiguous to igneous rocks, it is converted into plumbago; and hence such may have been the origin of the whole of it. In the Alps, plumbago is found in a clay slate that lies above the lias. It is also found in the coal series.

All the varieties of coal that have been described occur in the form of seams, or beds, interstratified with sandstones and shales; and most usually there are several seams of coal with rocks between them; the whole being arranged in the form of a basin. Fig. 52 is a sketch of the great coal basin of South Wales, in

Fig 52.

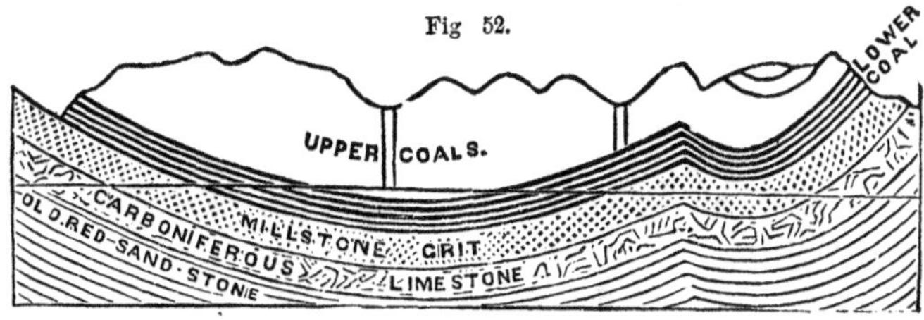

Great Britain; which contains twenty-three beds of coal; whose united thickness is ninety-three feet. When we consider how

much this arrangement facilitates the exploration and working of coal, we can hardly doubt but it is the result of Divine Benevolence.

The *Diamond*, which is pure crystalized carbon, has been found associated with a variety of New Red Sandstone, called itacolumite, at Golconda, India, and with talcose schist in Brazil. Both these rocks have been subject to high heat, and pressure, and hence perhaps the crystalization of the carbon. In general, the diamond is found in drift; having been removed from its original situation; and we may always presume that every mineral existing in the older rocks will be found also in Drift; because their detritus must contain them.

It has been inferred from the preceding facts that all the varieties of carbon, above described, had their origin in vegetable matter; and that heat and water have produced all the varieties which we now find.

Gems and Metals.—Almost all the precious stones, such as the sapphire, emerald, spinel, chrysoberyl, chrysophrase, topaz, iolite, garnet, tourmaline, etc., are found exclusively in the most crystalline rocks. Quartz in the various forms of rock crystal. chalcedony, carnelian, cacholong, sardonyx, jasper, etc., is found sometimes in the Mesozoic strata, and especially in the trap rocks associated with them.

Some of the metals, as platinum, gold, silver, mercury, copper, bismuth, etc., exist in the rocks in a pure, that is, metallic state; but usually they occur in the state of oxides, sulphurets, and carbonates, and are called ores. It is rare that any other ore is found in sufficient quantity to be an object of exploration on a large scale.

These ores occur in four modes: 1. In regular interstratified layers, or beds. 2. In veins or fissures, crossing the strata, and filled with ore united to foreign substances, forming a gangue or matrix. 3. In irregular masses. 4. Disseminated in small fragments through the rocks.

Iron is the only metal that is found in all the rock series in a workable quantity. Among its ores, only four are wrought for obtaining the metal: viz., the magnetic oxide, the specular or peroxide, the hydrated peroxide, and the protocarbonate.

Manganese occurs in the state of a peroxide and a hydrate, and is confined to the metamorphic rocks; except an unimportant ore called the earthy oxide, which exists in earthy deposits.

The most important ores of copper are the pyritous copper, and the carbonates. These are found in metamorphic rocks, in the Trias and the Tertiary. Immense quantities of native copper are mixed in the Lower Silurian and Huronian rocks about Lakes Superior and Huron.

The only ore of lead, of much importance, is the sulphuret. This is found in the Laurentian series, in both the stratified and unstratified rocks: in the Hudson River group, especially in the Western States; but it exists also in the Mesozoic system.

The deutoxide of tin is the principal ore of that metal. This is most commonly found in the oldest formations of gneiss, granite, and porphyry; also in the porphyries connected with red sandstone. It is found likewise in quantity sufficient to be wrought in Drift.

Of zinc the most abundant ore is the sulphuret, which is commonly associated with the sulphuret of lead, or galena. Other valuable ores are the carbonate, silicate, and the oxide, which occur in Mesozoic rocks.

The most common ore of antimony, the sulphuret, has hitherto been found chiefly in granite, gneiss, and mica schist.

The principal ore of mercury, the sulphuret, occurs chiefly in New Red sandstone: sometimes in a sort of mica schist.

Silver in its three forms of a sulphuret, a sulphuret of silver and antimony, and a chloride, has been found mostly in Hypozoic and Palæozoic slates; sometimes in a member of the New Red Sandstone series, and in one instance in Tertiary strata.

Cobalt, bismuth, arsenic, etc., are usually found associated with silver, or copper; and of course occur in the older rocks. The other metals, which, on account of their small economical value and minute quantity, it is unnecessary to particularize, are also found in the older strata; frequently only disseminated, or in small insulated masses.

Theory of the origin and distribution of Gold.—Gold in its original situation occurs mostly in veins of quartz that traverse the older Palæozoic slates and schists, frequently near their junction with eruptive rocks. Sometimes also it is found in the latter. Talcose schist is the most usual gold-bearing rock; the rocks containing it are metamorphic members of the Silurian, Devonian, and Carboniferous series—especially the first. In European Russia, for example, the Palæozoic rocks, scarcely even yet solidified, contain no auriferous bands; but by following the same strata into the Ural Mountains, where they have been lifted up, and metamorphosed in conjunction with the intrusion of porphyry, greenstone, syenite, and granite, gold is seen to abound.

But at what period was the gold introduced? In the Mesozoic and Tertiary strata none, or scarcely none, is found, and yet those rocks were derived from the more ancient Palæozoic members. Moreover, the loose deposits of gravel and sand, derived in part from the same Palæozoic strata, are the chief repository of gold. Hence the conclusion is irresistible, that the older schists were not impregnated with gold, while the Mesozoic and Tertiary strata were in a course of deposition, but after that time, the protrusion of the eruptive rocks produced the gold. Since then aqueous and atmospheric agencies have worn down the auriferous strata, carrying the metal into the lowest places, and thus bringing it within the reach of man.

Fig. 58.

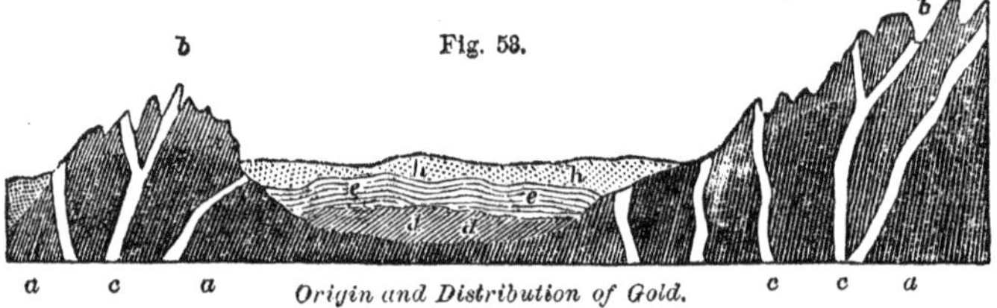

Origin and Distribution of Gold.

a, *a*, *a* represent the older slates, tilted up and metamorphosed by the intrusion of the veins, *c*, *c*, *c*, etc., which impregnated them with gold. Originally these slate mountains rose above *b*, *b*. By their erosion the Secondary deposits *d*, and the Tertiary deposits, *e*, were produced before the injection of the auriferous veins, *c*, *c*, *c*, etc. After their injection, the same erosion went on, reducing the mountains to the line *i*, *i*, and filling the low places with the deposits *h*, *h*, containing gold.

Thus it appears that gold was brought up from the earth's interior, a little time only (geologically speaking) before the appearance of man on the globe. Fishes and lizards, mollusks and crustaceans, did not need it; and therefore it was delayed till a being was about to be created who did.

The most important ranges of gold-bearing rocks are these: the Rocky Mountains, from Russian America through California and Central America, into parts of the Andes in South America; the Appalachian and associated ranges, from Canada to Alabama; the Uralian Mountains in Russia; and in Australia. The Californian ranges are the most productive. In 1854, 481,950 lbs. Troy of gold were mined in the whole world, of which the United States produced 200,000 lbs., Australia and adjacent islands, 150,000 lbs.; and the Russian Empire, 60,000 lbs.

With a few exceptions, working the original veins in which gold occurs has not proved remunerative, sanguine, as most gold-seekers are, that their fortune is made when they have discovered such a vein. But nature has done the work much better than man can, and collected in the lowest places gold in quantities, while in the rocks it is sparsely disseminated. Moreover, it is found that gold veins, unlike those of most other metals, diminish in richness as we descend.

It appears from the facts that have been detailed respecting the situation of the useful minerals that great assistance in searching for them may be derived from a knowledge of rocks and their order of superposition.

No geologist, for instance, would expect to find valuable beds of coal in the oldest crystalline rocks, but in the fossiliferous rocks alone; and even here he would have but feeble expectations in any rock except the coal formation. What a vast amount of unnecessary expense and labor would have been avoided, had men, who have searched for coal, been always acquainted with this principle, and able to distinguish the different rocks! Perpendicular strata of mica and talcose schists would never have been bored into at great expense, in search of coal; nor would black tourmaline have been mistaken for coal, as it has been.

By no mineral substance have men been more deceived than by iron pyrites: which is appropriately denominated *fool's gold*. When in a pure state, its resemblance to gold in color is often so great that it is no wonder those unacquainted with minerals should suppose it to be that metal. Yet the merest tyro in mineralogy can readily distinguish the two substances; since native gold is always malleable, but pyrites never. This latter mineral is also very liable to decomposition, and such changes are thereby wrought upon the rocks containing it as to lead the inexperienced observer to imagine that he has got the clue to a rich depository of mineral treasures; and probably nine out of ten of those numerous excavations that have been made in the rocks of this country, in search of the precious metals, had their origin in pyrites, and their termination in disappointment, if not poverty. This ore also, when decomposing, sometimes produces considerable heat, and causes masses of the rock to separate with an explosion. Hence the origin of the numerous legends that prevail respecting light seen, and sounds heard, in the mountain where the supposed treasure lies, and which so strongly confirm the ignorant in their expectation of finding mineral treasures. Now all this delusion would be dissipated in a moment were the eye of a geologist to rest on such spots, or were the elementary principles of geology more widely diffused in the community.

Another common delusion respects gypsum, which is as often sought among the hypozoic as in the secondary and tertiary rocks; although it is doubtful whether gypsum has ever been found in the former. A few years since, however, a farmer in this country supposed that he had discovered gypsum on his farm, and persuaded his neighbors that such was the case. They bought large quantities of it, and it was ground for agriculture, when accidentally it was discovered that it was only limestone: a fact that might have been determined in a moment at first, by a single drop of acid.

Caution.—It ought not to be inferred from all that has been said, that because a mineral substance has been found in only one rock, it exists in no other. But in many cases we may be certain that such and such formations can not contain such and such minerals. Of these cases, however, the practical geologist can alone judge with much correctness, and hence the importance of an extensive acquaintance with geology in the community. An amount of money much greater than is generally known has been expended in vain for the want of this knowledge.

The chemical changes which rocks have undergone since their deposition, as well as the operation of decomposing agents to which they are now exposed, properly belong to the chemistry of geology. But these points will be deferred to subsequent sections, because they will there be better understood.

SECTION III.

LITHOLOGICAL CHARACTERS OF THE ROCKS.

THE *lithological* character of a rock embraces its mineral composition and structure as well as its external aspect, in distinction from its zoological and botanical characters, which refer to its organic remains.

Rocks are deposited by water in two modes: first, as mere sediment, by its mechanical agency, in connection with gravity; secondly, as chemical precipitates from solution.

The first kind of rocks is called *mechanical* or *sedimentary* rocks; the second kind, *chemical deposits.*

As a general fact, the lower we descend into the rock series we meet with less and less of a mechanical and more and more of a chemical agency in their production.

In the fossiliferous rocks we sometimes find an alternation of mechanical and chemical deposits; but for the most part, these rocks exhibit evidence of both modes of deposit, acting simultaneously.

It is difficult to conceive how any rock can be consolidated without more or less of chemical agency, except perhaps in that imperfect consolidation which takes place in argillaceous mixtures by mere desiccation. Even in the coarsest conglomerate there must be more or less of chemical union between the cement and the pebbles.

The most common mechanical rocks are sandstones, conglomerates and shales.

When sand is cemented, the solid mass is called *sandstone;* rounded pebbles produce a *conglomerate*, or *pudding stone;* and angular fragments, a *breccia.*

Shale is regularly laminated clay, more or less indurated, and splitting into thin layers along the original lamination or planes of deposition.

Clay slate is a metamorphosed clay, differing from shale in having a superinduced tendency to split into thin plates, which may or may not coincide with the lamination of the rock.

Limestones embrace many varieties, as massive limestone, granular limestone, marbles, dolomite, oolite, chalk, and travertine. Most of these varieties are chemical deposits.

The time during which a number of rocks grouped into a formation is in the process of deposition, that is, until some important change takes place in the material or mode of production, is called a *geological period ;* and the point of time when the change occurs is called an *epoch.*

We learn much of the history of the world from the lithological characters of the stratified rocks. They indicate the mode of formation ; whether it was mechanical or chemical; whether the temperature was adapted to the existence of animals and plants ; and in connection with fossils, whether a deposit was marine or fresh water ; whether the deposition was made by a rough current or in placid waters ; and whether the water was deep or shallow.

We shall describe the lithological characters of each of the great systems in succession, beginning with the lowest stratified rock, and proceeding in an ascending order. In this way we shall incidentally read the history of the earth during the different periods. The unstratified rocks, sometimes associated with the sedimentary groups, will be described subsequently. According to our classification, the rocks are divided into the following systems :

I. STRATIFIED ROCKS.

I. Azoic,	II. Palæozoic,
III. Mesozoic,	IV. Cainozoic.

I. Azoic, Hypozoic (*Sedgwick*), Laurentian System (*Logan.*)

The rocks to be described (including granite, porphyry, etc.,) were formerly called *Primary*, because they were supposed to have been produced before the deposition of the fossiliferous strata ; whereas it now appears that several of these rocks have in some instances been formed at a later period. The term *Azoic* signifies *unfossiliferous*, and is the most satisfactory appellation for these crystalline rocks, which are not only the oldest rocks upon the globe, but are also found among the higher groups. The term *Hypozoic*, signifies that the rocks embraced in the system lie beneath those containing fossils. The term *Metamorphic*, which is sometimes applied to them, implies that they have been altered since their original production : but the same is true of some rocks containing fossils. The term Laurentian applies only to the lower part of the Azoic rocks, the upper part forming the Huronian system. It is a local name, derived from the Laurentine Mountains, in Canada, where this system is well developed. Prof. H. D. Rogers calls the Hypozoic or Laurentian rocks, Gneissic, and the Huronian, Azoic.

A subdivision of the Laurentian system has been proposed by Logan, which has not yet been carried out into details ; viz., into those rocks which contain lime, either as carbonate, or as a lime feldspar, and those which are destitute of lime in any form. There is no certain order of superposition among the different groups of this formation ; but we shall describe them in the order in which geologists have supposed them most commonly to occur.

1. *Gneiss.*—The essential ingredients in this rock are quartz, feldspar, and mica. The feldspars are both lime-feldspars, (Labradorite), and soda-feldspars, (Albite and Anorthite). Hornblende is sometimes present. These ingredients are arranged more or less in folia, and the rock is stratified. Where it passes into granite, however, (which is composed of the same ingredients), the stratification, as well as the foliation become exceedingly obscure; and it is impossible to draw a definite line between the two rocks. Gneiss, as well as mica schist, are remarkable in some places for tortuosities and irregularities exhibited by the strata and folia; while in other places these same rocks are equally distinguished for the regularity and evenness of the stratification, by which they are rendered excellent materials for economical purposes.

Gneiss sometimes contains crystals of feldspar, which give it a spotted appearance, and this is called *porphyritic gneiss.*

2. *Mica schist.*—This consists of successive layers of mica and quartz, the former predominating. It is not unusual to find small crystals of feldspar, disseminated through it. Garnets and staurotide are often so abundant in it, over extensive tracts, as properly to be regarded as constituents; hence the varieties, *garnetiferous*, and *staurotidiferous* mica schist.

3- *Saccharoid Azoic Limestone.*—The limestone connected with azoic rocks is generally white and highly crystalline, resembling loaf sugar so much as to be called *saccharoid.* In some situations it is dark colored, or it may receive bright colors from minerals disseminated through it. It is often highly magnesian. Many authors prefer the term *crystalline* to saccharoid; but many other limestones are crystalline.

4. *Talcose schist.*—This rock consists of successive layers of talc and quartz. Mica, calcite, feldspar, and hornblende, are frequently present. Often talc is replaced by talcite, or some mineral resembling talc. Talc is a hydrous silicate of magnesia, but the substituted minerals are silicates of alumina. Hence what is often called talcose schist is only an altered variety of clay slate.

Varieties.—In *chlorite schist* talc is replaced by chlorite, a hydrous silicate of alumina and magnesia. It is almost pulverulent and compact, of a green color, and the chlorite more abundant than the quartz. *Steatite* is often nothing but schistose talc, which is adherent enough to be wrought, and at other times it is somewhat granular and slightly indurated. This is the valuable stone so extensively used for furnaces, fire-places, aqueducts, etc., under the name of *soapstone* or *freestone*.

5. *Hornblende schist.*—Hornblende predominates in this rock, but its varieties contain feldspar, quartz, and mica. When it is pure hornblende, its stratification is often indistinct, and it passes, by taking feldspar into its composition, into a rock resembling greenstone. It occurs in every part of the Azoic system; but its most common associations are argillaceous slate, mica schist and gneiss; into which it passes by insensible gradations.

6. *Quartz rock.*—This rock is essentially composed of quartz, either granular or arenaceous. The varieties result from the intermixture of mica, feldspar, talc, hornblende, or clay slate. In these compound varieties the stratification is remarkably regular; but in pure granular quartz it is often difficult to discover the planes of stratification. It is interstratified with every one of the azoic rocks.

The arenaceous varieties of this rock form good *firestones*; that is stones capable of sustaining powerful heat. Some varieties of mica schist are still better. Gneiss of an arenaceous composition is also employed; as are several varieties of sandstone of different ages. The firestone of the English green sand is a fine siliceous sand cemented by limestone.

7. *Clay slate* or *argillaceous slate.*—This is a fine-grained fissile, highly indurated rock, splitting into plates by cleavage, altogether independent of the original laminæ. This superinduced structure may often be distinguished from the strata, by means of parallel bands of different colors and textures traversing the rock. It is generally a dull blue, grey, green, or black color, sometimes brick red, sometimes striped, sometimes mottled. This rock is best developed in the Cambrian series.

Novaculite or honestone is a compact variety of clay slate, which is highly prized for hones. It is less divided by cleavage planes, and has a very soft and smooth feel.

8. *Serpentine.* The description of this rock as a mineral embraces all that is needful to say of it in this place. We need, therefore, only refer to page 51.

II. PALÆOZOIC SYSTEM.

The Palæozoic rocks, or those in which the oldest forms of life are found, embrace deposits of vast thickness. They are

1. *The Cambrian*, 2. *The Silurian*, 3. *The Devonian*, or *Old Red Sandstone*, 4. *The Carboniferous* or *Coal Formation*, and 5. *The Permian System*.

1. *The Cambrian or Huronian Series.*

There has been much discussion among English geologists as to the upper limit of the Cambrian system. The most satisfactory classification makes of it a vast thickness of sandstones, schists, and slates underlying the Lingula flags in England, and the Potsdam sandstone in this country. Scarcely any organic remains are found in it in Europe, and none as yet in this country. Perhaps half of this group in Great Britain is clay slate. Its beds are there 26,000 feet thick. The term Cambrian is derived from the ancient name of Wales.

These rocks cover extensive areas in Great Britain, particularly in Wales, from which the well-known Welsh roofing slate is obtained; also in Ireland, Bohemia, and Scandinavia. They have been recognized in this country but recently. Logan has described a series of rocks about Lake Huron, referable to this group, which he has called the *Huronian Group*. The lowest member is a bluish slate, reposing unconformably upon the Laurentian rocks, succeeded upwards by various colored sandstones, slates, and an occasional band of limestone; the whole being 12,000 feet thick. Professor Rogers has described some rocks of that age in Pennsylvania.

2. *The Silurian Series.*

This system rests unconformably upon the Huronian series at Lake Huron, and elsewhere in this country upon the Laurentian group, and in Europe upon the Cambrian series. It is divided into the *Upper* and *Lower* Silurian, distinguished from each other by want of conformity and peculiar organic remains.

Fig. 54.

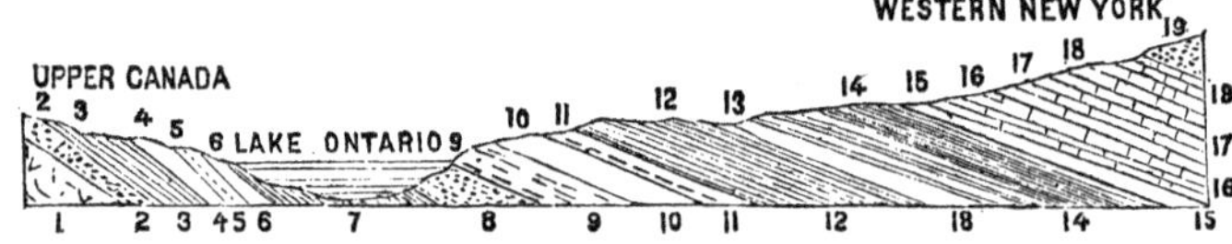

1. Laurentian system.

Lower Silurian.

2. Potsdam sandstone,
3. Calciferous sandrock,
4. Chazy limestone,
5. Trenton limestone,
6. Utica slate,
7. Lower Hudson river group,

Upper Silurian.

8. Oneida conglomerate,
9. Medina sandstone,
10. Clinton group,
11. Niagara group,
12. Onondaga salt group,
13. Helderberg series,

Devonian.

14 Hamilton group,
15. Tully limestone,
16. Portage group,
17. Chemung group,
18. Catskill red sandstone,

Coal Formation.

19. Conglomerate.

In Fig. 54 the position of the Silurian and Devonian rocks is shown as they occur in the Western part of New York. The Huronian group is wanting, as well as several of the other subdivisions, upon this section. It is very rare to find all the members of the series at one locality. For example, the Onondaga salt group is found only in Western New York and in British America. Elsewhere the Lower Helderberg limestone may succeed directly to the Niagara group.

1. *Lower Silurian.*—The *Potsdam sandstone* is a purely silicious sandstone. The *Calciferous sandrock* is a calcareous sandstone or an impure limestone; sometimes magnesian. The *Chazy and Trenton limestones* are black fossiliferous limestones. The *Utica slate* is a black shaly limestone. The *Lower Hudson river group* is mostly clay slate; but in the Western States its place is occupied by limestone; the upper part of the so called *cliff limestone.* Sometimes there is an unconformability between the Lower and Upper Silurian, as in England, and at the mouth of the St. Lawrence river in this country.

2. *Upper Silurian.*—The *Oneida conglomerate* is usually purely silicious, but passes insensibly into calcareous sandstone or dolomitic limestone in some districts. The *Medina sandstone* is a red sandstone, or shale. The *Upper Hudson river group* is partly clay slate, and partly talcose schist, with occasional beds of limestone. It has as yet been found only in Western New England or Eastern New York. The *Clinton group* is an alternation of shales, limestones, and iron ores or iron sandstones. The *Anticosti group* is an assemblage of argillaceous limestones occurring upon the island Anticosti in the Gulf of the St. Lawrence. It is probably equivalent to the formations between the Lower Hudson river group and the Clinton group.

The *Niagara group* is an alternation of limestones and shales; and sometimes the shales are wanting. The *Onondaga salt group* is an alternation of limestones and shales, the limestones predominating, from which issue salt springs. The *Lower Helderberg limestone* is a highly fossiliferous dark colored limestone, and is very persistent, while the previous member is most usually wanting.

The European members of the Silurian System are likewise composed of sandstones, limestones, and shales. The following figure represents the general order of these groups in Europe, with their names.

Fig. 55.

A. Hypozoic rocks.

Lower Silurian.

1. Lingula and Llandeilo flags,
2. Caradoc sandstone,

Upper Silurian.

3. Wenlock shale,
4. Wenlock limestone,
5. Lower Ludlow shale,
6. Aymestry limestone,
7. Upper Ludlow shale.

The Silurian rocks occupy large areas in Belgium, Germany, Scandinavia, and Russia, as well as in North and South America.

There has been much discussion among English geologists as to the limits of the Cambrian and Silurian series. Murchison regards them both as Silurian. Sedgwick divides the Cambrian into Lower, Middle, and Upper, and his Upper Cambrian is the same as what we have called Lower Silurian. The government surveyors of England have compromised these views, and describe these series as Cambrian, (Lower and Middle of Sedgwick), Lower or Cambro-Silurian, and Upper Silurian.

3. *The Devonian or Old Red Sandstone Series.*

The position of the different formations of the Devonian series has been already shown. The *Oriskany sandstone*, the *Cock-tail grit*, and the *Schoharie grit* are mostly silicious. They are succeeded by a persistent fossiliferous limestone, the *Upper Helderberg*. The *Marcellus shales* are composed of clay slate; the *Hamilton group* of thick bedded grits, used extensively for flagging stones, and slates; and the *Genesee slates* are also argillaceous. The *Portage and Chemung groups* are mostly grits and shales. The *Catskill Red Sandstone*, the upper member in this country, constitutes the Catskill Mountains in New York, where they are 3,000 feet thick. The whole thickness of the system in this country in 11,750 feet.

In Great Britain this system has long been known as the Old Red Sandstone, and was denominated Devonian by Sedgwick and Murchison, to designate the Old Red Sandstone as it was developed in Devonshire. In Scotland this formation is not less than 10,000 feet thick. In England it is divided into three groups: 1. *Tilestone*, or fissile beds used sometimes for tiles. 2. *Cornstone* and *Marl*, or argillaceous marly beds, alternating with sandstone, and sometimes with impure limestone. 3. *Old Red Conglomerate*, the uppermost division.

This formation is widely developed on the continent of Europe, as in Belgium and Westphalia, France and Spain. In Russia it covers more surface than the whole of Great Britain, not less than 150,000 square miles. In the United States it occupies extensive tracts.

4. *Carboniferous series, or Coal formation.*

This system derives its name from the great amount of coal found in it; it being the principal deposit from which coal is derived for economical purposes. In this country there are four general divisions: 1. *A conglomerate*, 2,660 feet thick in Pennsylvania. 2. *Carboniferous Limestone*, or Red Shales and Limestone, in Pennsylvania, 3,000 feet thick. This Limestone is gray and compact, traversed by veins of calcite, and is abundantly fossiliferous. When it is mostly made up of the remains of encrinites, it is called *Encrinal Limestone*. In England, where it forms the lowest member of the series, it is called *Mountain Limestone*. (See Fig. 52, on page 54). 3. *A conglomerate*, less than half the thickness of the lowest division. This is the *millstone grit* of Europe. 4. The true *coal measures*. These consist of irregularly interstratified beds of sandstone, shale, and coal. Frequently these are deposited in basin shaped cavities, but not always.

The following section of Carboniferous rocks, in Ohio, will illustrate the alternations of coal, shale, etc.

1. Sandstone,	
2. COAL,	6 feet.
3. Fine-grained slaty sandstone,	50 feet.
4. Silicious iron ore,	1.5 feet.
5. Argillaceous sandstone,	75 feet.
6. COAL,	3 feet.
7. Shale, containing vegetable impressions,	4 feet.
8. Sandstone,	80 feet.
9. Iron ore,	1 foot.
10. Argillaceous sandstone,	80 feet.

These rocks abound in faults produced by igneous agency; whereby the continuity of the beds of coal is interrupted, and the difficulty of exploring for coal increased in some respects; but in other respects facilitated; so that upon the whole, these faults are decidedly beneficial.

The thickness of the coal formation in Pennsylvania is about 7,000 feet; in Nova Scotia, 13,000 feet, in which there are 76 beds and seams of coal; in Great Britain 12,000 feet.

Coal has been found in nearly all parts of the world. Great Britain has 12,000 square miles of coal-fields, the continent of Europe about 10,000 square miles; the area of coal-fields in Nova Scotia and New Brunswick is 7,000 square miles, and in the United States there are more than 200,000 square miles underlaid by beds of coal.

Of particular coal fields in the United States the largest is the Appalachian, extending from Ohio and Northern Pennsylvania to Alabama, embracing 80,000 square miles. Others are the Indiana, Illinois, and Kentucky coal-field, covering an area of 50,000 square miles; the Iowa and Missouri coal-field, 60,000 square miles; the Michigan coal-field, 15,000 square miles, and the New England coal-field, 600 square miles. Still further west and south, the carboniferous rocks with coal are found, as in the southwest part of Nebraska, the east part of Kansas, and the north part of Texas. Still further west, along the eastern base of the Rocky Mountains, and extending north into the British possessions, are extensive deposits of lignite, either of Cretaceous or Tertiary age. They furnish a coal of much value, but not as good as that which is older. Those deposits have not yet been traced out; but will undoubtedly be found of great extent. *See H. Engleman's Report to Captain Simpson, appended to the latter's Report on Wagon Routes, etc., in Utah Territory, page* 49, 1858.

In McClintock's late Narrative in search of the remains of Sir John Franklin, (1860), is a Geological Map, which represents carboniferous sandstones with beds of coal extending from Lat. 72 to 77°, and Long. 92 to 125°. This is represented on the small Geological Map of North America, attached to Part V. of this work. Truly we do not yet know, by a great deal, the extent of the coal fields of this country.

Professor H. D. Rogers states approximately the amount of coal in the most important coal-fields in the world, as follows:

Belgium	36,000,000,000 tons.
France,	59,000,000,000 tons.
British Isles,	190,000,000,000 tons.
Pennsylvania,	316,400,000,000 tons.
Appalachian coal-field,	1.387,500,000,000 tons.
Indiana, Illinois, and Kentucky,	1,277,500,000,000 tons.
Iowa, Missouri, and Arkansas,	739,000,000,000 tons.
Total amount in North America,	4,000,000,000,000 tons.

In some parts of the world they are beginning to calculate how long their supplies of coal will last. In England not less than 6,000,000 tons of coal are yearly raised from the mines of Northumberland and Durham; at which rate they will be exhausted in about 250 years. In South Wales, however, is a coal field of 1,200 square miles, with 23 beds, whose total thickness is 95 feet; and this will supply coal for 2,000 years more. In North America there is twenty-one times as much coal as in Great Britain. Estimating our annual consumption of coal at twelve millions of tons, there is coal enough in North America to last 333,333 years.*

5. *The Permian Series.*

In Germany and England the Permian Series presents two very distinct groups of rocks; the lowest is made up of sandstones of various colors, with slates containing copper, and the highest is composed of magnesian limestones of various qualities.

The Permian system consists of numerous strata of great extent in Russia, (700 miles long and 400 broad), in the ancient kingdom of Permia, made up of common and magnesian limestones, with gypsum and rock salt conglomerates, red and green gritstones, shales, and copper ore, lying in a trough of the carboniferous strata, and below the Triassic system.

Until quite recently it was supposed that there were no Permian strata in North America. Quite extensive deposits, however, have been found in Kansas, Illinois, and Nebraska. Lithologically they are limestones, shales, and layers of clay.

Professor Emmons claims to have discovered Permian strata in North Carolina, and supposes that a large part of the sandstones along the Atlantic slope of the Appalachians are of the same age. It is probable that some of the lowest members of this deposit are of this age, but it needs positive proof.

III. MESOZOIC SYSTEM.

Under Mesozoic rocks are included all those from the top of the Permian to the base of the Tertiary. These are, 1. The Trias

* Great Britain mined in 1854, 64.661,401 tons of coal, and produces now about 67,000,000 tons annually. The United States produced in 1857, 10,500,000 tons. In the whole world it is estimated that about 100,000,000 tons of coal are annually consumed.

or New Red Sandstone, 2, the Jurassic series, and 3, the Cretaceous or chalk series.

1. *Triassic Series or New Red Sandstone.*

In Continental Europe the Triassic System is divided into three distinct groups, and hence its name. The lowest is the *Bunter Sandstein*, or *Gres Bigarre;* both terms meaning variegated sandstone. The colors are white, red, blue, and green. The composition of the rock is chiefly silicious and argillaceous, with occasional beds of gypsum, and rock salt. The second group is the *Muschelkalk*, a gray compact limestone, occasionally dolomitic. This member is wanting both in Great Britain and in this country; and hence there is great difficulty in distinguishing between the upper and lower divisions. The highest division is the *Variegated Marl*, or the *Keuper.* It consists of indurated clays of different colors, chiefly red, alternating with gray sandstone and yellowish magnesian limestone. Beds of gypsum and rock salt are common.

In this country this system is probably represented in a part of the Connecticut River sandstone.

In the Western parts of the United States, in the vicinity of the Rocky Mountains, there are deposits referable, it is said, to this series.

2. *The Jurassic Series.*

The Jurassic series, so called from its occurrence among the Jura mountains in Switzerland, embraces the Lias, the Oolite, and the Wealden formations of England.

Lias.—Lias is a rock usually of a bluish color like common clay; and it is indeed highly argillaceous, but at the same time generally calcareous. Bands of true argillaceous limestone do, indeed, occur in it, as well as of calcareous sand. It is widely diffused; is very marked in its characters, and contains peculiar and very interesting organic remains.

Oolite.—In many of the rocks of this series small calcareous globules are imbedded, which resemble the roe of a fish, and hence such a rock is called *roestone* or *Oolite.* But this structure extends through only a small part of this formation, and it occurs also in other rocks.

The Oolite series consist of interstratified layers of clay, sandstone, marl, and limestone. The Oolite proper, is divided into three groups, called the upper, middle, and lower, separated by clay or marl deposits.

The upper part of the Connecticut River sandstone, is probably of Jurassic age, as well as the long belt of Mesozoic rocks, or a part of them, along the Atlantic slope of the Appalachians. The fine coal-field near Richmond, Virginia, is also Jurassic. The Oolite in England, contains a large number of reptilian remains: and in this country the Connecticut River sandstone contains the remarkable fossil footmarks or *ichnites*.

Wealden Formation.—This formation embraces, 1. *The Weald Clay;* 2. *Hastings Sand;* 3. *Purbeck Strata.* They were first described in the South-east of England, chiefly in the *wealds* or woods of Sussex and Kent, and are composed of beds of limestone, conglomerate, sandstone, and clay, which abound in the remains of fresh water and terrestrial animals, and appear to have been deposited in an estuary that once occupied that part of England. Similar beds occur in Scotland, and in a few places on the European Continent. Some place the Wealden under the cretaceous formation, as below.

3. *Cretaceous Series.*

In Europe and Asia, this series is usually characterized by the presence of chalk in the upper part, and sands and sandstones in the lower. In North America the chalk is wanting.

Fig. 56 represents the succession of strata at Lulworth Cove in England, and their connection with the Wealden formation.

Fig. 56.

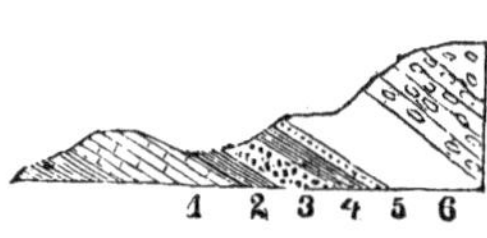

Section of the English Cretaceous.

1. Hastings sand, } Wealden.
2. Purbeck strata, }
3. Lower greensand, } Greensand,
4. Gault, }
5. Upper greensand, }
6. Marly and lower, } Chalk.
7. Upper, }

Section of the Cretaceous in Nebraska.

1. Yellowish sandstone.
2. Dark gray clay.
3. Lead gray marl.
4. Bluish plastic clay.
5. Arenaceous clays.

Section of the Cretaceous in New Jersey.

1. Fine clay and Potters clay.
2. Dark-colored clay.
3. " with greensand. } Nos. 4 and 5 of Nebraska section.
4. Greensand, 1st. or lower bed. }
5. Ferruginous sand. }
6. Greensand, 2d. bed. }
7. Quartzose sand.
8. Greensand, 3d. or upper bed.

The principal cretaceous formations in the United States are in New Jersey and in the Southwestern and Western Territories. The latter deposits cover many thousand square miles. We have given above sections of these two deposits, parts of each containing the same fossils. The exact correspondence of the European and American strata is not yet ascertained.

Chalk is a pulverulent carbonate of lime, and its varieties have resulted from the impurities that were deposited with it. The upper beds are remarkable for the great quantity of flints dispersed through them, generally in parallel position. The famous Dover cliffs in England are composed of chalk.

Greensand is a mixture of arenaceous matter, with a peculiar green substance greatly resembling chlorite, or green earth. It has been used extensively as a fertilizer, as it contains large quantities of silicate of iron and potassa.

Gault or Galt, is a provincial name for blue clay or marl, forming an interstratified bed in the greensand of England.

IV. CAINOZOIC SYSTEM.

These include all the more recent formations, viz.: 1. the Tertiary, and 2. the Alluvium.

1. *Tertiary Series.*

The Tertiary rocks have been divided into three distinct groups of marine strata, distinguished by important peculiarities in their organic remains, and separated from one another by strata which contain fresh water and terrestrial remains.

Lyell has given names to these groups which are generally adopted: 1. the *Eocene*, signifying *the dawn*, or commencement of the existing types of organic life, and containing about four per cent. of shells identical with living species. This is divided into three parts: the *Lower* Eocene, embracing the Nummulitic formation of the Alps and the London clay; the *Middle* Eocene, embracing portions of the Paris basin, etc., and the *Upper* Eocene, embracing the upper marine beds of the Paris basin, etc. 2. The *Miocene* (*less recent*), containing about twenty-five per cent. of shells identical with living species. 3. The *Pliocene* (*more recent*) or *Newer Tertiary*, containing shells, of which about two-thirds are identical with living species.

In Europe and Asia, the rocks of this period are found principally in basins, apparently deposited in lakes and estuaries of limited extent. London, Paris, and Vienna, are each situated upon Tertiary basins. Fig. 57 represents the London basin, lying in a trough of the Chalk series.

In North America the Tertiary deposits are found chiefly upon the Atlantic seaboard, and upon the Gulf of Mexico, running northerly into the Territories. There is a remarkable deposit of this age in Nebraska, upon the *Mauvais Terres or Bad Lands* upon White river. Local names have been given to the different deposits, whose relation to the European divisions is as follows: the *Claiborne Period* corresponds to the Lower Eocene; the *Vicksburg Period* to the Upper Eocene; and the *Yorktown Period* to the Miocene and Pliocene.

Fig. 57.

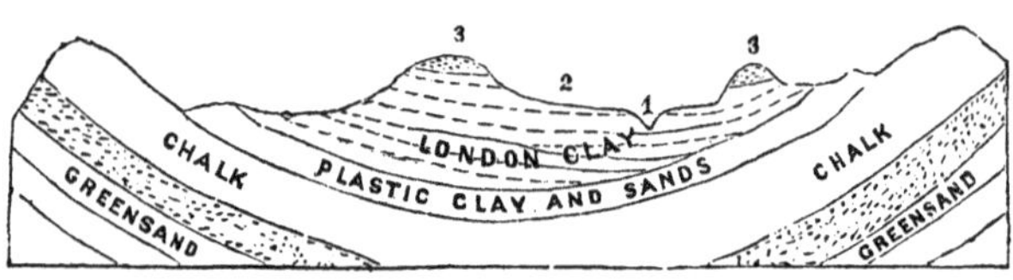

1. River Thames. 2. London. 3. Marine sands.

The Tertiary rocks are in general distinctly stratified, and the strata are usually horizontal. But in some cases, as at the Isle of Wight and upon Martha's Vineyard, they are inclined at a large angle.

The Tertiary rocks are mostly of mechanical origin: nevertheless, several beds are the result of chemical precipitation; as gypsum, limestone, and rock salt.

The varieties of rocks composing the Tertiary strata are concretionary, tufaceous, argillaceous, and silicious; or limestone, marl, plastic clay, silicious and calcareous sands, green sands, gypsum, lignite, rock salt, and buhrstone.

2. *Alluvium.*

Much of this deposit consists of materials which have resulted from the comminuting, rounding, and sorting action of water, and hence the name from *alluvio*, an inundation, or *alluo*, to wash.

Alluvium is divided lithologically into two sections; *Drift*, and *Modified Drift*, or Alluvium proper. Chronologically, it is divided into four periods: 1. The *Drift Period;* 2. The *Beach Period;* 3. The *Terrace Period;* and 4. The *Historic Period*, or the present system of life and action.

1. *Drift.*

Owing to the diversity of opinion that has prevailed respecting the origin of this deposit, it has received various names, such as *diluvium, bowlder formation, erratic block group*, etc. The term diluvium is objectionable, because it implies that its origin was the deluge, or some deluge, the very point to be proved. Drift is a better term, because short and free from hypothetical allusions.

A bowlder is a loose block of stone larger than a pebble, and either rounded or angular.

Drift is a mixture of abraded materials—bowlders, gravel, and sand, blended confusedly together, and driven mechanically forward by some force behind. Yet in some places there are marks of stratification or lamination, as if water had been concerned in the work of deposition.

Drift is distinguished from the Tertiary by lying always above it; and by the peculiarities of the organic remains; and from modified drift, by always lying beneath it, and being less comminuted.

Modified drift is not only stratified and laminated, but sorted also; the size of the fragments thus selected depending upon the force of the current that did the work. Drift is usually not sorted; sometimes it is so in particular places.

It is difficult to draw the precise line between drift and unmodified drift, because they blend into each other. Modified drift is always stratified, while the drift is generally a heterogeneous mixture without arrangement, yet large bowlders are sometimes imbedded in sand, as if there were sometimes a combination of forces in accumulating the same pile.

The bowlders so characteristic of drift are sometimes seen insulated upon other rocks, and so equally poised that a small force will make them oscillate, though weighing many tons. They are called *Rocking Stones.*

Fig. 59 represents a rocking stone in Fall River, Massachusetts, poised upon

Fig. 58.

Rocking Stone, Barre, Massachusetts.

granite, and weighing 160 tons. Fig. 58 shows another, a double one, in Barre, Massachusetts.

Many of the most valuable of the precious stones and metals are found in drift; such as the diamond, the sapphire, the topaz,

Fig. 59.

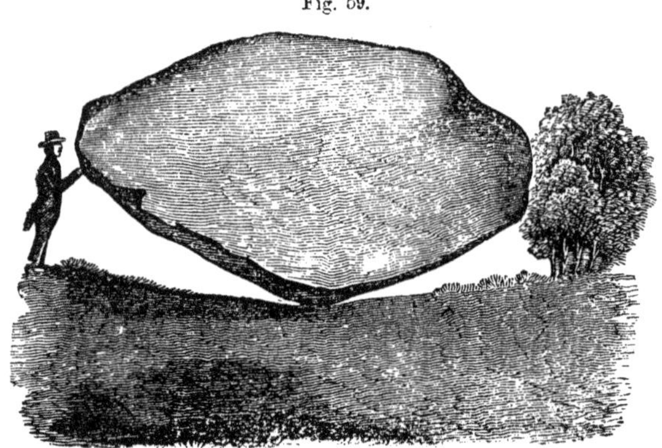

Rocking Stone, Fall River, Massachusetts.

the ruby, and the zircon; as well as platinum, gold, and tin. Platinum, gold and the diamond, are explored almost exclusively in this formation.

2. *Alluvium Proper, or Modified Drift.*

Considered lithologically, Alluvium embraces the following deposits:

Soil.	Silicious Marl, or deposits of the skeletons of Infusoria.
Clay.	Bitumen.
Sand.	Sulphate of Lime.
Peat.	Hydrate of Iron.
Marl.	Bog Manganese.
Calcareous Tufa, or Travertin.	Chloride of Sodium (Sea Salt).
Coral Reef.	Sandstones, Conglomerates, and Breccias.
Silicious Sinter.	

Soil is disintegrated and decomposed rock, with such a mixture of vegetable and animal matter that plants will grow in it.

Clay differs from sand in two respects: 1. The materials have been reduced to a much finer state than sand; 2. It contains much more alumina.

The vast accumulations of *sand*, the result of alluvial agency, occur not merely in the bed of the ocean and in lakes, but also upon the dry land, where they are called *dunes* or *downs*. These are composed almost entirely of silica; and being destitute of organic matter, can not sustain vegetation.

Peat, when perfectly formed, is destitute of a fibrous structure, and when wet is a fine black mud; and when dry, a powder. It consists chiefly of the decomposed organic matter called *geine* or *humic acid*, with crenic and apocrenic acids, phosphates, etc., part of which are soluble, and a part insoluble, in water. These deposits of peat are sometimes 30 or 40 feet thick; but they are not formed in tropical climates on account of the too rapid decomposition of the organic matter.

Alluvial marl is usually a fine powder, consisting of carbonate of lime, clay, and soluble and insoluble geine; and is found usually beneath peat in limestone countries; sometimes at the bottom of ponds. It is produced partly by the decay of shells of molluscous animals, and partly by the deposition of carbonate of lime from solution. It contains numerous small fresh water shells, and has hence received the name of *shell marl*.

Other kinds of Marl.—Several other substances, that contain no carbonate of lime, have often been denominated marl, by agriculturists, and not without reason; for they have produced effects analogous to calcareous marl. But it seems very desirable that terms should not be applied too loosely, and we propose the following designations for these substances:

Calcareous Marl: that which contains carbonate of lime in any quantity.

Silicious Marl: that in which silica predominates, and no calcareous matter is present.

Aluminous Marl: that in which clay predominates, and no calcareous matter is present.

Greensand Marl: that which contains greensand. This is the substance that has been of late employed with signal success as a fertilizer of land in New Jersey, Virginia, Delaware, etc.

Calcareous tufa or *travertin* is a deposit of carbonate of lime, made by springs containing that substance in solution. It forms a solid limestone, sometimes even crystalline, and of considerable extent, so as to be used for architectual purposes. Thermal waters produce it most abundantly, as in central France, Hungary, Tuscany, and Campagna di Roma; but it is also deposited by springs of the ordinary temperature, as at Saratoga and in the Apennines. Travertin is also precipitated by rivers, as in Tuscany.

Very similar to travertin are the concretionary calcareous deposites formed in caves: those depending from the roof, like icicles, are called *stalactites*, and those on the floor, *stalagmites*. Often a stalactite and a stalagmite unite and form a column.

Coral reefs are extensive deposits of carbonate of lime, etc., formed by myriads of coral animals in shallow water, in tropical seas. They form the habitations of these animals, and of course are organic in their structure.

Silicious sinter, or tufa, is a deposit of silica, made by water of thermal springs, which sometimes hold that earth in solution. Successive layers of sinter and clay frequently occur, and these are sometimes broken up and re-cemented so as to form a breccia.

Silicious marl, or the *fossil shields of microscopic plants and animals.* Beneath the beds of peat and mud in the primary regions of this country, a deposit often occurs from a few inches to several feet thick, which almost exactly resembles the calcareous marl that is found in the same situation. When pure, it is white and nearly as light as the carbonate of magnesia; but it is usually more or less mixed with clay. It is found by analysis to be nearly pure silica; and it turns out to be almost entirely composed of silicious shields, or skeletons, of those microscopic animals called *infusoria*, or of plants which have lived and died in countless numbers in the ponds at the bottom of which this substance has been deposited.

Some springs produce large quantities of *bitumen* in the form of naptha and asphaltum.

Although *sulphate of lime* very generally exists in the waters of springs, yet it is rarely deposited. One or two examples only are mentioned, where a deposit of this salt has been made; as at the baths of San Philippo, in France.

The Hydrated peroxide of iron or bog ore is a common and abundant deposit from waters that are capable of holding it in solution; and it appears, also, that this ore is often made of the shields of infusoria, which are often ferruginous.

Chloride of Sodium, or rock salt, is sometimes deposited in inland seas or salt water lagoons. The salt is precipitated only when the water is completely saturated; or in small lagoons, the water might be evaporated, leaving the salt behind.

The waters of Lake Elton, in Asiatic Russia, and of other lakes adjoining the Caspian Sea, have deposited thick beds of rock salt at their bottom. The same is true of Lake Indersk, on the steppes of Siberia; of Lake Bakr Amal, in Ethiopia; in Patagonia; and in a lagoon adjoining Lake Oroomiah, in Persia. The bottom is covered by an incrustation of salt from three to five inches thick.

Alluvial sandstone, conglomerate, and breccia, are formed by the cementation of sand, rounded pebbles, or angular fragments, by iron, or carbonate of lime, which is infiltrated through the mass in a state of solution. They are not very common, nor on a very extended scale.

Thickness of Strata.—If all the stratified rocks have been deposited from water, as we suppose, the layers must have been originally nearly horizontal. Rocks are not usually inclined more than ten degrees by original deposition, but there may be cases where the angle would be as large as 25° or 30° over limited areas. Hence if we get the perpendicular thickness of a series of strata we ascertain the character of the crust of the globe to that depth.

If we measure the breadth of a series of upturned strata, on a line at right angles to their strike, and ascertain their dip, we have given the hypothenuse and angles of a right-angled triangle to find the perpendicular, which is the thickness of the strata. If the strata are perpendicular, a horizontal line across their edges gives their thickness.

In calculating the thickness of rocks in any given district, we must be careful not to measure the same strata more than once. For when strata are folded over, the upper beds will be folded beneath the lower and dip at the same angle. Especially if the crest of the fold has been denuded is there danger of mistake. Without regard to this principle, most enormous thicknesses might be calculated.

By measurements and calculations of this sort, it has been ascertained that the total thickness of the fossiliferous strata in Europe is not less than 15 miles. In this country, as has been already shown, the total thickness of the fossiliferous strata is nearly ten miles.

We see from these statements how groundless is the opinion, that geologists are able to ascertain the structure of the earth only to the depth that excavations have been made, which is less than a mile; especially when we recollect that the unstratified rocks are uniformly found beneath the stratified; and since their igneous or aqueo-igneous origin is now generally admitted, it can hardly be doubted that they came from very great depths; so that probably the essential composition of the globe is known almost to its center.

UNSTRATIFIED OR IGNEOUS ROCKS.

The differences among the unstratified rocks result from two causes. 1. A difference in chemical composition. 2. The diversity of circumstances under which they were produced.

All the varieties of these rocks pass into one another by insensible gradations, even in the same mountain mass; giving rise to endless varieties, which can not be described minutely in a treatise like the present.

The two predominant and characteristic minerals in the unstratified rocks are feldspar and pyroxene, or hornblende.

Pyroxene and hornblende have very nearly the same chemical constitution, and seem capable of being converted into each other.

It is now agreed among geologists, that the unstratified rocks have resulted from the action of heat, either as dry heat or from aqueo-igneous fusion. Hence the best writers denominate them Igneous Rocks, and divide them, as in our classification in Section I., into three groups: 1. Granitic; 2. Trappean; 3. Volcanic.

The following table shows to what divisions the different igneous rocks belong.

GRANITIC ROCKS.

Quartz, feldspar, and mica, or hornblende, etc.		*Quartz and feldspar.*
Granite.	Tabular granite.	Pegmatite.
Syenite.	Concretionary granite.	Eurite.
Protogine.		

TRAPPEAN ROCKS.

Silicco—feldspathic.	*Feldspar and hornblende, etc.*
Porphyry.	Greenstone or Diorite.
Felstone.	Melaphyre.
Pitchstone.	Andesite.
Clinkstone.	Hypersthene rock.
Hornstone.	Hornblende rock, etc.
Cornean.	

VOLCANIC ROCKS.

Essentially feldspathic.	*Feldspar and augite.*
Trachyte.	Basalt or Dolerite.
Trachytic porphyry.	Amygdaloid.
Domite.	Peperino.
Pearlstone.	
Obsidian.	
Pumice.	
Tuff.	

The melted matter that is ejected from a volcano, or remains within it, is called *lava*. Hence it is not improper to apply the term to any rock that is proved to have been in a melted state. But it is usual to confine it to the more modern unstratified rocks, such as have been ejected from a crater.

Lava cooled rapidly, and not under pressure, forms glass, or scoria; but cooled slowly, and under pressure, it becomes sometimes crystalline. Now the older unstratified rocks, such as granite, syenite, porphyry, and greenstone, are more or less crystalline; whereas basalt, trachyte, and other igneous rocks are compact or cellular. Hence it is inferred, that the first were cooled under a vast pressure of the ocean and its subjacent beds; and hence they are called *plutonic rocks*; whereas the latter are denominated *volcanic rocks*.

There is strong reason to believe that in some instances, as in Saxony, and in Sutherlandshire, and Arran in Scotland, granite has been protruded through the strata after it became solid. Solid basalt was protruded in a similar manner, according to Von Buch, in the year 1820, in the island of Banda, in great quantities.

The most important of the unstratified rocks will now be described in the order of our classification, as nearly chronologically (beginning with the oldest), as the present state of knowledge will admit.

I. GRANITIC ROCKS.

The essential ingredients of granite are quartz, feldspar, and mica. Its prevailing colors are white and flesh-colored. In some cases the materials are very coarse, the crystalline fragments being a foot or more in diameter. In other cases, they are so fine as to be scarcely visible to the naked eye; and between these extremes there exists an almost infinite variety. The fine-grained varieties are best for economical uses; but the coarser varieties abound most in interesting simple minerals.

Varieties. Graphic granite or Pegmatite, is composed of quartz and feldspar, in which the former has an arrangement which makes the surface of the rock exhibit the appearance of letters, as in Fig. 60.

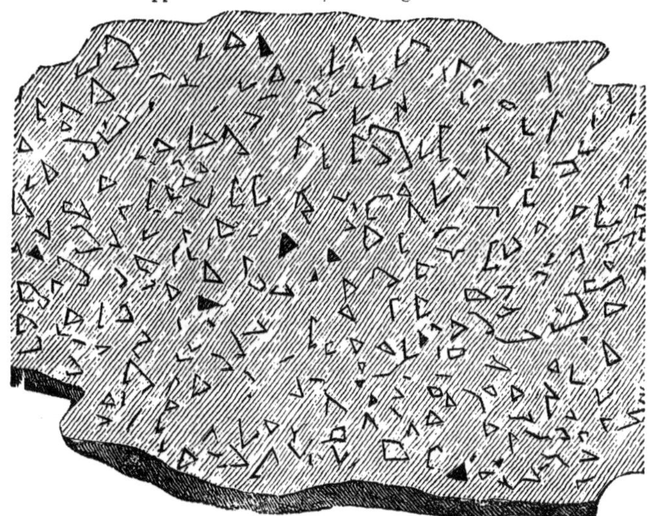

Graphic Granite.

When granite contains distinct crystals of feldspar, as in Fig. 61, it is called *porphyritic granite.* When the ingredients are blended into a finely granular mass, with imbedded crystals of quartz and mica, it has been called *Eurite.*

Fig. 61.

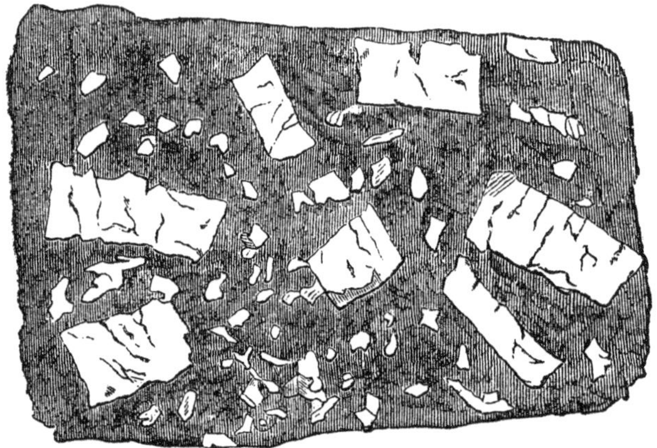

Tabular or regularly jointed granite is distinguished by the regularity of its forms, which result from two causes, joints and interstratification with slates. The granite is divided into large tables and prisms of various forms and sizes, as in Fig. 62. A more remarkable variety of granite is the *concretionary.*

Fig. 62.

It is a fine-grained white granite, containing black mica, which often is agglomerated into spherical or elongated nodules, from half an inch to five inches in length. The larger ones have their longer axes parallel. Fig. 63 represents these nodules. Some of them so much resemble one of our common nuts, as to have received the name of *petrified butternuts.* The best locality of this variety is in Craftsbury, Vermont.

There may be subdivisions of all these varieties into others, which have not as yet received distinct names, according to the particular species of feldspar or mica present. Thus, granite may contain either orthoclase, albite, labradorite, etc., as its feldspar, and either muscovite, biotite, or phlogopite, for its mica.

Fig. 63.

Syenite.—Syenite is composed essentially of feldspar, quartz, and hornblende, the first predominating. When mica is also present, the compound is frequently denominated *Syenitic granite*. Syenite may be found passing into granite on the one hand, and into the trap family, on the other.

Conglomerate syenite is common syenite containing numerous rounded and angular pebbles of other rocks. This is an important variety to be examined in the study of the origin of these rocks.

When it was ascertained that the famous rock from Syene, in Upper Egypt, (so much employed in ancient monuments), from which the name of syenite was derived, was nothing but granite with black mica, and also that Mount Sinai in Arabia was composed of genuine syenite, a French geologist proposed to substitute *Sinaite* for syenite: but the suggestion, which was certainly a good one, has not been adopted.

Most of the syenite so famous in New England for architectural purposes, as that from Quincy and Cape Ann, is composed of feldspar, quartz, and hornblende, the latter frequently disappearing.

3. *Protogine.*

Protogine, called also talcose granite, is a mixture of quartz, feldspar, and talc. In Europe it abounds among the Alps, and in Cornwall, in England. In this country it occurs among the Laurentian rocks of New York, and in the metamorphic granite of Devonian age in New England.

II. TRAPPEAN ROCKS.

4. *Porphyry.*

Rocks with a homogeneous, compact, or earthy base, through which are disseminated crystalline masses of some other mineral of contemporaneous origin with the base, are denominated *porphyry*. True classical porphyry, such as was most commonly employed by the ancients, has a base of compact feldspar, with imbedded crystals of feldspar.

Fig. 64 was copied from a pebble corresponding to the beautiful green porphyry of the ancients, found on the beach in Scituate, Massachusetts.

Fig. 64.

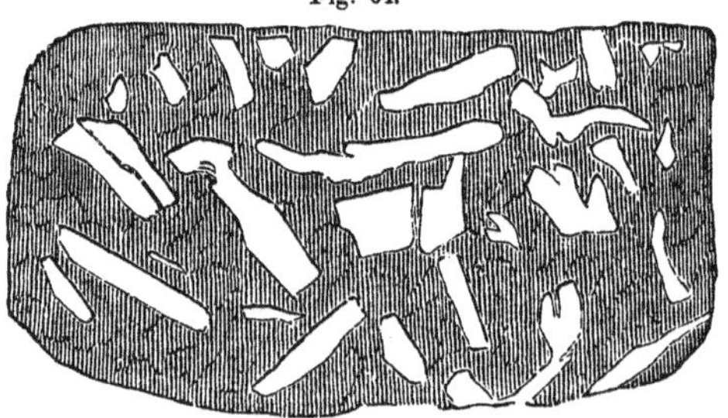

When the base is greenstone, pitchstone, trachyte, or basalt, the porphyry is said to be greenstone porphyry, pitchstone porphyry, trachytic porphyry, and basaltic porphyry. The base is sometimes clinkstone, and the imbedded crystals may be feldspar, augite, olivine, etc.

Hence the term porphyry designates only a certain form of rock, but does not refer to any particular kind. When porphyry is spoken of in general terms, however, feldspar porphyry is usually meant.

The name porphyry signifies *purple*, *πορφυρα*, such having been the most usual color of the ancient porphyries; but this rock exhibits almost every variety of color. It is the hardest of all the rocks and when polished, is probably the most enduring.

Compact feldspar, or *Felstone* sometimes called *petrosilex*, is a hard compact stone of various colors; fusible before the common blowpipe, and often translucent on the edges like hornstone. Its predominant ingredient appears to be feldspar, (*clinkstone* or *phonolite*;) or fissile *petrosilex*, a greenish or grayish rock, dividing into slabs or columns, ringing under the hammer, and apparently a variety of compact feldspar. *Hornstone* is a compact mineral, often translucent like a horn; of various colors; in hardness and fracture approaching flint; infusible before the blow pipe; and hence composed chiefly of silica. *Cornean* is between hornstone and compact feldspar, compact and homogeneous; supposed to consist of feldspar, quartz, and hornblende. All these substances form the basis of porphyry; and hence we have clinkstone porphyry, hornstone porphyry, etc. When black augite forms the base of porphyry, it is called *melaphyre*.

5. *Greenstone*.

Several unstratified rocks, whose principal ingredients are feldspar and hornblende or augite, are called *Trap Rocks*, from the Swedish word *trappa*, a stair; because they are often arranged in the form of stairs or steps.

Although the term trap is loosely applied, most writers limit it to the varieties of rock called greenstone, syenitic greenstone, basalt, compact feldspar, clinkstone, pitchstone, wacke, amygdaloid, augite rock, hypersthene rock, trap porphyry, pitchstone porphyry, and tufa. Macculloch includes claystone and syenite.

Greenstone, or diorite, is ordinarily composed of hornblende and feldspar, orthoclase and albite, both compact and common, the former in the greatest quantity.

Dr. Macculloch calls those varieties of greenstone which have a green color, *augite rocks*; because augite is the predominant ingredient. When hypersthene takes the place of hornblende, he calls the compound *hypersthene rock*. When greenstone is composed almost entirely of hornblende, the rock is denominated *hornblende rock*. When the grains of feldspar and hornblende are quite coarse, it is called *syenitic greenstone*, which often takes quartz into its composition, and passes into granite. All the above rocks are frequently porphyritic; and hence we have augitic, or pyroxenic porphyry, dioritic porphyry, etc. *Euphotide*, is a rock composed of compact feldspar and diallage.

III. VOLCANIC ROCKS.

6. *Basalt, or Dolerite*.

This rock appears to be composed of augite, feldspar, (labradorite), and titaniferous iron; and sometimes olivine in distinct grains. Its color is black, bluish, or grayish; and its texture compact and uniform; more so than greenstone. Augite is the predominant ingredient.

It is probable that most of the trap rocks of our Atlantic States are greenstones or diorites. Many of them have an impalpable texture like basalt; and chemical analysis alone can decide their composition in such cases. Probably the trap-rocks of the Rocky Mountains are basalt, and as recent as the basalt of Europe.

7. *Amygdaloid*.

This term, like porphyry, is not confined to any one sort of rock, but indicates a certain form, which extends through all the trap family. Amygdaloid abounds in rounded cavities, like the scoriæ and pumice of modern lavas, and these are often filled with calcite, quartz, chalcedony, zeolites, and other minerals, which have taken the shape of the cavity; so that the rock appears as if filled with almonds, and hence the name from the Latin, *amygdala*, an almond. These cavities, however, have sometimes been lengthened by the flowing of the matter, while melted, so that cylinders are found several inches long. When they are not filled, the rock is said to be *vesicular*.

Fig. 65 represents a fragment of lava, partly vesicular, and partly amygdaloidal; the white kernels being composed of carbonate of lime.

Fig. 65.

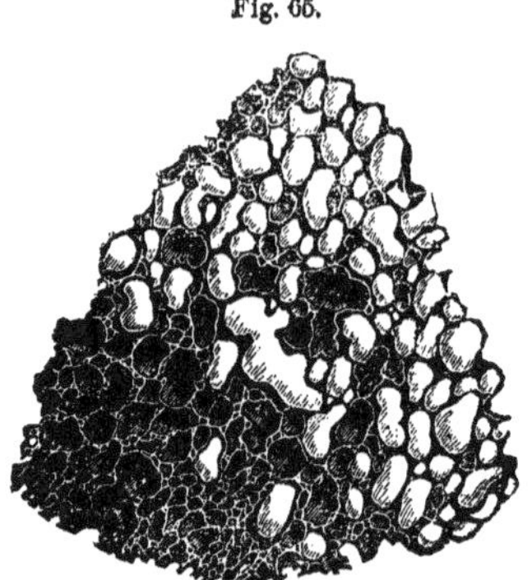

Amygdaloid.

A soft variety of trap rock, resembling indurated clay, is called *wacke*, which may or may not be vesicular. From its resemblance to the toad, probably, it is called in Derbyshire, *toadstone*.

8. *Trachyte.*

Trachyte is of a whitish or grayish color, usually porphyritic by feldspar crystals, and essentially composed of glassy feldspar, with some hornblende, mica, titaniferous iron, and sometimes augite. Its name is derived from the Greek, *τραχυς*, *rough*, from its harshness to the touch. It was an abundant product of volcanic action during the tertiary period, and usually appears to be older than basalt, although trachytic lavas have continued to be ejected down to the present day. Trachyte occurs in Auvergne and Hungary, and in vast quantities in South America. It constitutes the loftiest summits of the Cordilleras.

Trachyte in an earthy condition, as it occurs in the Pays de Dome, in Auvergne, is called *domite.* Trachyte is usually porphyritic, and hence we have *trachytic porphyry.*

9. *Lava.*

Lava, as remarked in another place, embraces all the melted matter ejected from volcanos; and the two minerals, feldspar and augite, constitute almost the entire mass of these products. When the former predominates, light-colored lavas are the result, when the latter, the dark varieties. The former are called *feldspathic* or *trachytic*, and the latter, *augitic* or *basaltic* lavas.

Other simple minerals occur in lava. Thus, in the product of Vesuvius alone, not less than 100 species have been detected; but they form so inconsiderable a part of the whole mass, as not to deserve consideration in a general view like the present.

Trachytic lava corresponds in most of its characters to the trachyte of the older igneous rocks. When cooled under pressure, solid rock results; but when cooled in the air, it is porous, fibrous, and light enough to swim on water, as is the case with pumice, large masses of which are found sometimes in the midst of the ocean. Sometimes it is porphyritic, like the older trachytes.

In like manner the basaltic or augitic lavas exceedingly resemble the more ancient basalt; and are, in fact, the same thing, produced under circumstances a little different. When cooled under pressure, compact basalt is the result; but cooled in the open air, they are scoriaceous or vesicular, and are usually called scoriæ.

Graystone lava is a lead gray or greenish rock, intermediate in composition between basaltic and trachytic lavas; but the feldspar predominates, being more than 75 per cent.

Vitreous lava has a fracture like glass. *Obsidian* seems to be merely melted glass. *Pitchstone* is less glassy, with an aspect more like pitch. It is usually composed of feldspar and augite, and often passes into basalt. Its composition however varies.

The small angular fragments and dust of pumice, (which is vesicular trachytic lava), and of scoriæ, (which is vesicular basaltic lava), which are produced by an eruption, falling into the sea, or on dry land, and mixing with sand, gravel, shells, etc., and hardened by the infiltration of carbonate of lime or other cement, constitute the substance denominated *tuff* or *tufa.* When this rock occurs with trap, it is called *trap tuff;* and when with modern lava, *volcanic tuff.* If it contains large and angular fragments, it is called *volcanic breccia.* When the fragments are much rolled, the rock is a *tufaceous conglomerate.* The basaltic tuffs are denominated by the Italian geologists, *peperino.* A kind of mud is poured out of some volcanic craters, which forms what is called *trass.*

Sometimes, especially at the great volcano of *Kilauea,* on the Sandwich Islands, when lava is thrown into the air, the wind spins it out into threads, resembling flax, and drives it against the sides of the crater. This is called *volcanic glass;* and by the natives of Sandwich Islands, *Pele's hair;* Pele having formerly been regarded as the presiding divinity of the volcano of Kilauea.

Other substances ejected from volcanos are fragments of granite and other rocks, scarcely altered; cinders and ashes of various degrees of fineness, which are sometimes converted into mud by the water that accompanies them; also sulphur in a pure state; various salts and acids; and several gases, among which are the hydrochloric, sulphurous, and sulphuric acids; alum, gypsum, sulphates of iron and magnesia, chlorides of sodium and potassium, of iron, copper, and cobalt; chlorine, nitrogen, sulphuretted hydrogen, etc.

Prismatic or Columnar Structure.—One of the most remarkable characteristics of the trap rocks is their columnar structure. This consists in the occasional division of their substance into regular prisms, with sides varying in number from three to eight, usually five or six, whose length is sometimes several hundred feet. They are often jointed, that is, divided crosswise into blocks from one to several feet in length, whose extremities are more or less convex or concave, the one fitting into the other. Frequently these columns stand nearly perpendicular, and when worn away on the side they present naked walls, which appear like the work of art. They stand so closely compacted together, that though perfectly separable, there is no perceptible space between them. The diameter of the column varies from a few inches to more than five feet.

The columnar and trappose forms of basalt and greenstone have produced some of the most remarkable scenery on the globe. Fingal's Cave, in the island of Staffa, (one of the Western Islands of Scotland,) and the Giant's Causeway, in the north of Ireland, are almost too well known to need description. Staffa is composed entirely of basalt, with a thin soil, and its shores are for the most part a steep cliff, 70 feet high, formed of columns. The cave is a chasm (in these columns), 42 feet wide, and 227 feet long, formed by the action of the waves. The following sketch, Fig. 66, will convey an idea of the situation of the cave and of the general structure of the island.

The Giant's Causeway consists of an irregular group of pentagonal columns, from one to five feet thick, and from 20 to 200 feet high, jointed as usual. Where the sea has had access to them, their upper portions are worn away, while the lower part remains extending an unknown distance beneath the waves, and seeming the ruin of some ancient work of art, too mighty for man, and therefore referred to the giants. Here also is a cave of considerable extent.

Fig. 66.

Fingal's Cave, Staffa.

Fig. 67 shows the appearance of a few columns, having the concave extremity uppermost at the Giant's Causeway.

Fig. 67.

Fig. 68 represents an overhanging group of greenstone columns, at Mt. Holyoke, in Massachusetts, which is called Titan's Piazza. The lower end of the columns, several rows of which project over the observer's head, are exfoliated in such a manner as to present a convex and even attenuated surface downwards.

Fig. 68.

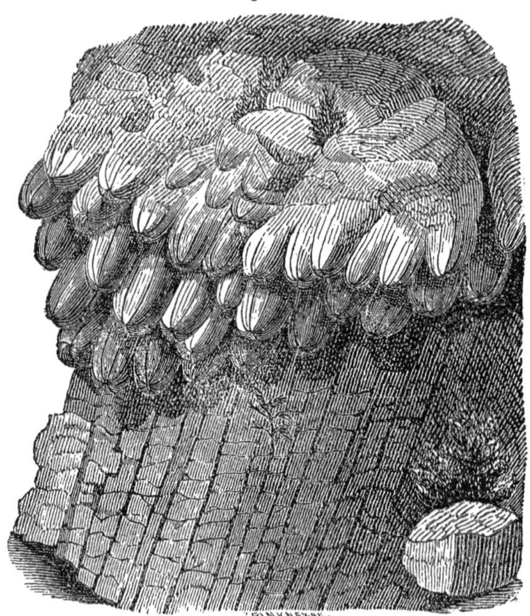

Titan's Piazza, Mt. Holyoke.

Fig. 69 shows a mass of trap columns on the north shore of Lake Superior.

Fig. 69.

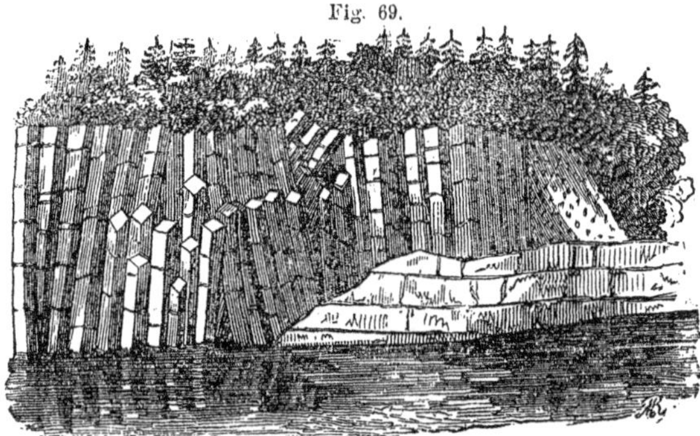

Basaltic Columns, Lake Superior; N. Shore.

When a trap vein, or dyke, is columnar, the columns often lie horizontal, or rather perpendicular to the sides of the vein; and thus is produced a wall of stones regularly fitted to one another and laid up, apparently by man; while often a decomposition of the surfaces of the blocks produces a powder resembling disintegrated mortar. Such a wall occurs in Rowan county, North Carolina. Fig. 70 shows a similar example on Lake Superior.

Fig. 70.

Trap Dyke, Lake Superior.

Greenstone columns are quite common in North America, either standing upright or leaning a few degrees. The Palisadoes upon the Hudson river, a few columns upon Penobscot river, in Maine, and others at *Titan's Pier*, near Mt. Holyoke, on Connecticut river, are well-known examples in the eastern part of the continent. But the most extensive development of them is in Oregon and Washington Territories, especially upon Columbia river. The banks are from 400 to 1000 feet high, and are made up of columns of trap or basalt in successive rows, superimposed upon one another, and separated by a few feet of amygdaloid, conglomerate, and breccia.

We suppose that the columnar structure of trap rooks has resulted from a sort of crystallization, while they were cooling under pressure from a melted state, for two reasons: *First*, Precisely similar columns are found in recent lavas; and *secondly*, from experiment. Mr. Gregory Watt melted 700 pounds of basalt, and caused it to cool slowly, when globular masses were formed, which enlarged and pressed against one another until regular columns were the result.

Magnetism of Rocks.—Many unstratified and metamorphic rocks sensibly affect a magnet. When there is a large amount of magnetic iron present, the magnetic needle is affected at a considerable distance from the ledge. But there are many cases where the magnetizer is present in such small quantities, as to be unappreciated by the needle, except when placed in immediate proximity to the rock.

Baron Humboldt first pointed out the magnetism of rocks, in a hill of serpentine, which he supposed to have only two poles, that is, to constitute one magnet. In 1815, Dr. John MacCulloch described a similar character in such rocks as granite, porphyry, syenite, serpentine, several kinds of slates, and especially of trap. He discovered that several magnets existed in the same mass of rocks.

We have found remarkable examples of the same kind, and some others still more complex, upon the trap rocks of New England. Upon examining a continuous surface, at certain localities, of only a few square feet, we frequently find several distinct magnetic poles, either north or south, and sometimes both, within a few feet or inches of each other. This fact is discovered by moving a pocket-compass over the surface. Wherever a pole exists, the opposite pole of the needle will point to it, as the compass is moved about. Fig. 71 represents a surface where these different poles were found. The nature of the pole is indicated by the words N. Pole and S. Pole. The size of the area shown in the figure is six by eight feet, the squares representing feet on Mt. Holyoke.

Fig. 71.

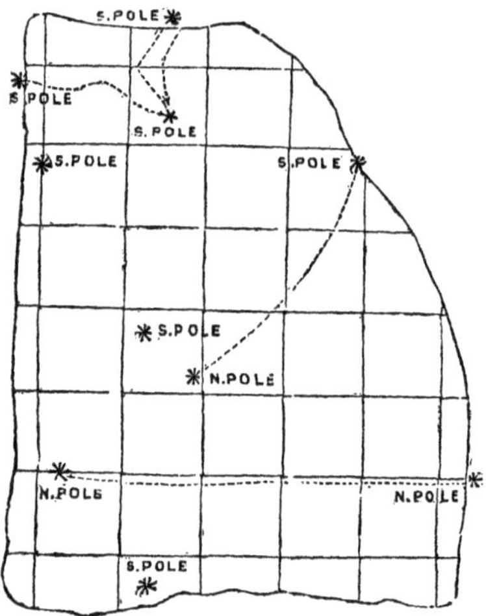

Another remarkable fact is shown upon the figure. Dotted lines are seen connecting some of these poles, along which the needle continues reversed, just as it does at their extremities, or when it approaches an isolated pole; the north end of the needle pointing to the pole if it be a south one, and the south end doing the same if it be a north one. In other words, along these dotted lines there appears to be an infinite number of poles, which are called *Lines of Polarity.* They are easily traced out by moving the compass over the surface.

Hand specimens of lava from Mts. Etna and Vesuvius, and of obsidian, from Mt. Ararat, in our possession, show magnetism and polarity. Observers have noted the same upon the lava of Mt. Ararat, 7,280 feet above the ocean. Bischoff, in his "Chemical and Physical Geology," has pointed out a large number of cases; but nowhere, save in the cases that have fallen under our own observation, have we seen described the phenomena of opposite poles in close proximity, and of lines of poles.

Theoretical Suggestions.—The existence of magnetic iron ore in so many rocks, and the fact that it often possesses polarity, leads naturally to the conclusion that this must be the cause of the magnetism and polarity of rock, and probably it is so. Yet some of the phenomena seem not fully explained on this supposition. Bischoff says, that in basalt "the magnetic polarity bears no definite proportion to the density of the rock, and consequently to the greater or less amount of magnetic iron ore, or hornblende; and further, that when the surface of the rock is decomposed, and the magnetic iron ore converted into hydrated peroxide of iron, the magnetic action of the rock is not weakened," although the peroxide is not magnetic. Nor does this explanation account for the production of opposite poles on the same surface, confusedly mixed together; nor for lines of polarity, extending over many feet of surface. But if the phenomena are not produced by magnetic iron ore, we have no theory to offer.

Relative Age of the Unstratified Rocks.—In the stratified fossiliferous rocks the relative age of the different groups is determined by their position; that is, the oldest are at the bottom, and the newest at the top; unless we can prove disturbance and inversion subsequent to their original formation. It has been usual to regard the unstratified rocks as lying in a reverse order; that is, the highest is the oldest, and they become newer as we go downward. This is undoubtedly true, if we suppose these rocks to be the result of melted matter, which has continued to cool deeper and deeper, so as to form a thicker crust, and some of which has been thrust up through the fissures in the stratified rocks. The particular geological period when an unstratified rock has been thus pushed upward, can be known by finding how high it has reached among the strata. Hence, Sir Charles Lyell has divided the unstratified rocks into the *Primary Plutonic*, *a*, *a*, Fig. 72, the *Secondary Plutonic*, *b*, *b*, the *Tertiary Plutonic*, *c*, *c*, and the *Recent Plutonic*, *d*, *d*, according to the period in which they have been erupted. But in fact we do not find that particular igneous rocks were produced only during particular fossiliferous periods; for most of them have been formed during nearly all these periods. The Granitic rocks, however, were most abundant during the older periods, but they have been found high in the tertiary, as in Catalonia; but lava is still poured out so as to cover alluvium. The crystalline unstratified rocks were most abundantly produced in the earlier periods; while the trappean and volcanic varieties have been most abundant at later periods.

Fig. 72.

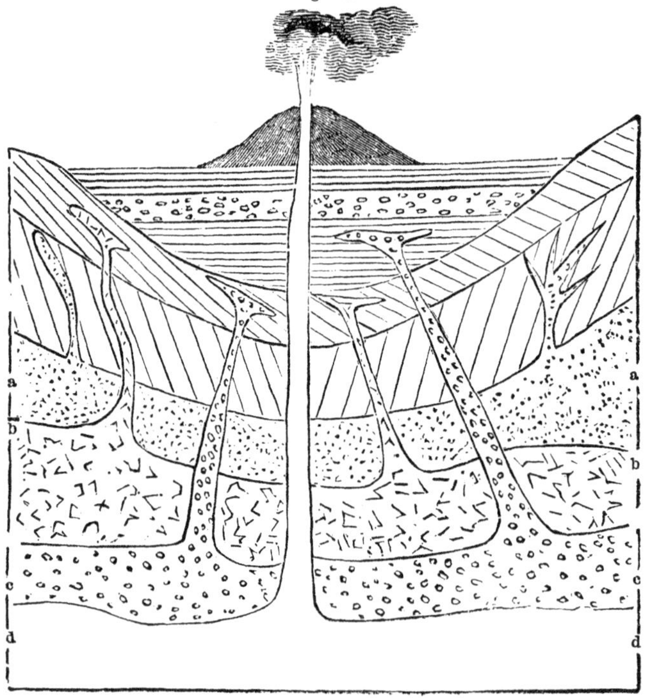

Section of the Relative Age of the Unstratified Rocks.

Fig. 36, as well as the above, is intended to show the relative age of the igneous rocks; Granite, and Syenite are marked as the oldest; next Porphyry and Trap; next the Volcanic. But the supposition on which this order is made out, that all unstratified rocks are melted matter ejected from the interior of the globe, is probably not true. No small part of them are probably metamorphosed stratified rocks, as we shall endeavor to show in our Section on Metamorphism. As to the age of such, it can be determined only when we can fix upon the period of the metamorphism; as we often can with much probability. But mere position will not determine the age.

Origin of the different varieties of Unstratified Rocks.—If the unstratified rocks were all derived from the same melted mass in the earth's interior, we should suppose they would not differ from one another at any period of their eruption. But, in fact, they do so differ as to show, first, that the ingredients from which they were derived were different; and secondly, that the circumstances under which they were formed, as to temperature, fusion, and pressure, were different. The following average analysis of orthoclase granite, greenstone, (feldspar and hornblende), and doleritic lava from Etna, will give an idea of the different chemical composition of the granite and more recent igneous rocks.

	Granite.	*Greenstone.*		*Lava.*
Silica, . . .	74.84	54.86		48.83
Alumina, . . .	12.80	15.56		16.15
Potash, . . .	7.48	6.83		0.77
Soda, . . .				3.45
Lime, . . .	0.37	7.29		9.31
Magnesia, . . .	0.99	9.39		4.58
Oxide of Iron, . .	1.93	4.03	*Prot.*	1[illegible]3[illegible]
Oxide of Manganese, .	0.12	0.11		0.54
Hydrofluoric Acid, .	0.21	0.75		

It will be seen that silica was more abundant in the granite than in the later rocks, and that the lime, magnesia and iron, were in much greater quantity in the latter than the former. The consequence is, that the silicates of lime, magnesia and iron, exist largely in the trappean and volcanic rocks, but scarcely at all in the granitic. Now these silicates act as fluxes, and hence the later rocks are much more fusible than the granitic. Indeed, the later are almost infusible, and would require a much higher temperature, or rather a more complete solution, either simply igneous or aqueo-igneous to form them, than the traps or the lavas. Hence the latter might be produced in such a state of a menstruum as would not produce granite, even though all the essential ingredients were present. And that state of the menstruum in which granite would form, might prevent that play of affinities requisite to produce the trappean and volcanic products. In this way we might show, both how all the varieties of unstratified rocks might be produced from the same fluid mass, and how one period might be more favorable to the formation of the granites, another to the traps, and another to the lavas. Still we incline to the opinion that in very many instances the varieties of unstratified rocks have resulted from the metamorphism of different kinds of stratified rocks.

TABLE OF THE COMPOSITION OF ROCKS.

Rocks	Silica.	Alumina.	Protoxide of Iron.	Peroxide of Iron.	Oxide of Manganese.	Lime.	Magnesia.	Potash.	Soda.	Organic Matter.	Water.	Sulphuric Acid.	Carbonic Acid.	Specific Gravity.
Clay	62.8	25.5		1.25		0.36	0.47	2.51			6.65			
Shale	61.9	21.7	4.73			0.09	0.59	3.16	0.25	0.70	6.73			
Clay Slate	67.4	18.2	4.71	1.02	0.30		4.0	2.65			1.74			2.8
Gypsum						32.6					20.9	46.5		2.3
Chalk and Oolite						56.1							4.3	2.5
Dolomite						52*	45*							2.7
Hydraulic Limestone	15.4	9.1		2.25		25.5	12.4						34.2	
Silurian Limestone / Winooski Marble	10.3	12.2				35.3*	42.2*							
Carrara Limestone	99.3*		0.25				0.28*							2.70
Sandstone (Quader)	98.6	0 08		0.44	0.03	0.02		trace	trace		0.83			2.67
Greensand (N. Jersey)	49.3	7.71	24.7			5 08		10.0			6.00			2.55
Gray Wacke	87.0	6.4		4.7		0.19	0.37	0.53	0.34		0.37			2.67
Mica Schist	55 1	12.6		16.9			11.0	2.2	1.24		2.13			2.69
Talcose Schist	57.8	7.06		9.45			25.6							
Talcose Schist	69.9	20.0				1.51	1.80	1.45	3.33		2.40			2.7
Chlorite Schist	31.5	5.44		0.18			41.5				9.32			
Hornblende Schist	48.6	16.4	4.69	18.6	0.48	7.16	2.32	0.50	0.89		0.21			
Gneiss	71.	15.2												2.72
Serpentine	38.7	30.6	12.8		0.9	4.51	30.0	0.29	0.11		9.17			2.58
Granite	67.3	16.1		1.9		0.6		13.3			0.8			2.74
Syenite	70 0	12.4	1.51	0.17		3.45	3.46	3.13	3.44		0.56			2.74
Feldspar Porphyry	70.5	13.5		5.5		0.25	0.40	5.50	3.55		0.77			2.62
Greenstone, or Diorite	60.0	12.3		12.3	8 1	1.4		9.6	1.1		0.2			2.85
Melaphyr	53.2	19.8	8.56		0.51	3.87	4.96		7.02		2.14			
Basalt or Dolerite	48 5	30.2				11 9	6.9	0.65	1.96					2 86
Trachyte	76 7	14.2				1.44	0.28	3.20	4.18					2.68
Lava, top of Mt Ararat	69.5	15.0	3.35			4.7	0.98	1.46	4.46		0.35			2.60
Lava, Trachytic	57.8	17.6	Prot. & per 2.09	4.64	0.82	5.46	2.76	1.42	6 82					2.75
Obsidian, Ararat	77 8	11 8		2.35		1.81		2.44	4 15		0.51			2.36

* Carbonate of Lime and Magnesia.

Chemical Composition of the Rocks.—It is only quite recently (1860), that we have had many reliable analyses of the rocks, and even now the table which we present is not complete. But the subject has become one of great interest, especially in its bearings upon metamorphism, which is now a most important part of the science. The analyses in the following table have been derived chiefly from Bischoff's Chemical Geology, though some have been taken from American analysts.

The column of specific gravities has been added chiefly to enable any one to determine the weight of rocks, a problem which every one has often occasion to solve. The specific gravity shows us how much heavier a given quantity, say a cubic foot, of the rock is than the same quantity of distilled water. Now since a cubic foot of water weighs 1000 ounces avoirdupois, multiply the number of cubic feet in the mass, whose weight you seek, and this product by the specific gravity, and you will obtain the weight in ounces, which divided by 16 will give it in pounds. Thus, suppose a mass of Greenstone measures 20 cubic feet, then $20 \times 1000 \times 2.85 \div 16 = 3562.5$ pounds.

This table might perhaps more logically have been placed at the end of Section II.; but there is an advantage in knowing something about the characters of the rocks before studying their chemical composition.

SECTION IV.

OPERATION OF ATMOSPHERIC AND AQUEOUS AGENCIES IN PRODUCING GEOLOGICAL CHANGES.

The basis of nearly all correct reasoning in geology, is the analogy between the phenomena of nature in all periods of the world's history: in other words, similar effects are supposed to be the result of similar causes at all times.

This principle is founded on a belief in the constancy of nature; or that natural operations are the result of only one general system, which is regulated by invariable laws. Every other branch of physical science, equally with geology, depends upon this principle; and if it be given up, all reasoning in respect to past natural phenomena is at an end.

It does not follow from this principle that the causes of geological change have always operated with equal intensity, nor with entire uniformity. How great has been the irregularity of their action is a subject of debate among geologists.

It is important to ascertain the *true dynamics* of existing causes of geological change; that is, the amount of change which they are now producing. For until this is done, we can not determine whether those causes are sufficient to account for all the changes which the earth has undergone.

The elements of these atmospheric and aqueous agencies are

heat, cold, and water in its manifold states, liquid, vaporous, and solid. In their action they may be combined, or act separately. The atmospheric agencies are oxygen, nitrogen, carbonic acid, vapor, and winds. The aqueous agencies are frost, snow, ice, glaciers, icebergs, springs, rivers, lakes and oceans. We shall not attempt to designate the separate action of these agents, but merely point out their most important combined effects.

All these agencies, where they act mechanically, remove materials from higher to lower levels. Mountains diminish in size, valleys receive accumulations, but only to assist the fragments in their progress towards the ocean. Thus the general tendency is to transport the continents into the ocean.

Disintegration of Rocks.—Water acts upon rocks and soils both chemically and mechanically; chemically, it dissolves some of the substances which they contain, and thus renders the mass loose and porous; mechanically, it gets between the particles and forces them asunder; so that they are more easily worn away when a current passes over them. Congelation still more effectually separates the fragments and grains, and thus renders it easy for rains and gravity to remove them to a lower level. In a single year the influence of these causes may be feeble; but as they are repeated from year to year, they become, in fact, some of the most powerful agencies in operation to level the surface of continents.

As rain in falling through the air absorbs carbonic acid, it acts with greater energy in the decomposition of rocks, especially those which are calcareous. It also penetrates into their pores and crevices, and initiates the process of metamorphism, by changing the mineral structure of rocks. Easily decomposing crystals, as pyrites or calcite, may be entirely removed, and their cavities be filled with some other mineral, which will assume the exact form of the original crystal, and thus be a *pseudomorph*.

It is a fact established by the experiments of Professors W. B. and R. E. Rogers, that pure water will dissolve more or less every variety of rock except quartz, on which nothing will act but hydrofluoric acid, unless it be converted into silicates. Much more decided will be the action of water if it contain, as it commonly does, carbonic acid. Other energetic chemical agents are produced which, along with carbonic acid, are carried we know not how deep into the earth, not only disintegrating the surface, but effecting important changes even in rocks that show no signs of alteration. This subject will be dwelt upon more in detail in our Section on Metamorphism.

Detritus, or Debris of Ledges.—It is chiefly by the action of frost and gravity, that those extensive accumulations of angular fragments of rocks are made that often form a *talus*, or slope, at the foot of naked ledges, and even high up their faces. In some cases, though not generally, this detritus has reached the top of the ledge, and no further additions are made to the fragments, which usually slope at an angle not far from 40°. Examples of this detritus are usually most striking along the mural faces of ledges, especially where the upper layers are the hardest.

Fig. 73.

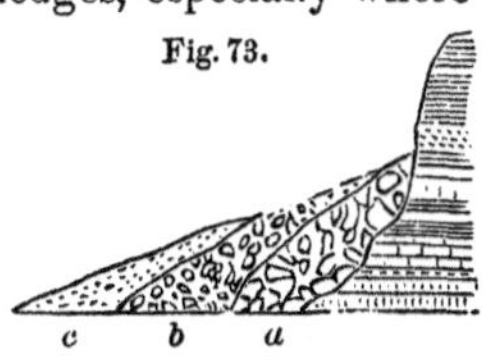

Fig. 73 represents a cliff, at the base of which there is produced a talus of large angular blocks, *a*. These are further acted upon at *b*, and *c*, until the rains or tides remove the finer portions in the form of mud.

These slopes of *debris* show that the earth can not have existed in its present state an immense period of time; because the process of disintegration is not yet complete. Had these agencies been at work upon the earth in its present form from eternity, the earth would have become a vast plain, and the land have been all swallowed up and covered by the ocean.

Dunes or Downs.—The sand which is driven upon the shore by the waves is often carried so far inland as to be beyond the reach of the returning wave; and thus an accumulation takes place, which is the origin of most of those moving sand hills, known by the names of *dunes* or *downs*. When the sand becomes dry, the sea breezes drive it further and further inward, the land breezes not having equal power to force it back; and at length it becomes a formidable enemy, by overwhelming the fertile fields, filling up rivers and burying villages. Sometimes these dunes occur in the interior of a country.

Every one is familiar with the history of these dunes in Egypt. The westerly winds have brought in the sands from the Lybian desert, and all the west side of the Nile, with the exception of a few sheltered spots, has been converted into an arid waste. In Upper Egypt especially, the remains of ancient temples, palaces, cities, and villages, are numerous among the drifting sands. In Europe, around the Bay of Biscay, a similar destructive process is going on. A great number of villages have been entirely destroyed; and no less than ten are now imminently threatened by sand hills, which advance at the rate of 60 or even 72 feet annually. On the coast of Cornwall, in England, similar effects have taken place. These dunes are common on the coast of the United States, especially on Cape Cod, in Massachusetts, where strenuous efforts have been made to arrest their progress, and to prevent the destruction of villages and harbors that are threatened. Upon the south shores of Lake Superior they are also found.

Aqueous agencies act mechanically and chemically. Mechanically, in the form of glaciers, avalanches, icebergs, frost, snow, ice, landslips, rivers, tides, waves, and oceanic currents, they are continually abrading the ledges and re-arranging the materials, particularly under water. Chemically, water dissolves portions of rocks, and deposits them again by evaporation or precipitation.

GLACIERS.

Glaciers are rivers of ice, descending from the regions of perpetual frost, to levels below the usual snow line. They are inclosed in valleys, or are suspended upon the flanks of mountains. They are properly streams filled with the overflow or *waste* of the vast snow fields occupying the higher regions. These may be aptly compared to a sheet of ice descending from a tin-covered roof. In the Alps the glaciers terminate sometimes as high as 7,000 or 8,000 feet above the ocean; but some descend to 3,400 feet, while the usual snow line does not descend below 8,600 feet.

Glaciers are found among the Himalayahs, the Caucasus, and Altai mountains, in Asia; among the Alps, where they have been most studied, in Norway, Iceland, and Spitzbergen, in Europe; in Patagonia in South America, and within both frigid zones. There are 400 glaciers among the Alps, covering about 1,400 square miles of surface. They are divided into two groups—the glaciers of Mont Blanc, and of the Finster Aarhorn districts. The most important glaciers have received distinct names, as Bossons, Aletsch, and Viesch, among the Alps, and the great Humboldt glacier in Greenland, described by Dr. Kane, and figured in our *Frontispiece.*

The elevated crests and plateaux above glaciers are more or less covered with snow. These vast fields of powdery and crystalline snow are termed *mers de glace*, or seas of ice. In its descent this snow becomes more granular, and forms the *névé* (French), or *firn* (German). Additions are made to the névé, or the upper part of the glacier, every year, so that the mass is stratified.

The névé gradually changes into the *blue compact ice*, which forms the true glacier. The latter is permeated by a delicate structure, similar to the cleavage of slaty rocks, or vertical ribbons

5

of blue ice parallel to the course of the glacier, alternating with bands of white ice, from $\frac{1}{48}$ of an inch to several inches wide.

The névé has a concave even surface, but the glacier proper is convex, and is generally riddled by fissures or crevasses, often hundreds of yards long, and hundreds of feet deep, crossing in every direction. These are produced by the unequal temperature of the mass. The glacier is cut up by the fissures into masses of various shapes. Sometimes the masses are square or trapezoidal; or constitute needle shaped pyramids, called *Aiguilles.* Where the fissures are wanting, numerous little streamlets flow over the surface in small channels. These may terminate in circular or elliptical holes, called *puits,* often of great depth. Sometimes the puits are only a few feet deep, and are filled with water—like pot-holes near waterfalls. When the water freezes in them, the ice is formed in beautiful concentric layers, which are called *glacier stars.* The *meridian holes* are semi-circular hollows, invariably having the arc turned to the north, and the chord to the south; thus serving as compasses and sundials to travelers. They are produced by the heat of the sun's rays upon small accumulations of gravel. The pebbles absorb more heat than the ice, and hence sink into the ice where the sun's rays act the longest, and with the greatest intensity.

Glaciers are of two kinds. The first class lie in valleys of greater or less depth, and have a declivity or slope from 3° to 10°. The second class are generally small, resting upon the declivities of mountains, with a slope varying from 15° to 50° and upwards.

Of the first class there are three varieties: the *canal shaped,* of uniform width with scarcely any branches; the *oval shaped,* and the *basin shaped,* both of which are contained in deep valleys among mountains, so that the width of the outlet is a fraction of the diameter of the glacier itself. They may be compared with expansions of rivers into lakes.

Fig. 74 is a view of a canal shaped glacier in its upper part, as it procedes from the distant *mer de glace,* and winds through the long valley. It is one of the Alpine glaciers.

The greater part of the Alpine glaciers are from 18 to 21 miles long, between a mile-and-a-half and three miles wide, and from 100 to 600 feet thick. The thickness of the upper is always greater than that of the lower end. Glaciers have been described among the Himalayahs several thousand feet in thickness. The great Humboldt glacier, in Greenland, is 300 feet thick, 45 miles wide, as it flows into Peabody Bay, and .200 miles long; the largest glacier by far yet described. (See Frontispiece.*)

In warm seasons the lower ends of glaciers are gradually melting, and inequalities are being produced over their surfaces. This superficial waste is called *ablation.* When large flat blocks of stone lie upon the ice, by ablation the phenomenon of *glacier tables* is produced. The rock protects the ice beneath itself from melting, but the ice of the glacier around it gradually disappears, until the block is poised upon a single pedestal, like a table. Prof. Forbes describes one of these tables in the Alps, 23 feet by 17, and 3½ feet thick. Other objects of interest are the *gravel cones.* These are cones of ice, usually about a foot in height, covered with gravel. They are formed by ablation. Like the tables of stone, a mass of gravel is elevated upon a pedestal, and the gradual waste of the ice beneath allows the outside

* We are indebted to the liberality of Childs & Peterson for permission to copy this drawing from their splendid work, *Arctic Explorations,* by Dr. Kane, Philadelphia. 1856.

Fig. 74.

View of the Glacier of Viesch.—(Upper part).

pebbles to fall down and cover the cone. Prof. Agassiz describes some of these cones in the Alps 13 feet high, and 13 feet broad.

As glaciers advance they break off masses of rocks from the sides and bottoms of the valleys, and crowd along whatever is movable, so as to form large accumulations of detritus in front and along their sides. When the glacier melts away, these ridges remain, and are called *moraines.* Agassiz describes three kinds: 1. The *terminal moraine*, or that at the extremity of the glacier; 2. *Lateral moraines*, or those ridges of detritus formed along the flanks of the glacier; 3. *Medial moraines*, or those longitudinal trains of blocks which sometimes accumulate upon the top of the glacier, especially where glaciers unite from two valleys, and crowd the detritus between them upon their tops. The moraines are sometimes 30 or 40 feet high.

At the lower extremity of the glacier there is a vault from which issues, especially in the summer, a stream of water, which ramifies upward beneath the ice like rivers in general. This stream, continuing from generation to generation, wears out a channel in the rocks as it descends from the glacier.

Fig. 75 shows the lower extremity of the glacier of Viesch, with a distinct terminal moraine, which at the sides is connected with lateral moraines. A stream of water is seen issuing from the glacier, which has worn a channel in the rocks. On Fig. 74 are shown both lateral and medial moraines; the latter considerably scattered. Fig. 78 exhibits a fine example of a medial moraine.

Although the inferior surface of the glacier is pure smooth ice, yet it is usually thickly set with fragments of rock, pebbles, and coarse sand, firmly frozen into it, which makes it a huge rasp; and when it moves forward, these projecting masses, pressed down by the enormous weight of the glacier, wear down and scratch the solid rocks; or when the materials in the ice are very fine, they smooth and even polish the surface beneath. The movable materials beneath the ice are crushed and rounded, and often worn into sand or mud. The rocks in place, against which the glacier presses, are also smoothed and striated upon their sides. These striæ, wherever found, are perfectly parallel to one another, because the materials producing them are fixed in the bottom of the ice. But as the glacier advances from year to year, new sets of scratches will be produced, which sometimes cross those previously made, at a small angle.

Fig. 75.

Glacier of Viesch.—(Lower part).

Fig. 76 shows a specimen of schistose serpentine smoothed and scratched beneath a glacier in the Alps.

Fig. 76.

Rocks striated by Glaciers.

Fig. 77 is a similar specimen, exhibiting two sets of striæ crossing one another at a considerable angle.

Fig. 77.

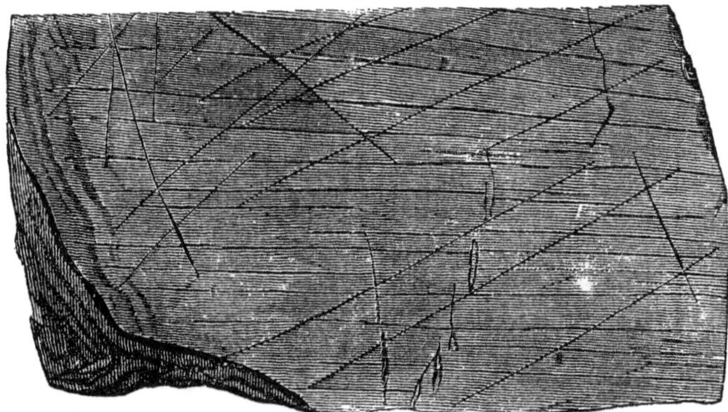

Rocks striated by Glaciers.

When the ledges beneath the glaciers are uneven, and exhibit many angular projections, the angles are worn off, and the surfaces assume that peculiar rounded and undulating appearance denominated by Saussure *roches moutonees*, or embossed rocks. They are shown poorly in Figs. 74 and 75.

Currents of water sometimes conspire with the movements of the glacier, and form grooves or troughs of considerable depth and width on the top of precipitous rocks, to which currents of water could have no access were not the space around them filled with ice. Such furrows are called, in Switzerland, *lapiaz* or *lapiz.*

Motion of Glaciers.—Professor J. D. Forbes supposes he has proved that the ice comports itself precisely like a stream of water in the following particulars: 1. The motion of the Mer de Glace, during summer and autumn, is as great as four feet in a day in some places, and only eight or nine inches in others. 2. Each portion of the glacier moves continuously, and not by fits and starts. 3. The glacier, like a stream, has its pools and its rapids. 4. The motion of the ice is favored by an increase of temperature. 5. Yet the motion does not entirely cease in the winter. 6. The centre of the glacier moves faster than the sides; and the top faster than the bottom. 7. The maximum velocity is not always in the middle of the glacier, but may be upon either side. 8. The changes in the velocity of the ice take place gradually by the yielding of the entire mass, not by the justling of fragments, or the formation of rents.

The Mer de Glace moved 16,500 feet in 44 years, or upon an average, 375 feet annually. Sometimes glaciers advance lower and lower for several years, in consequence of low temperature; or they may retreat during a succession of unusually warm seasons. As they advance and retreat they produce and leave successive moraines, especially terminal ones.

In the view of Prof. Forbes, a glacier is an imperfect fluid, or a viscous body, which is urged down slopes, like a river, by the mutual pressure of its parts. The ice is not inflexible, but more or less plastic, in consequence of having its minute pores and fissures permeated with water. As much of the water freezes in cold seasons, the motion of the glacier is retarded in winter and accelerated in summer. The decrease of the glacier in summer by ablation, and by the attenuation and collapse of the parts which move most rapidly, is repaired during the winter, when, the higher regions of the glacier moving relatively faster than the lower, the yielding mass of ice is pressed upwards.

Recent experiments show that ice has a property of plasticity called *regelation*. With or without pressure, fragments of ice will cohere at 32°, and by pressure they may be molded into any form. Hence the glacial valley is a mold, in which the ice is pressed by its own gravity. Also, when separate glacier branches unite themselves into a single trunk, regelation cements all the parts into one whole; and thus the combination of two glaciers is like the confluence of two rivers. The ribboned structure of the ice is thought by some to have been produced by pressure, like the slaty cleavage of rocks.

The theory of Agassiz imputed the onward movement of glaciers to the expansion of the water by freezing, which during the day in mild weather is constantly infiltrated into the cracks and pores of the ice. This view is ably defended by its learned author in his fascinating and splendid work, *Etudes sur les Glaciers.*

In that work he gives a curious example of the progress of a glacier. A certain explorer fastened a pole to a large block in the moraine of a glacier, high up towards its source, and mentioned the fact in his book. Some ten years afterwards another explorer started to find the block, and was agreeably surprised to meet it some eight or ten miles nearer the lower end of the glacier. This block with many others is shown on Fig. 78, which exhibits also several glacier tables.

Fig. 78.

We introduce Fig. 79 to give an idea of the stupendous chain of mountains in the Alps, called Mont Blanc, as seen from the summit of the Breven, which is 8.500 feet high; *a*, is Chamouny in the valley; *b*, Mont Blanc; *c*, Mer de Glace; *d*, Boissons Glacier; *e*, Augille verte; *f*, Dome du Goute; *g*, Montanvert. Any one who has been there will recognize these spots with great distinctness.

Avalanches, Icebergs, etc.—When the slope down which a glacier descends is very steep, or it is crowded to the edge of a precipice, huge masses sometimes are broken off by gravity, and tumbling down the mountain produce great havoc. In the Alps

Fig. 79.

these are called *avalanches* ; and so large are they sometimes, that from one to five villages, with thousands of inhabitants, have been destroyed in a moment.

Landslips are a somewhat similar occurrence, happening in countries where perpetual frost does not exist. They frequently occur in the spring, when the frost leaves the soil, and the great weight of snow and ice drags along with it trees, soil, and rocks, down the mountain's side. Sometimes these slides take place in the summer, after powerful rains; as that in the White Mountains in 1826, by which a family were destroyed. Marks of ancient slides are visible on the sides of the Hopper on Saddle Mountain in Massachusetts, and upon the west side of Mansfield Mountain, in Vermont. Arctic voyagers say that similar phenomena are common in the Polar regions.

When a glacier discharges itself into the ocean, great masses of ice, perhaps loaded with detritus, quietly separate from it, and are floated off by currents. Or if the glacier is crowded off a steep shore, the masses will be precipitated into the water. Two systems of fissures, dividing the ice into square blocks, facilitate the separation. These blocks are *icebergs*.

The representation of the Humboldt Glacier (*Frontispiece*), shows the origin of many of the icebergs in the Northern Atlantic. These bergs separate from this glacier in lines parallel to the shore. A representation of a single berg, with the ship of Captain Parry in front of it, is given in Fig. 80.

Fig. 80.

Fig. 81 exhibits a remarkable berg, figured by Captain McClintock, standing on a narrow pedestal, which, however, seems to be connected with a broad mass of ice mostly beneath the surface of the sea. It must be large compared with the berg to prevent the latter from toppling over.

These bergs from the north frequently float as far south as 40° north latitude before they are all melted, and they have been the occasion of many shipwrecks between the United States and Europe.

When the ice in high latitudes breaks up in the summer, vast surfaces of it, called *Field Ice*, are borne hither and thither by the winds and currents, often

Fig. 81.

imprisoning vessels, *nipping* them severely, and sometimes raising them entirely out of the water.

A portion of an ice-field is a *floe*. The *ice-belt* is a continued margin of ice adhering to the coast above the ordinary level of the sea, in high latitudes. The *ice-foot* is a limited ice-belt. Land-ice is field-ice adhering to the coast, or included between headlands. An *ice-raft* is a mass of ice transporting foreign matter. Icebergs are frozen from fresh, and floes from salt water.

Icebergs and floes may be of great size and wide surface. Icebergs rise sometimes from 250 to 300 feet above the water, and as every cubic foot above the surface implies eight cubic feet below it, they must descend over 2,000 feet. The floes are often one, two, five, and even thirteen miles long; and one northern voyager relates that a party traveled northerly upon one of them for days, supposing it either fixed to the shore or covering the land, and knew not their mistake, till an observation for latitude showed them that the ice was moving southerly as fast as they moved northerly.

The bergs are sometimes loaded with detritus of boulders, sand and gravel. Capt. Scoresby conjectures that some which he saw contained from 50,000 to 100,000 tons of such materials. On a large berg, observed in 1839, in S. lat. 61°, a boulder was observed frozen in, six feet by twelve in diameter, which had been carried 1,400 miles, that being the distance to the nearest land.

Dr. Kane has given several fine drawings to illustrate this subject: one of which we have copied in Fig. 82. This raft is loaded with masses of slate. Dr. Kane says, "I have found masses that have been detached in this way floating may miles out to sea—long symmetrical tables, 200 feet long, by 80 broad, covered with large angular blocks and bowlders, and seemingly impregnated throughout with detrited matter. These rafts in Marshall Bay were so numerous, that, could they have melted as I saw them, the bottom of the sea would have presented a more curious study for the geologist than the bowlder-covered lines of our middle latitudes." He speaks of an ice-belt (which in summer is detached and floats off,) as "covered with its millions of tons of rubbish, greenstones, limestones, chlorite slates, rounded and angular, massive, and ground to powder."

Fig. 82.

Icebergs are often stranded, even in deep water. The collect upon the loose materials at the bottom, and even on the projecting rocks of such a mass, with stones frozen into its bottom, and moving at the rate of several miles an hour, when it strikes the surface, must be very great, and such as we shall find has been the result of some agency in high latitudes when we come to describe the phenomena of drift.

Ice islands sometimes get stranded upon the top of some rock that rises in the ocean, and then frozen to it, so that when by winds and waves the icy cap is loosened, it tears off more or less of the rock beneath, and bearing it away in the direction of the current, drops the attached fragments upon the bottom.

If this operation be often repeated, it might produce a train of boulders on the bottom of the sea.

When the sea is rough, an iceberg may be lifted up by the waves, and as they retire be allowed to fall; so that, if the water be shallow, it will come down like a mighty maul, and with a force which even solid rock could scarcely resist.

RIVERS.

Rivers produce geological changes in four modes: 1. By excavating some parts of their beds. 2. By filling up other parts. 3. By forming deposits along their banks. 4. By forming deposits, called *deltas*, at their mouths.

Most of the larger rivers, especially where they flow through a level country, are filling up their channels; but where smaller streams pass through a mountainous region, the process of excavation is still going on; and it is accomplished in a good measure by means of *ice freshets*.

Valleys of Denudation.—In many instances it can be shown that the present beds of rivers were only in a small part produced by their own erosions, and that previous agencies in great part prepared the valleys through which they run.

The general features of continents, and the larger valleys, are due to the position of the underlying strata. If these have the shape of a trough or basin, the depression is a natural valley; and as it is produced by the sinking of the strata, it is called a *valley of subsidence*, as in Fig. 83. Such valleys may be enlarged by erosion.

But often these depressions are actually excavated from strata inclined at various angles. These are *valleys of denudation*. There are two general varieties of them. First, when the crest of an anticlinal ridge has been worn away, constituting a *valley of elevation*, as in Fig. 84. Secondly, when a valley has been excavated out of vertical or inclined strata, as in Fig. 85. These are *valleys of erosion*. Ravines, gorges, and cañons (pronounced *canyōns*), are narrow valleys chiefly along the present beds of rivers, excavated by the streams alone without foreign assistance. In many instances one is surprised at the magnitude of the excavations.

Fig. 83.

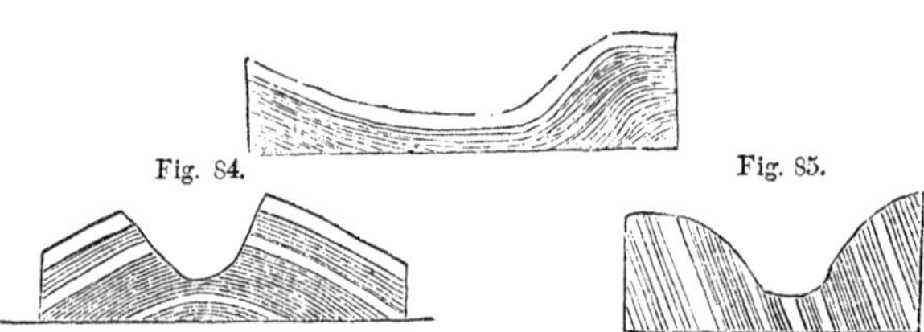

Fig. 84. Fig. 85.

Out of a multitude of examples we select only a few to illustrate the first mode in which rivers produce geological changes.

1. The gulf, seven miles long and 150 feet deep, between Niagara Falls and Lake Ontario, has long excited the attention of geologists, and some of them have imagined other agencies beside the river to make the erosion. But we see a work now going on there from year to year, which needed only time enough to have excavated the whole gorge, and time is an element for which a multitude of geological facts make an almost unlimited demand. At Niagara Falls 670,000 tons of water are precipitated into the gulf every minute. The upper shelf of rock is quite hard, but the layers of strata beneath are worn out by the dripping water, and then the weight causes the hard crust to break off from time to time in large masses. The rate of retrocession has been loosely estimated to average from one foot to one yard per year. But this rate, if correct, would not be what it was when the fall was nearer Lake Ontario, nor what it will be as it approaches Lake Erie, because in both cases the rocks are different.

2. On Genesee river, in New York, we find very striking evidence of erosion. Between its mouth and Rochester, seven miles, are three cataracts,

some miles apart; and some of these falls have receded further than others, because there are three kinds of rock crossed, which are worn away with different degrees of rapidity. South of Rochester we find a gorge worn 14 miles long, from Mount Morris to Portage, sometimes 350 feet deep, with three distinct falls, originating in the same cause as that above mentioned, and which proves beyond question that the river has done the work.

3. On the Potomac, ten miles west of Washington, the Great Falls have worn out a gorge from 60 to 70 feet deep, and four miles long, in hard mica schist.

4. *Cañons.*—In our southwestern States, New Mexico, Arkansas, etc., where for days the traveler finds no trees, the rivers have cut long and deep gulfs into the horizontal strata. Their existence is not suspected till a person finds himself arrested on the brink of a precipice, often hundreds of feet high, at the foot of which, and bounded by an opposite wall, the stream runs. These gulfs are called *Cañons*, and often they are so long that for days the traveler can find no place to cross, nor to water his animals. They extend also up the tributaries; a conclusive proof that the streams themselves, and not convulsions have produced them. The Grand Cañon on the Canadian river is 250 feet deep, and 50 miles long. The Cañon of Chelly, in New Mexico, has walls from 100 to 800 feet high, and 25 miles long. Captain Marcy describes a Cañon on Red River, in Texas, 70 miles long, and from 500 to 800 feet deep. The annexed sketch of a *Cañon*, on *Psuc-see-que Creek*, in Oregon, will give a good idea of this class of phenomena. (Fig. 86.)

Lieut. Ives statements in his Report of 1858, respecting the gorges upon the Colorado river, in California, are almost incredible, and are certainly without a parallel. At the head of navigation, the deep and narrow current of the river flows between massive walls of rock which rise sheer from the water for over a thousand feet, seeming almost to meet in the dizzy height above. The sun rarely penetrates the depths of this "Black Cañon," 25 miles in length. Above this there is a vast table land, 8000 feet above the ocean, and hundreds of miles in breadth, extending eastward to the mountains of the Sierra Madre, and stretching far north into Utah. The Colorado and its tributaries, flowing to the south-west, have cut their way through this immense plateau, making cañons which are in some places *more than a mile in depth.* The streams form a labyrinth of yawning abysses, generally inaccessible.

So numerous and so closely interlaced are the cañons, that often they leave only scattered remnants of the original plateau—narrow walls, isolated ridges, and slender, seemingly tottering spires, shooting up to an enormous height from the vaults below.

5. Upon the eastern continent we would refer to the *Wadys* of Syria and Palestine; to the Via Mala, on the upper part of the Rhine, 1600 feet deep, 4 miles long, and only 20 feet wide; to the Defile of Karzan, on the Danube, 200 yards wide, and 2000 feet deep; a gorge on the Sutlej river, among the Himalayahs, 1500 feet deep, and a mile long; to a gorge in Australia, on Cox river, 800 feet high, and 2200 yards wide; and to a multitude of other examples.

Rivers accumulate materials in parts of their channels, or along their banks. *Terraced valleys* and *levees* are the results of this agency. The terraces are objects of great importance in our reasonings; the levees are akin to, and connected with deltas. The large rivers do not carry all their detritus to the delta, but deposit some of it along their sides. In times of freshets these deposits

Fig. 86.

are made upon the banks, above the ordinary level of the water. Thus great embankments will be formed upon each side. These levees are augmented by the agency of man, for the protection of his property from overflow. The river Po, in Italy, is restrained within proper bounds by levees higher than the roofs of the houses of the cities protected by them. The Mississippi is confined by levees for a considerable distance above and below New Orleans. A serious breach or *crevasse* in the dykes would inundate the city and vicinity.

Deltas of Rivers.—The delta of the Mississippi, the father of waters, has formed most of the lower part of Louisiana, and has advanced several leagues since New Orleans was built. All large rivers enter the ocean by several mouths like this. The delta of the Ganges commences 220 miles from the sea, and has a base 200 miles long, and the waters of the ocean at its mouth are muddy 60 miles from the shore. Since the year 1243 the delta of the Nile has advanced a mile at Damietta; and the same at Foah since the 15th century. In 2,000 years the gain of the land at the mouth of the Po has been 18 miles, for 100 miles along the coast. The delta of the Niger extends into the interior 170 miles, and along the coast 300 miles, so as to form an area of 25,000 square miles.

The Delta of the Ganges with its numerous mouths is represented in Fig. 87.

Fig. 87.

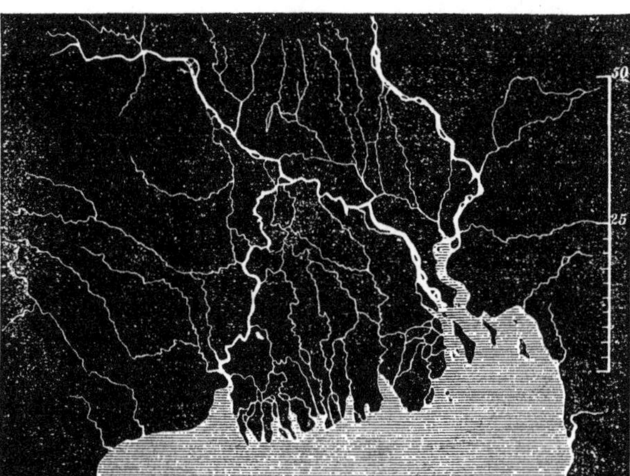

An immense alluvial deposit is forming at the mouth of the Amazon and Orinoco rivers, most of which is swept northerly by the Gulf Stream. The waters of the Amazon are not entirely mixed with those of the ocean at the distance of 300 miles from the coast. The quantity of sediment annually brought down by the Ganges amounts to 6,368,077,440 tons, or sixty times more than the weight of the great pyramid in Egypt. The Mississippi annually discharges 14,883,360,000,000 cubit feet, or 101.1 cubic miles of water, into the Gulf of Mexico; and deposits at the same time 28,188,000,000 cubic feet, or two billion tons, of solid matter. It is estimated that the whole delta contains 2,700 cubic miles of solid matter, and that 14,000 years would be required for its formation, at its present rate of growth. In Massachusetts, the matter carried down by the Merrimac has been estimated to be 840,000 tons per annum.

The extensive deposits thus forming daily by rivers need only consolidation to become rocks of the same character as the shales, sandstones, and conglomerates of the secondary series.

The delta of the Rhone, on the shores of the Mediterranean, is said to be mostly solid calcareous and even crystalline rock.

Bursting of Lakes.—A few examples have occurred in which a lake, or a large body of water long confined, has broken through its barrier and inundated the adjacent country. An interesting example of this kind occurred in 1810, in the town of Glover, in Vermont, in which two lakes, one of them a mile and a half long and three-fourths of a mile wide, and in some places 150 feet deep; and the other, three-fourths of a mile long, and a half a mile wide, were let out by human labor, and being drained in a few minutes, the waters urged their way down the channel of Barton river, at least 20 miles to Lake Memphremagog, mostly through a forest, cutting a ravine from 20 to 40 rods wide, and from 50 to 60 feet deep; inundating the low lands, and depositing thereon vast quantities of timber.

In 1818, the waters of the Dranse, in Switzerland, having been long obstructed by ice, burst their barrier and produced still greater desolation, because the country was more thickly settled than the borders of the lake above named. The enterprise of an engineer averted part of the desolation by tunneling the barrier; but not sufficient to prevent the destruction of 400 houses and many pleasure grounds.

It has been supposed, that should the falls of Niagara ever recede to Lake Erie, a terrible inundation of the region eastward would be the result; but De la Beche has proved satisfactorily that the only effect would be a gradual draining of Lake Erie, with only a slight increase of Niagara river.

Pond and Lake Ramparts.—These have not yet been described in any works on geology. Around the borders of some not very deep lakes and ponds in high latitudes, ridges or embankments of bowlders have been formed, the outer being the steepest side. So perfect are the walls thus produced, that many have supposed them to have been the work of aborigines. In Wright county, Iowa, there is a rampart ten feet high, composed of boulders from fifty pounds to three tons in weight, surrounding a lake 1,900 acres in extent, and from two to twenty-five feet deep. But there are no scattered bowlders in the water or in the vicinity of the lake upon the shore. Several lakes and ponds in Vermont, also, have these ramparts about parts of their shores; as at the north part of Willoughby Lake, at Averill and Franklin ponds.

As this phenomenon has no connection with glaciers or drift, we venture to propose a theory for it here. We think that the bowlders composing the ramparts have been brought from the bottoms of the lakes, and pushed upon the shore by the outward expansion of the water in freezing. Instances are on record where large stones of tons weight have been moved several feet in a single season. And if a few inches progress only be made in a single winter, a hundred winters might witness the removal of all the blocks in shallow water to the shore, and the crowding of them into a ridge having the form of a rampart. A similar phenomenon on the shores of Lake Onega, in Russia, is described by Sir Roderick Murchison, and explained in an analogous manner.

AGENCY OF THE OCEAN.

The ocean produces geological changes in three modes: 1. By its waves; 2. By its tides; 3. By its currents. Their effect is twofold: 1. To wear away the land; 2. To accumulate detritus so as to form new land.

The action of waves or breakers upon abrupt coasts, composed of rather soft materials, is very powerful in wearing them down, and preparing the detritus to be carried into the ocean by tides and currents. During storms, masses of rocks, weighing from ten to thirty tons, are torn from the ledges, and driven several rods inland, even up a surface sloping with a considerable dip towards the ocean.

In the 13th century, a strait half as wide as the channel between England and France, was excavated in 100 years in the north part of Holland; but its width afterwards did not increase. The English channel also is supposed to have been formed in a similar manner. In England, several villages have entirely disappeared by the encroachments of the sea. At Cape May, on the north side of Delaware Bay, the sea has advanced upon the land at the rate of about 9 feet in a year; and at Sullivan's Island, near Charleston, South Carolina, it advanced a quarter of a mile in three years. But perhaps the coast of Nova Scotia and New England exhibits the most striking examples of the powerful wasting agency of the waves, whose force there is often tremendous, especially during violent northeast storms. Where the coast is rocky, insulated masses of rocks, called *Drongs*, are left on the shore, giving a wild and picturesque effect to the scenery, as in the following sketch, Fig. 88, which was taken upon Jewell's Island, in Casco Bay, Maine.

Fig. 88.

Drongs on Jewell's Island, Casco Bay.

Fig. 89 shows Drongs, or Needles, on the coast of England. Fig. 90 exhibits the famous Cheesewring, near the Lizard, in Cornwall, England. Remarkable as such columns are, we might fill pages with sketches of similar ones, and many of them in our own country.

Fig. 89.

Purgatories.—When the rocks exposed to the waves are divided by fissures, running perpendicular to the coast, the mass between two fissures is sometimes removed by the water, thus leaving a chasm, often of great size, into which the waves rush during a storm with great noise and violence. Such fissures have been called *purgatories*, in New England, when they are quite narrow. They are only examples, on a small scale, of the Fiords of Norway, Greenland, and the northeastern coasts of North America. Well known examples of these Purgatories occur in the vicinity of Newport, Rhode Island. Similar fissures exist in the interior, as in Sutton and Great Barrington, Massachusetts.

Beaches of Shingle and Sand.—The shingle, or perfectly water-worn pebbles of a coast, and sand, are sometimes driven upon the shore by the waves, so as to form beaches; and sometimes even large bowlders are thus urged inland by powerful storms, so as to lie in a row on the shore. In some cases of this sort, after the beaches have been formed, the waves rather protect the coast than encroach upon it.

Fig. 90.

WAVES AND TIDES.

The effects of waves upon shores is very great. Those of the ordinary size will form wave ridges, or *ripple marks*, at the depth of 70 feet, upon a sandy bottom; and in violent gales the bottom may be disturbed at the depth of 400 feet.

A bowlder containing more than 200 cubic feet was hurried up an acclivity to the distance of 150 feet, upon one of the British islands. Near Edinburgh it was ascertained, by careful experiment, that the average force exerted by waves in summer was 611 pounds every square foot; and in the winter 2,086 pounds for every square foot. At the Bell Rock Lighthouse, in the German Ocean, during a violent storm, the pressure exerted was nearly three tons to every square foot. When we reflect that the weight of bodies in water is but little more than half their weight in air, we shall see that great effects may be caused even by ordinary waves.

In large inland bodies of water, such as the Mediterranean, Black and

Caspian Seas, and Lake Superior, tides are scarcely perceptible; never exceeding a few inches; and in the open ocean they are very small; not exceeding two or three feet; but in narrow bays, estuaries, and friths, favorably situated for accumulating the waters, the tides rise from 10 to 40 feet; and in one instance even 60 or 70 feet on the European coasts; and in the Bay of Fundy, in Nova Scotia, 70 feet. In such cases, especially where wind and tide conspire, the effect is considerable upon limited portions of coast, both in wearing away and filling up.

When tides enter rivers, the water is forced to rise suddenly, in consequence of the contraction of the channel. This produces a wave as high as the tide, called "the Bore," which rushes up the channel with great rapidity, and acts powerfully as a denuding agent. Upon Calcutta river, it is called the "*Eagre*," and its approach is much dreaded by ship owners.

Earthquake Waves, or Waves of Translation, are powerful agents of erosion. They constitute one of the phenomena of earthquakes. Their effects are unusually disastrous, because the water itself is moved along bodily. They have been known to attain the height of sixty feet, and to move at the rate of twenty miles in a minute. One of these huge waves rose and fell eighteen times upon the coast of Africa in 1755.

OCEANIC CURRENTS.

Oceanic currents are produced chiefly by winds. Modern researches have revealed the existence of a great number of these ocean rivers. The most extensive of them is the Gulf Stream. An equatorial current from the Southern Atlantic empties into the Caribbean sea, receiving the waters of the Amazon and Orinoco on its way, and is thus a feeder of the Gulf Stream. It properly commences in the Gulf of Mexico, whence it issues by the Strait of Florida, and one part of it stretches northeasterly, passing along the coast of the Atlantic States, and extending beyond Norway and Spitzbergen; it is common to find tropical fruits and pieces of wood transported by this current from the West Indies to the Hebrides. The other branch passes from Florida across the Atlantic towards Madeira, uniting with a current down the west coast of Africa. This is a warm current, but is divided into alternate warm and cold portions. Its velocity is variable; but may be stated as from one to three and even four miles per hour; its mean rate being 1.5 mile. Its velocity decreases towards the northeastern extremity. Return currents originate about the North Pole, or come through Behring's Straits, and pass south, partly through Baffin's Bay, and partly east of Greenland, uniting on the Labrador coast and passing along the coast of British America and the United States to Florida. The latter is a cold current. In the Indian and Pacific Oceans there are great currents, particularly a cold current setting out from Cape Horn, which continues along the coast to Central America, then crosses the Pacific towards Borneo, and loses itself south-westerly in the Antarctic regions.

A constant current sets into the Mediterranean through the Straits of Gibraltar, at less than half a mile per hour. It has been conjectured, but not proved, that an under current sets outwards through the same strait, at the bottom of the ocean. Lyell also suggests that the constant evaporation going on in that sea may so concentrate the waters holding chloride of sodium in solution, that a deposit may now be forming at the bottom. But the deepest soundings yet made there, (5,880 feet), brought up only mud, sand, and shells. Numerous other currents of less extent exist in the ocean, which it is unnecessary to describe. They form, in fact, vast rivers in the ocean, whose velocity is usually greater than that of the larger streams upon the lands.

The ordinary velocity of the great oceanic currents is from one to three miles per hour; but when they are driven through narrow straits, especially with converging shores, and the tides conspire with the current, the velocity becomes much greater, rising to eight, ten, and even in one instance, to fourteen miles per hour. The depth to which currents extend has not been accurately determined. Experiments indicate that they may sometimes reach to the depth of more than 500 feet. It ought to be remembered, however, that the friction of water against the bottom greatly retards the lower portion of the current; so that the actual denuding and transporting power in these currents is far less than the velocity at the surface would indicate.

Alike uncertain are the data yet obtained for determining what velocities of water at the bottom are requisite for removing mud, sand, gravel, and bowlders. It has been stated, however, (and these are the best results yet obtained,) that 6 inches per second will raise fine sand on a horizontal surface; 8 inches, sand as coarse as linseed; 12 inches, fine gravel; 24 inches per second, will roll along rounded gravel an inch in diameter; and 36 inches will move angular fragments of the size of an egg. The velocity necessary for the removal of large bowlders has not been measured. A velocity of 6 feet per second would be 4 miles per hour; of 8 feet per second, 5.4 miles per hour; of 12 feet per second, 8.2 miles per hour; of 24 inches per second, 16.4 miles per hour; of 36 feet per second, 24.6 miles per hour. Fine mud will remain suspended in water that has a very slight velocity, and often will not sink more than a foot in an hour; so that before it reached the depth of 500 feet it might be transported, by a current of 3 miles per hour, to the distance of 1,500 miles.

It hence appears that most rivers, in some part of their course, especially when swollen by rains, possess velocity of current sufficient to remove sand and pebbles; as do also some tidal currents around particular coasts; but large rivers, and most oceanic currents, can remove only the finest ingredients; and as to large bowlders, it would seem that only the most violent waves and mountain streams can tear them up, and roll them along.

Oceanic currents have the power greatly to modify the situation of the materials brought to the sea by rivers and tides, and to spread them over surfaces of great extent.

Thus the waters of the Amazon, still retaining fine sediment, are found on the surface of the ocean 300 miles from the coast, where they are met by the equatorial current, which runs there at the rate of four miles per hour. Thus are these waters carried northerly along the coast of Guiana, where an extensive deposit of mud has been formed, which extends an unknown distance into the ocean. In like manner the muddy waters of the Orinoco and other rivers are swept northerly. Scoresby counted 500 icebergs starting from the frozen regions, at one time, for the south. Doubtless a great part of the Banks of Newfoundland is produced by the deposition of the materials from these bergs.

Of the above agents of erosion the ocean has, without doubt, been by far the most potent. It must be borne in mind, that our present continents, certainly North America, have been several times submerged beneath the ocean, and again elevated above it by slow vertical movements; so that every part of these countries has been again and again subjected to the long-continued action of the waves and currents; in other words, every portion of the surface has been repeatedly the shore of the ocean, against which its waves, tides, and currents, have impinged as fiercely as they now do. During the Silurian and Devonian periods the surface, composed of rocks of that age, must have been beneath the ocean. But during the Carboniferous period, large portions at least must have been above the waters, to furnish the gigantic vegetation which was converted into coal. Subsequently that same surface, in some countries certainly, must have gone down to receive the thick marine beds of the Oolite and Chalk. During the Tertiary period, there appears to have been sometimes an alternation of salt and fresh water deposits. But subsequently it seems the whole of our western continent was submerged, and then again raised essentially to its present height.

AMOUNT OF DENUDATION

The great amount of denudation that has been the result of these several agencies may be learned by the following facts:

1. The great abundance of loose materials, often hundreds of feet thick, that are spread over the surface almost everywhere.

2. The evidence of the action of existing agencies. Upon sea coasts, cliffs are rapidly worn away; along rivers, deep gorges and valleys are excavated. In districts where the strata are but slightly inclined, outlying precipices and isolated hills were once continuous with each other. So vast has been the denudation, that we must call in the aid of the ocean in addition to rivers for its accomplishment. Such a process has gone on, for example, in the Palæozoic rocks in the Appallachian coal basin, to form valleys for the Ohio river and its tributaries; also in the Mesozoic rocks of the Connecticut river valley; where probably 1,000 square miles of surface have been denuded of sandstone to the depth of at least 1,000 feet.

3. All the fossiliferous consolidated rocks, six or seven miles thick in some places, are formed of materials eroded from older strata, stratified or unstratified; and probably, also, all the stratified unfossiliferous rocks, whose thickness is of equal amount.

4. The most striking evidence of the enormous extent of erosion is found in the vast amount of materials that must be supplied to fill up deficiencies in the strata. We never doubt but that a gorge in horizontal strata, (as B in Fig. 91), was once filled with sedi-

Fig. 91.

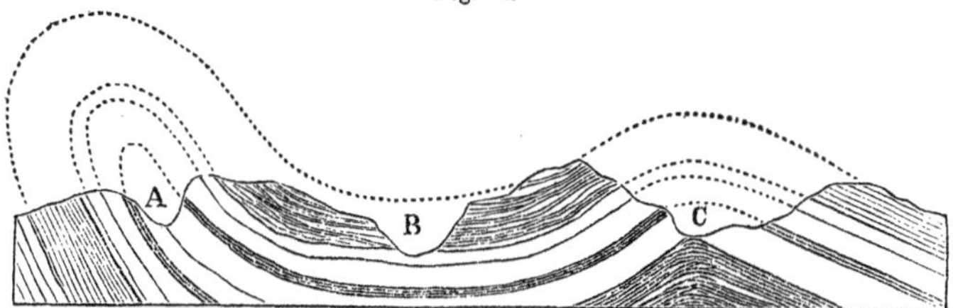

ments connecting the two sides. So when we find the same strata upon both sides of a valley of elevation, dipping in opposite directions, (as C in the same figure), we conclude that they once were joined together; and upon geological sections, such former extension is usually represented by dotted lines above the present surface. Another case is illustrated at A. A valley has been excavated in nearly perpendicular strata. As this is not their normal position, we endeavor to ascertain their former extent.

The amount of eroded material cannot be ascertained as accurately as at C, because the height of the original fold above the present surface is often a matter of conjecture. Hence we estimate not merely the amount sufficient to fill the valley, but the probable extent of the contiguous strata before their removal.

Upon these principles English geologists have ascertained that in South Wales, and the adjacent English counties, a mass of rock from 3,000 to 10,000 feet thick has disappeared; in other words, the country was two miles higher than it is now. A few measurements of this kind have been made in the United States. The junior author of this book has found that 5,000 feet of strata have been removed from an anticlinal valley in Brattleboro, Vermont; and that nearly 10,000 feet of vertical thickness have disappeared from the surface at Shelburne Falls, Massachusetts.—*See Final Report on the Geology of Vermont.*

From these investigations it may be inferred that the matter torn from the present surface was far greater than all which still remains above the level of the ocean.

CHEMICAL DEPOSITS FROM WATER.

Calcareous Tufa, or Travertin.—In certain circumstances water holds in solution a quantity of carbonate of lime, which is readily deposited when those circumstances change. The deposit is called *travertin*, or *calcareous tufa.*

At Clermont, in France, a single thermal spring has deposited a mass of travertin 240 feet long, 16 feet high, and 12 feet wide. At San Vignone, in Tuscany, a mass has been formed upon the side of a hill, half a mile long and of various thickness, even up to 200 feet. At San Filippo, in the same country, a spring has deposited a mass 30 feet thick in 20 years. And a mass is found there, 1.25 mile in length, one-third of a mile wide, and in some places 250 feet thick. In the vicinity of Rome, some of the travertin can hardly be distinguished from statuary marble; and that which is constantly forming near Tabreez, in Persia, is a most beautiful variety of semi-transparent marble, or alabaster. At Tivoli, in Italy, the beds are sometimes from 400 to 900 feet thick, and the rock of a spheroidal structure.

Marl.—The only kind of marl now in the course of formation, is that deposited at the bottom of ponds, lakes, and salt water, known by the name of *shell marl;* and which consists of carbonate of lime, clay, and peaty matter; as has been described in a preceding section. The marls in the tertiary strata are frequently indurated, and go by the name of rock marl. Much of the marl used in Virginia, and other Southern States, is composed mostly of fossil marine shells; and this is a true *shell marl.* But that usually so called contains only a small proportion of shells; the remainder being pulverulent carbonate of lime, except the clay and peaty matter, mixed with the carbonate. These beds of marl often cover hundreds of acres, and are several feet thick. In Ireland they contain bones of a large extinct species of elk, as well as shells of *Cypris*, *Lymnæa*, *Valvata*, *Cyclas*, *Planorbis*, *Ancyclus*, etc. The marls of this country contain shells of *Planorbis*, *Lymnæa*, *Cyclas*, and other small fresh-water molluscs.

A part of these marls is probably a chemical deposit. Carbonate of lime is scarcely soluble in pure water, but is abundantly soluble in water impregnated with carbonic acid. Yet the excess of acid is easily expelled, and then the salt will be deposited in a pulverulent form, unless there be some reason why it should be crystalline. As marl beds chiefly occur in the vicinity of limestone, it is easy to surmise the origin of the carbonate of lime. The tributaries convey it in solution from the ledges into the pond. There the constant evaporation of the water causes the dissolved portion to fall to the bottom. Molluscs add their shells to the mass, and at length a thick deposit will be formed. When the pond is drained or dries up, the marl may be gathered. This process may suggest the origin of many of the limestones of the older series, as the marls need only induration to resemble them completely.

Silicious Sinter.—Thermal waters alone contain silica in solution to any important amount. The most noted of these are the Geysers in Iceland, where a silicious deposit about a mile in diameter, and 12 feet thick, occurs; and those of the Azores, where elevations of silicious matter are found 30 feet high. The stems and leaves of the frailest plants are converted into sinter, or covered with it. Thermal springs, also, not in volcanic regions, as on the Washita river, in this country, and in India, deposit a copious sediment of silica, iron, and lime.

Bog Ores.—The numerous deposits of the hydrated peroxide of iron, or bog iron ore, so widely diffused, may originate from springs, from the fossil shields of animalcula, or from the decomposition of beds of iron ore or pyrites. A popular theory of the origin of bog ore is this: Waters containing organic matter from vegetable decay, reduce to the state of protoxide the peroxide of iron disseminated through sediments, and thus dissolve it. The oxygen of the air then peroxidizing the iron, it is precipitated from the water as the hydrated peroxide. Under various circumstances of change, it may become iron ore of various densities and compositions; and thus is explained the origin of iron ore of every age.

Bog manganese, or *Wad*, is almost as widely diffused through the rocks as the peroxide of iron; but its quantity is so small that it exerts but a slight influence in producing geological changes, and will therefore be passed over without description. The same remark will apply to sulphate of lime, carbonate of magnesia, chloride of calcium, etc., which occur in most natural waters, and sometimes form deposits of small extent.

Petroleum, Asphaltum, etc.—The great amount of bituminous matter with which certain springs are impregnated renders them deserving of notice as existing causes of geological change, capable of explaining certain appearances in the older rocks; many of which are highly bituminous. In the Burman Empire, a group of springs or wells at one locality yielded annually 400,000 hogsheads of petroleum. It is also found in Persia, Palestine, Italy, and the United States. In this country it has the name of Seneca Oil, from having been early observed on the surface of springs at Seneca, in New York. It is thrown up in considerable abundance, also, at the salt borings on the Kenawha river, in Ohio; where a few years ago a large quantity of it, floating on the surface of a small stream, took fire, and the river for half a mile in extent appeared a sheet of flame. A deposit of coal oil near Titusville, Crawford county, Pennsylvania, upon Oil Creek, is now (1860), attracting much attention.

In Palestine the Dead Sea is called the lake Asphaltites, from the asphaltum which formerly abounded there. But the most remarkable locality of bituminous matter is the Pitch Lake, in the island of Trinidad, in the West Indies. It is three miles in circumference, and of unknown thickness. It is sufficiently hard to sustain men and quadrupeds; though at some seasons of the year it is soft.

Mineral pitch was a principal ingredient in the cement used in constructing the ancient walls of Babylon, and of the temple in Jerusalem. It has lately been employed in a similar manner, and it is said very successfully, to form a composition for paving the streets of cities.

The various bitumens are produced from vegetables by the processes by which these are converted into coal in the earth.

Hence the bitumens that rise to the surface of springs, or form inspissated masses on the earth's surface, or between the layers of rocks, are supposed to be produced from vegetable matters buried in the earth; and to be driven to the surface by internal heat; and the fact that such deposits usually occur in the vicinity of active or extinct volcanos, gives probability to this theory.

Phenomena of Springs.—Water is very unequally distributed among the different strata; some of them, as the argillaceous, being almost impervious to it; and others, as the arenaceous, admitting it to percolate through them with great facility. Hence, when the former lie beneath the latter in a nearly horizontal position, the lower portions of the latter will become reservoirs of this fluid.

Hence, if a valley of denudation cuts through these pervious and impervious strata, we may expect springs along their junction.

In Fig. 92, if B, B, be the pervious, and C, C, the impervious strata, and a valley of denudation has been excavated in them, we may expect springs at E, E. If a fault occurs where pervious and impervious strata join each other, the water will be accumulated in the lowest portion of the pervious strata, and we may expect to find a spring at L.

And vice versa, the geologist can sometimes discover the line of a fault by the occurrence of mineral springs, where nothing else indicates its existence at the surface.

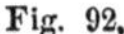

Fig. 92.

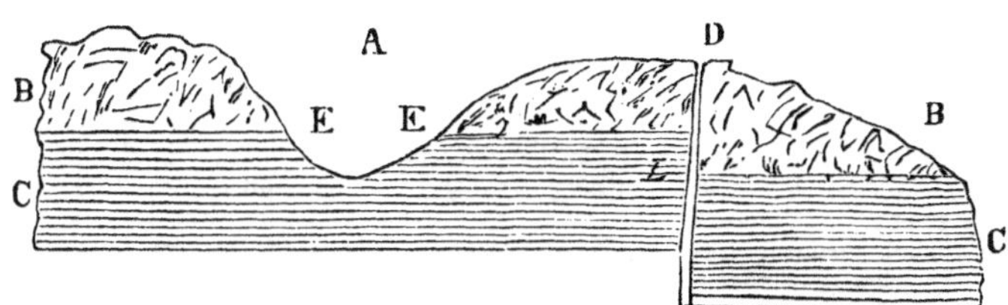

In many parts of the world if the strata be penetrated to a considerable depth by boring, water will rise sometimes with great force to the surface, and continue to flow uninterruptedly. Such examples are called *Artesian Wells*, from having been first discovered in the province of Artois, in France, the ancient Artesium.

The theory of these wells is simple. In Fig. 93, suppose the stratum M, M, to be pervious to water, while the rocks above and below, A, A, and B, B, are impervious. The result is, that all the water which accumulates in the stratum M, M, will press toward the lower part of the basin. If now an opening be made at any place lower than the outcrop of the stratum, the water will be forced above the surface in a jet, by hydrostatic pressure, and an artesian well will be the consequence.

Fig. 93.

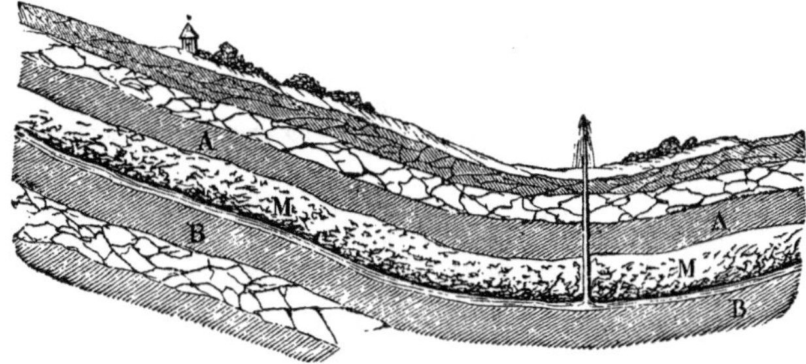

We infer from this theory: 1. If any water bearing stratum, passing under a place where boring is attempted, rises higher at any point of its prolongation than the surface where the boring is made, the water will rise above that surface, and it will fall as much below that surface as is the level of the highest part of the pervious stratum.

2. Hence borings of this sort may fail; first, because no water bearing stratum is reached; and, secondly, because that stratum does not rise high enough above the place to bring the water to the surface.

3. These explorations have proved that subterranean streams of water exist; some of which have a communication with the water at the surface.

EXAMPLES.—At St. Ouen, in France, at the depth of 150 feet, the borer suddenly fell a foot, and a stream of water rushed up. At Tours the water brought up from the depth of 374 feet fine sand, vegetable matter, and shells of species living in the vicinity, which must have been carried to that depth within a few months preceding. In Westphalia the water brought up several small fish, although no river existed at the surface within several leagues. The borings in the United States prove that cavities containing water exist even in granite.

Depth of the Borings.—One of the deepest wells in this country is at Louisville, Kentucky. It is 2,086 feet deep. It discharges 330,000 gallons of water every twenty-four hours, which rises to the height of 170 feet above the surface. The aperture is three inches in diameter. The water is much warmer than the average of the surface water, being 76½°, and it is unaffected by the external temperature. In Columbus, Ohio, one of these wells was 1,858 feet deep in December, 1858, and is said to have been carried several hundred feet lower since that time. In Paris there is an artesian well at Grenelle, 1,800 feet deep, and capable of producing 14,000,000 of gallons of water daily. At Niondorf, in Germany, there is one 2,247 feet deep. In the Duchy of Luxembourg, an excavation was made several years ago, to reach a stratum of salt water, which had been carried to the depth of 2,336 feet, in 1847.

Natural deserts may sometimes be changed into regions of fertility by these wells. Several wells have been bored in the Llano Estacado, in Texas, but we believe without much success. Upon the great desert of Sahara, in Africa, five of these wells, called "Wells of Gratitude," have been excavated, to the great relief of the nomadic tribes roaming there, as well as of travelers.

Thermal Springs will be considered elsewhere.

Mineral Springs.—All waters found naturally in the earth contain more or less of saline matter; but unless its quantity is so great as to render them unfit for common domestic purposes, they are not called mineral waters.

The ingredients found in mineral waters are the sulphates of ammonia, soda, lime, magnesia, alumina, iron, zinc, and copper; the nitrates of potassa, lime, and magnesia; the chlorides of potassium, sodium, ammonium, barium, calcium, magnesium, iron, and manganese; the carbonates of potassa, soda, ammonia, lime, magnesia, alumina, and iron; the silicate of iron; silica, strontia, lithia, iodine, bromine, and organic matter; the phosphoric, fluoric, muriatic, sulphurous, sulphuric, boracic, formic, acetic, carbonic, crenic, and apocrenic acids; also oxygen, nitrogen, hydrogen, sulphureted hydrogen, and carbureted hydrogen.

Theory.—Many of the above ingredients are taken up into a

state of solution from the strata through which the water percolates; others are produced by the chemical changes going on in the earth, by the aid of water and internal heat; and others are evolved by the direct agency of volcanic heat.

Salt Springs.—The most important mineral springs in an economical point of view are those which produce common salt. These are called salines, or rather such is the name of the region through which the springs issue. They occur in various parts of the world; and the water is extensively evaporated to obtain table salt. They contain also other salts; nearly the same, in fact, as the ocean.

Some of these springs contain less, but usually they contain more salt, than the waters of the ocean. Some of the Cheshire springs in England yield 25 per cent.; whereas sea water rarely contains more than 4 per cent. In the United States they contain from 10 to 25 per cent. They are found in New York, Ohio, Virginia, Pennsylvania, Illinois, Michigan, Missouri, Arkansas, and Upper Canada. 450 gallons of the water at Boon's Lick, in Missouri, yield a bushel of salt; 300 gallons at Conemaugh, Penn.; 280 at Shawnee town, Ill.; 120 at St. Catharine's, U. C.; 75 at Kenawha, Va.; 80 at Grand River, Arkansas; 50 at Muskingum, Ohio; and 41 to 45 at Onondaga, N. Y; 350 gallons of sea water yield a bushel at Nantucket. In Ohio 1,300,000 bushels of salt were manufactured from these springs in 1855. The springs in New York yield annually about six millions of bushels, and those in Virginia three and a half millions. In all these places deep borings are necessary, sometimes even as deep as 1,000 feet; and usually the brine becomes stronger the deeper the excavation.

Origin of Salt Springs.—In many parts of Europe, salt springs are found rising directly from beds of rock salt, so that their origin is certain. In this country, beds of rock salt have been found in Virginia, and they doubtless exist wherever salt springs occur. The springs in this country issue almost invariably from the Silurian rocks.

Gas Springs.—Carbonic acid, and carbureted hydrogen, are the most abundant gases given off by springs. They sometimes escape from the soil around the springs, over a considerable extent of surface, and produce geological changes of some importance. Carbonic acid, for example, has the power of dissolving calcareous rocks, and of rendering oxide of iron soluble in water. It contributes powerfully also to the decomposition of those rocks that contain feldspar. Carbureted hydrogen is sometimes produced so abundantly from springs that it is employed, as at Fredonia, in New York, in supplying a village with gas lights. At Charlestown, in Western Virginia, and in Pomeroy, Ohio, it is so abundant that it has been employed for boiling down the salt water that is driven up by it with great force. In almost all the States west of New England this gas rises from springs in greater or less abundance, generally from salt springs.

Origin of these Gases.—Some of these gases, as carbonic acid, are given off most abundantly from springs in the vicinity of volcanos; and in such a case there can be no question but they are produced by decompositions from volcanic heat. When they proceed from thermal springs, there is a good deal of reason for believing that internal heat may have produced them. But where they rise from springs of the common temperature, they must generally be imputed to those chemical decompositions and recompositions that often occur in the earth without an elevated temperature. Although carbureted hydrogen may sometimes proceed from beds of coal, it may also proceed from other forms of carbonaceous matter; as from bitumen disseminated through the rocks.

SURFACE GEOLOGY.

By *Surface Geology* is meant the history of the superficial deposits which have accumulated upon the earth since the tertiary period. It is the geology of the Alluvial Period. We have already described the agencies which have produced the effects, as existing causes are adequate for the work. The facts are first stated, and then the theories.

The most general division of the superficial deposits is into Drift and Modified Drift. These may be subdivided into four periods, viz.:

I. *Drift.*	II. *Modified Drift.*
1. The Drift Period,	2. The Beach and Sea Bottom Period,
	3. The Terrace Period,
	4. The Historic Period.

The agencies which produced all the different forms of Alluvium were at work in each of these periods, and are still in operation. The second period, for example, embraces the time when most of the deposits formed were beaches and sea bottoms. But these accumulations have also been made in the third and fourth periods, though not so abundantly as terraces. Hence each period receives its name from the predominant form of the deposit then made.

There is a great diversity of views in relation to Surface Geology among geologists. We present the subject in the light which, after much study and observation, appears to us most probable.

I. DRIFT.

Unaltered or unmodified drift is a mixture of abraded materials, such as bowlders, gravel, and sand, blended confusedly together, and piled up by some mechanical force that has pushed it along over the surface. Yet in some places the materials are somewhat stratified and laminated, as if by water. In other cases we find more or less of an alternation of finer materials, such as sand and gravel, with the coarser unstratified accumulations mentioned above.

All the great bowlders scattered over the surface belong to the unmodified variety. Some of the bowlders are scratched, thus showing collision with other rocks.

A common variety of drift is the *bowlder clay*. This is a heterogenous mixture of a stiff, dark-bluish clay, with rounded and striated pebbles and bowlders of all sizes. It is very common to find it exposed on the banks of streams which are so precipitous as to prevent the growth of vegetation upon them. Marine shells are found in this clay in Scotland and England. Generally the shells are crushed to fragments, which are more or less comminuted. About thirty species have been found, most of which live in the vicinity at the present time, but a few of them are more boreal in their character, being adapted to the climate of Iceland or Greenland.

The coarse drift lies upon some older formation, though sometimes deposits of clay or sand intervene. It is usually succeeded upward by regular stratified deposits of the same materials, which have been reduced to a finer state, sorted into finer or coarser layers, and deposited in more and more delicate layers as we ascend. These deposits, mainly horizontal, may be called *Modified Drift.*

Fig. 94.

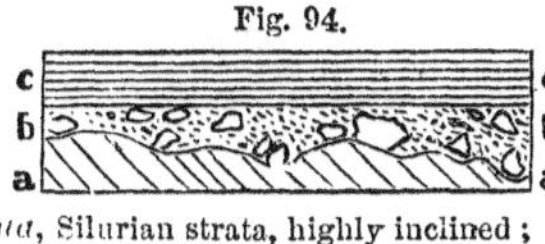

aa, Silurian strata, highly inclined;
bb, Drift;
cc, Modified Drift.

Fig. 94 illustrates the position of the unmodified drift; *e. g.*, lying unconformably upon Silurian rocks, and overlaid by modified drift.

Drift is easily distinguished from the subjacent tertiary strata, by superposition, by the marks of a much more powerful mechanical agency in its production, and by the absence of organic remains; for probably in most cases where organic remains have been reported in drift they have been derived from modified drift.

We can see from the preceding remarks that it is not easy to say precisely where is the line between drift and modified drift; but it is easy to distinguish between the coarse irregular beds of bowlders, gravel, and sand, lying immediately upon the older rock, and the fine stratified deposits of clay, sand, and loam, that lie much higher, and frequently form the banks of rivers. We can see that the latter have been produced from the former by the comminuting, sorting, and re-depositing power of water, as the chief agent; whereas we can hardly account for the formation of the coarse drift without the aid of ice in some form.

Dispersion of Drift.—It is a characteristic of drift, by which it is distinguished from disintegrated rock, that it has been removed from its original position, it may be only a few rods, but more often a great many miles. And by the bowlders and trains of gravel and sand wh... it has left along the way we can trace it back to its origin.

In the dispersion of drift we find the evidence of two distinct phases of action, which may, however, have been the result of the same general cause, operating in different circumstances. In the first case, the drift has been carried out from the summits and axes of particular mountains along the valleys, and spread over the neighboring plains.

An example of this mode of dispersion exists in the Alps. The bowlders there have usually been carried down the valleys, and they exist in the greatest abundance opposite the lower opening of those valleys.

Northern Scandinavia is another example of a centre of dispersion for drift. Norwegian bowlders are found in a southwest direction, in England and Denmark; in a southerly direction they are found in Prussia; east and northeast, in Russia, and northerly in Russian Lapland.

In the second phase of this action the force seems to have operated on a wider scale, having driven the materials in a southerly direction, over most of the northern part of the American continent, and over a part of Europe. It is probable, however, that if we could learn more of the drift in high latitudes, where the ground is covered with snow most of the summer, we should find a point beyond which the bowlders took a north direction. Indeed, in McClintock's explorations in search of Sir John Franklin, from 1857 to 1859, he found several examples in North Lat. 74°, where bowlders had been transported from 100 to 200 miles north and northwest of the parent rock. If we could be sure that there is no mistake as to these facts, it would settle the question as to the northern direction of the drift on this continent. At any rate, we have reason to suppose that some of our high mountain chains may have been centres or axes from which glaciers, as in the Alps, have proceeded outwardly. We attempt elsewhere to prove that the range of the Green and Hoosic Mountains in New England, once formed such an axis. The White Mountains, in New Hampshire, and the mountains of Essex county, in New York, also, may be found in future to have been such centres of dispersion.

There are three general directions in which bowlders have been transported, in this country: to the southwest, to the south, and to the southeast. Those from the northeast to the southwest are the least common; hence it is supposed that these were transported the earliest. Those from the northwest to the southeast are the most common. This course carried the bowlders very obliquely across the precipitous ridges of the Green and Hoosic mountains, in New England, for example, without deviating from a right line. The largest blocks usually lie nearest to the bed from which they were derived, and they continue to decrease in size and quantity for several miles; sometimes as many as 50 or 60, and not unfrequently even 100 miles, though usually the sea-coast is reached short of that distance. The islands off the coast are covered with detritus derived from the mainland.

In the Western States large bowlders of Azoic rocks are found scattered over Silurian and Devonian strata; and are significantly called *lost rocks.* About Lake Superior, the bowlders have been driven in a southwesterly direction. Around the Lake of the Woods the course is nearly from north to south.

The distance to which bowlders have been driven from their native beds in this country is very great. In New England they have been traced rarely more than from 100 to 200 miles. In Ohio and Michigan, Azoic bowlders are very common, which have been transported from the region of the great lakes. This would make their longest transit from 400 to 600 miles.

Hence the dispersion of bowlders may be of great service to the geologist. For if fragments of a peculiar kind of rock are found in any district, and it is wished to know their source, by following the direction of the drift current, as indicated by striated rocks in the vicinity, the parent ledge will be found.

In passing to the eastern continent, we find, as already stated, that on the eastern coast of England the drift came from Scandinavia and from Scotland. On the west side of England, the bowlders were carried from the northwest to the southeast. The dispersion of blocks from several local centres, as Wales, Ben Cruachen, and Ben Nime, seems to be independent of that more general force, apparently marine, that swept southeasterly over the whole island, and also over Ireland.

The drift of Scandinavia reaches as far east as the Uralian mountains. Siberia is said to be mostly free from it. In northern Syria drift phenomena have been observed. Bowlders of greenstone have been traced southerly sixty miles from their source, as far as Beirut, about 32° North Lat. In South America, beyond 41° South Lat., transported bowlders show themselves in Chile and Patagonia, where they seem to have traveled in an easterly and westerly direction. Enormous transported blocks have also been found in British Guiana.

Size and Amount of Erratic Bowlders.—Frequently the surface is almost entirely covered for many square miles with large transported blocks of stone, which are but little rounded.

The hilly parts of New England illustrate this statement. Also in eastern Massachusetts, near the coast, unmodified drift is unusually common. Fig. 95 will give some idea of a desolate landscape near Squam, in Gloucester.

Fig. 95.

View in Gloucester, Mass.

The size of single bowlders is sometimes enormous. The Needle mountain in Dauphiny, said to be a bowlder, is 3,000 feet in circumference at the bottom, and 6,000 at the top. Fig. 96, represents the block called *Pierre à Bot*,

Fig. 96.

Pierre à Bot, Switzerland.

containing 40,000 cubic feet, near Neufchatel, on the Jura. It has been transported from near Martigny, more than 60 miles, across the great valley of Switzerland. Prof. Forbes describes a bowlder in the Alps 62 feet in diameter, containing 244,000 cubic feet.

In this country bowlders occur of equal dimensions. In Danvers, Mass., there is one called the *Ship Rock*, shown in Fig. 97, and which is the property of the Essex Society of Natural History.

Fig. 97.

Ship Rock.

On the top of Hoosic Mountain, in North Adams, Massachusetts, is a bowlder of granite, called the *Vermonter*, which weighs 510 tons. It has been transported from Oak Hill, across a valley 1,300 feet deep. The *Green Mountain Giant* is represented in Fig. 98. It is 40 feet long, 36 feet wide, and 27 feet high, and it weighs 3,400 tons. This has also been brought across a valley 1,000 feet deep, from the crest of the Green Mountains. It is now located upon a mountain in Whitingham, Vermont.

At Fall River, Massachusetts, there was a bowlder of conglomerate, which originally weighed 5,400 tons, or 10,800,000 pounds; but it is all used up for building purposes.

Effects of the Drift agency upon Ledges.—One of the most remarkable effects of drift action is the smoothing, rounding, scratching, and furrowing of the surface of rocks in place. Ledges

Fig. 98.

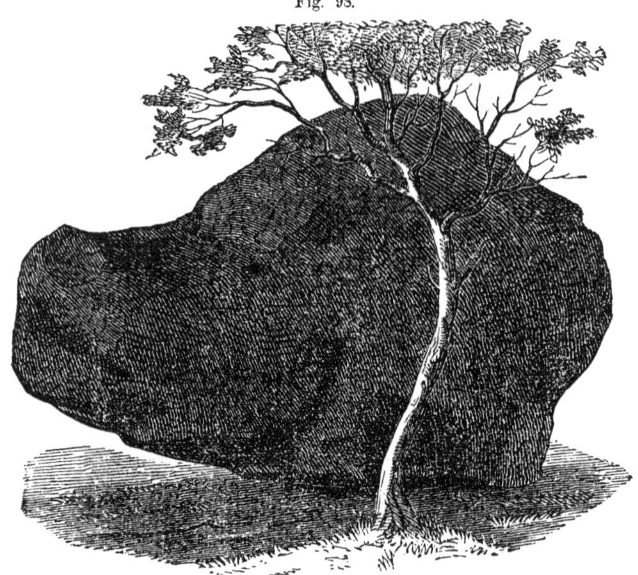

Green Mountain Giant.

that are susceptible of polish are sometimes as smooth as polished marble. Universally the ledges over which the drift materials have passed are more or less smoothed and rounded.

A careful examination of the mountains of New England shows that their north, northeastern and northwestern sides are worn and rounded throughout. An interesting example is Mount Monadnoc, in New Hampshire; which is the more striking, because it is mostly naked rock. The surface of the mountain is very uneven; but the protuberances are nearly all rounded, and few are left angular, except on the southeastern side. The axes of the intervening hollows usually correspond nearly to the direction of the striæ; so that the surface appears like the swell of the ocean after a storm. Seen in a certain direction these swells appear like domes. Fig. 99 will give some idea of a spot on the southwest part of this mountain about five rods square. This appearance corresponds precisely with that in the Alps, denominated by Saussure *roches moutonnees*, produced by glaciers.

When rocks or mountains have been thus acted upon, we can easily see which side has been struck by the denuding force, because that side is rounded or embossed. In Sweden this is called the *stoss* or *struck* side. The other is called the *lee* side.

Unless these smoothed and rounded ledges have been decom-

Fig. 99.

Embossed Rocks (Roches moutonnees), Monadnoc.

posed upon their surfaces, they are covered with *striæ*, usually parallel to one another, and indicating the exact course of the drift agency. They are rarely met with on pure limestone, unless the rock has been protected by soil; on account of its great liability to disintegration. Most of the coarse granites and conglomerates, as well as gneiss, are so much decomposed at the surface as to have lost all traces of these markings. Greenstone, syenite, and porphyry are frequently rounded and smoothed; but the markings are usually faint on account of the great hardness of the rocks. Ledges of talcose, micaceous, and argillaceous rocks retain the striæ most distinctly. Were the rocks of the Northern States to be laid bare, nearly half of the surface would show marks of this scarification. In New England the proportion would be much greater.

If we find embossed rocks, with no striæ upon them, we can determine the direction of the force by which they have been rounded, by ascertaining which is the *stoss*, and which the *lee* side. The bosses can hardly lose their form by the ordinary natural agents, because they act upon the whole surface equally. Drift action is chiefly distinguished from aqueous action upon rocks by the great evenness and uniformity of its erosion. Water will smooth rocks, but not uniformly over so great surfaces.

Care must be taken by the observer not to confound drift furrows and striæ with those grooves on the surface of rocks produced in the direction of the cleavage planes,.or the planes of stratification, by the unequal disintegration of the harder and softer parts; nor with the furrows between the veins of segregation, produced in the same manner.

The drift striæ vary in direction from northeast to southwest and northwest to southeast. Multitudes of examples may be found all over the country directed to every conceivable point between these two courses. Of these the first are probably the oldest, and the second the most recent. In New England the first set are found principally upon elevated peaks. Those from north to south are found at all altitudes.

In general, these striæ do not alter their course for any topographical feature of the country. They cross valleys at every conceivable angle, and even if the striæ run in a valley for some distance, when the valley curves the striæ will leave it, and ascend hills and mountains, even thousands of feet high. But these striæ are never found upon the south sides of mountains, unless for a part of the way where the slope is small. Mt. Monadnoc, of New Hampshire, is an illustration of these statements. It is a naked mass of mica schist, 3,250 feet high, rising like a cone out of an undulating country. And from top to bottom it has been scarified on its northern and western sides, indicated by striæ running up the mountain, at first southeasterly, and at the top at S. 10° E. There are deep furrows and other phenomena upon the summit, and the striæ continue a short distance upon the southern slope of the mountain.

Rarely do the striæ appear to have been influenced in their course by the general features of the country. In general, in great north and south valleys, they correspond to the axis of the valley; as, for example, the valley of the Connecticut, where most of the striæ run north and south. Upon the valleys of the Lamoille, Winooski, and Missisco rivers, in Vermont, the deflection from the usual course is quite marked. These rivers cross the Green mountains nearly at right angl s, running, therefore, about east and west. Upon the elevated land, averaging about 2,000 feet above the valleys, the striæ have a general southerly direction, but at the bottoms of the valleys they have an easterly direction, running up the stream. It is as if, when the highest peaks of the Green mountains were islands in the glacial ocean, a great iceberg was accidentally caught in one of these valleys, and was forced onward in an unusual direction.

Sometimes there are several sets of striæ crossing one another at a small angle, the lines of each set preserving their parallelism. Cases where two and three sets cross each other are quite common. The angle of intersection is sometimes as great as 45°. Upon Isle La Motte, in Lake Champlain, there are eight distinct directions of the striæ; the two most widely separated running S. 8° W., and S. 65° E.

Fig 100.

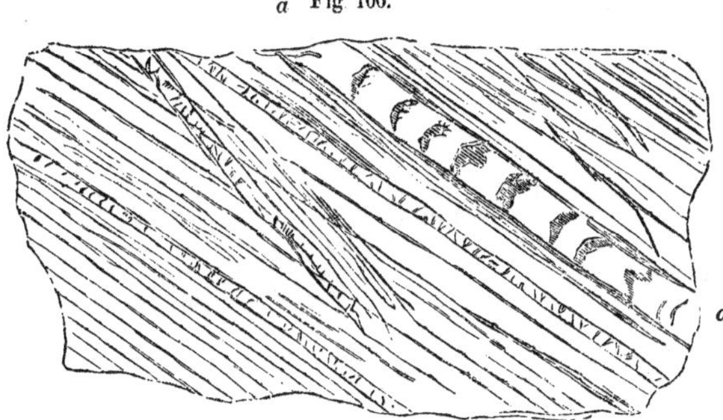

Fig. 100 represents drift striæ upon a slab of Trenton limestone from Shoreham, Vermont. This shows two facts of much interest: first, we have here a broad furrow, *a*, *a*, *flaked up* every inch or two, as could have been done only by a very heavy body moving with some friction; secondly, we have broad scratches, deviating from the common course, and at length terminating, just as would be done by a loose pebble waddling to one side and finally completely crushed beneath the heavy graver.

The summit of Mount Holyoke, in Massachusetts, which has been very much abraded by the agency under consideration, sometimes presents insulated hummocks of greenstone, resembling the "sacks of wool," described by Sefstroom, as shown in Fig. 101.

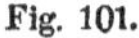
Fig. 101.

Hummock on Holyoke.

Sometimes, instead of striæ, we find the summit of a mountain ploughed into deep furrows, which enlarge so as to form deep parallel valleys.

A most remarkable example of this kind is the summit of Mount Holyoke, mentioned above. This is a narrow, very precipitous ridge of greenstone, rising 700 or 800 feet above the valley of the Connecticut, and lying in the curvilinear direction shown in Fig. 102, where the line N S represents the meridian, and corresponds to the direction taken there by the drift, which struck the mountain from the north. On that side the mountain is a nearly perpendicular wall of rock. Yet the summit is intersected with numerous grooves and valleys in the direction of the lines A, A, A, A, N S, from a few inches to several hundred feet deep. And not only do we see the marks of abrasion in the bottoms and on the sides of these valleys, but the fact that they preserve their parallelism so perfectly, although the mountain curves so much, shows that they were produced by some abrading agency rather than by the original structure or elevation of the mountain. For had they resulted from the latter causes, we might expect them to change their course to the lines B, B, B, as the mountain continued to curve more and more.

These furrows and valleys must be imputed to the joint action of ice and water. If water alone were concerned, the valleys could not have so nearly preserved their parallelism. Indeed, unless the large valley around the mountain had been filled with ice, it is difficult to see how streams of water could have flowed over its summit so as to produce these valleys. Ice alone, moving over the top, might have begun the work, (and this would explain

Fig. 102.

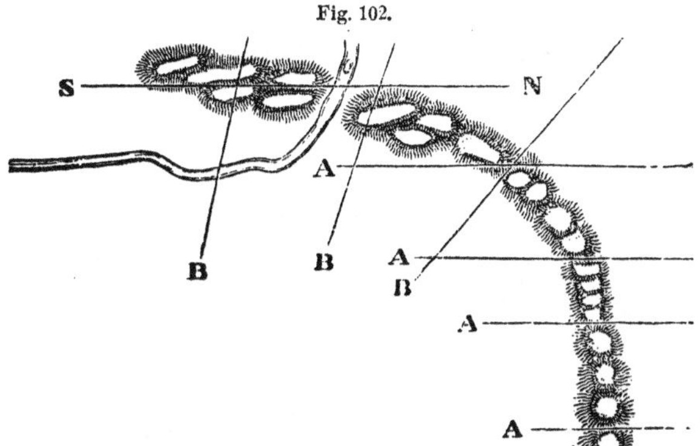

the parallelism of the valleys), but could not have made so deep erosions without wearing down the intervening ridges.

It appears that in all cases the striæ, furrows and valleys, that have been described upon the surface of rocks, correspond in direction to the course taken by the drift, and thus the two classes of phenomena are proved to have resulted from the same general cause.

Transport of Drift from Lower to Higher Levels.—The embossed and striated rocks show that in some instances the drift has been transported from lower to higher levels. On the northern slopes of mountains the striæ run from the bottom to the top, the course being shown by the stoss side, without essentially changing their parallelism. The slope up which the force has carried materials may be as great as 60°, as illustrated upon Mt. Monadnoc. The bowlders which have been carried up to the tops of these mountains will remain to attest this truth. We need only refer to the Green Mountain Giant and the Vermonter to confirm this statement.

Ledges Fractured by Glacial Action.—Sometimes the end or side of a ledge of a rock bears evidence of having been subjected to a crushing force, which has broken the strata into numberless fragments. Many quarries of building stones and roofing slate show this action, which, of course, has greatly injured their value. Fig. 103 represents one of these fractured ledges, where the crushing force must have come from the east, in Guilford, Vermont. The thickness of the crushed fragments is twenty feet.

Similar cases are found elsewhere in Vermont; near Niagara Falls, in New York; at Middlefield, and Lowell, in Massachusetts; at Newark, New Jersey; in Wales and Scotland, etc.

Fig. 103.

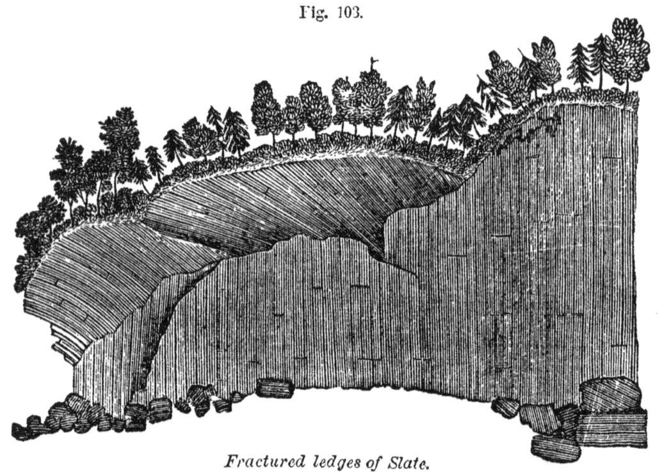

Fractured ledges of Slate.

These fractured ledges are difficult to explain. Where the strike of the cleavage is at right angles to the direction of the valley, it may be supposed that a glacier formerly descended the valley, breaking the strata, and pushing them downwards. In other cases it might be explained by the joint action of frost and gravity. If we suppose that water percolates into crevices, and freezes, it might separate the layers; and if a heavy weight of snow and ice had accumulated upon it, gravity might produce a slide. But this will not explain all the phenomena. A more probable theory is that huge icebergs or glaciers of great weight crowded along the surface might crush and displace the strata to a considerable depth.

Trains of Bowlders.—Rarely the bowlders derived from a single locality are arranged in a line or in several lines streaming off in the direction in which the drift agency operated. Such bowlders are not much rounded, and they lie upon the surface of the common drift, not being mixed with it.

Fig. 104 represents one of these trains in Berkshire County, Massachusetts. The mountains from which the angular blocks of hard talcose slate have been torn off, lies in Canaan, New York; and from thence they lie in trains, running for a few miles S. 56° E., and then changing to S. 34° E., and extending yet further, making in the whole distance not less than fifteen or twenty miles; at least one of them extends that distance, passing obliquely over mountain ridges some 600 or 800 feet high. Its width is not more than thirty or forty rods. The blocks are of all sizes, from two or three feet in diameter to those containing 16,000 cubic feet, and weighing nearly 1,400

Fig. 104.

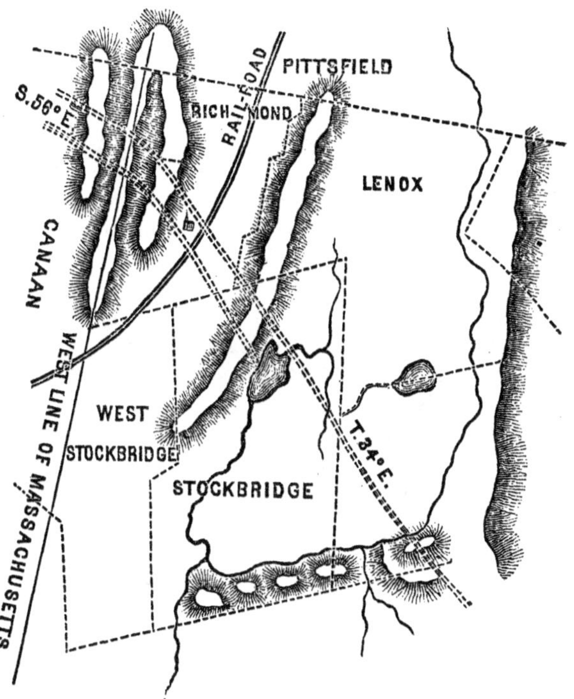

tons; and in some places they almost cover the surface. The trains lie upon the surface of the common drift, and are not mixed with it. An analogous case, one mile long, is in Huntington, Vermont.

We incline, upon the whole, to regard these trains as *Osars*, to be described under Modified Drift.

Vertical and Horizontal Limits of the Drift Agency.—The drift agency was mainly confined to the colder regions of the globe. In America it extended to the fortieth degree of latitude from both poles. Upon the eastern continent the southern limit is variable, reaching, in one case, to the thirty-second degree of north latitude. In the southern parts of both Asia and Africa there is no drift except where glaciers exist or have existed, as in the Himalayahs.

The upper limit of the drift is a little over 5,000 feet in this country. All the mountain peaks east of the Rocky Mountains are covered by its relics, except several hundred feet of the conical summit of Mount Washington, in New Hampshire. This summit is covered with angular fragments of rock which have never been removed, except by frost.

In some parts of Europe the drift agency did not extend to the tops of high mountains, nor was that upper limit horizontal. In the Alps this upper limit varies from 3,000 to 8,000 feet, and its inclination is never quite three degrees.

But the lowest level of drift agency is unknown. The striæ left by it are seen descending beneath the ocean, where it is impossible to trace them any further. The detritus from icebergs may cover the bottom of the present northern ocean, several thousand feet below the surface.

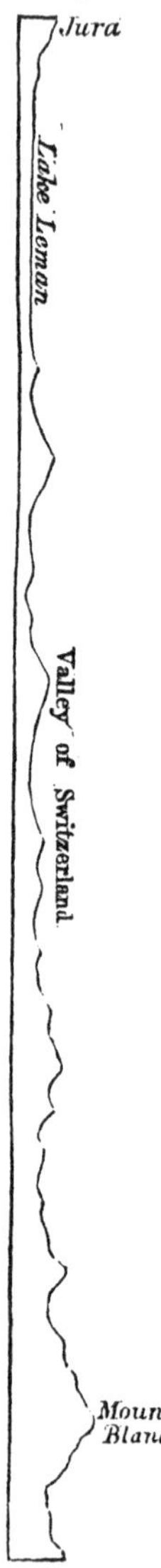

FORMER EXTENT OF GLACIERS.

The researches of Venetz, Charpentier, Agassiz, Guyot, Forbes and others, have brought to light marks of ancient glaciers in the Alps at a much lower level than those now existing, and in advance of them. The evidence consists of moraines, insulated blocks, and especially of smoothed, striated, and rounded rocks in place, produced by a force crowding down the valleys that descend from the summits of the Alps.

The theory of Charpentier, Agassiz and others, is, that the great valley of Switzerland was once filled with ice, and the blocks were carried by its motion from the Alps to the Jura. Fig. 105 will show how small must have been the declivity, much less than is now sufficient to cause a glacier to move,—none of them making much progress where the slope is not over 3°. Hence, Sir Charles Lyell and Mr. Darwin suppose that when the great valley of Switzerland was beneath the ocean, and the Alps were raised above it, and the Jura formed an island,

Fig. 105.

the glaciers, descending from the former into the ocean, sent off icebergs, loaded with blocks, which stranded on the Jura. But both theories admit the former wide extension of the Alpine glaciers.

Mt Snowden, the highest peak in Wales or England, was a center from which glaciers formerly radiated. Prof. Ramsey proves that these glaciers existed previous to the drift period, which has left its deposits to the height of 2,300 feet above the ocean, and that others have existed since that time. The high mountains of Scotland, Ben Cruachen, Ben Nime, Schiehallien, the Grampians and Ben Nevis, were evidently once the seat of glaciers. There is also evidence to prove the former extension of the glaciers of the Himalayahs, in India, far beyond their present limits.

The traces of former glaciers in the United States have been found upon the Green Mountain range in Massachusetts and Vermont. The eastern slope of this range is twenty miles wide, while its western slope is much more precipitous. Across this eastern slope several rivers have cut deep valleys, opening into the valley of Connecticut river. These streams run nearly east, while the high hills through which they pass show on their summits the striæ and other phenomena of the drift agency. The direction of these striæ is nearly north and south, deflected often toward the east from the south, and to the west from the north, a few degrees. But on the steep sides of the east and west valleys, is another set of striæ, running nearly east and west, formed by a force directed down the valleys, as is proved by the stoss side of the ledges. These could in no possible way have been produced by the drift agency, but they are precisely the effect that would be produced by glaciers sliding down the valleys towards Connecticut river from the crests of the range. The examples in Massachusetts are these: on Westfield river, in Russell; near Huntington; also in Russell, on Westfield Little river; at Sodom Mountain, in Granville; and on Deerfield river and some of its branches. In Vermont these ancient glaciers existed on the headwaters of Deerfield river, in Searsburg; at Windham and Grafton, on Saxton's river; on a branch of West river, in Jamaica; on the Otta Queechee, in Plymouth and Bridgewater; on White river and its branches; at Hancock, on the west side of the range, and elsewhere. It is probable that this range formed a crest from which glaciers descended on both sides, principally before the drift period.

Traces of glaciers in earlier periods have been supposed to exist. In England, striated blocks which can not be distinguished from those marked by modern glaciers have been found in deposits of the Permian period; and geologists have traced out the course of this ancient glacier, and find that its outline agrees with that of modern glaciers, and that its greatest length was fourteen miles.

In this country striæ have been found upon Trenton limestone, in the valley of Lake Champlain, and at Copenhagen, Lewis county, New York, which appear to have been made during the deposition of the rock itself. We should suspect also, from the great size of the fragments, that some of our Mesozoic conglomerates were produced by something like drift agency.

Distinctions between the marks of Drift and of Glaciers.—There may be no perceptible difference between the marks of drift and of ancient glaciers in many cases. But generally they may be distinguished from each other; and the following are the most important distinctions:

1. Glacier striæ differ often widely in direction from drift striæ. The drift striæ may be referred to three general directions—to the south, to the south-

east, and to the southwest—while the glacier directions are exceedingly various, sometimes coinciding with, and often crossing those left by the drift.

2. Glacier striæ occur only in valleys, while the drift striæ overtop mountains; or, when found in valleys, may cross them obliquely

3. Glacier striæ descend from higher to lower levels, except in limited spots, where they may be horizontal. Drift striæ as frequently ascend mountains hundreds of feet, and rarely descend to lower levels.

4. Drift is spread promiscuously over the surface, and the blocks are a good deal rounded. The detritus of glaciers more or less blocks up the valleys, and the fragments are frequently quite angular. These, however, are in part covered with other materials, which have descended from the mountains.

II. MODIFIED DRIFT.

Whenever there is evidence that the coarse drift has been acted upon by waves, or currents, subsequent to its production, whereby the fragments have been rounded, comminuted, their striæ removed, and those of different sizes sorted and arranged in different layers, we call the mass *Modified Drift.* This term embraces what some authors call Pleistocene.

It should be understood, that not unfrequently, especially near the outer limits of drift action, we find beds of modified and rearranged stratified materials, beneath, and in the midst of coarse drift; nor is it possible in going upward, to draw a definite line between modified and unmodified drift. We can only say, that usually the coarse drift lies lowest, and shows less effect from water than the materials lying higher in the series. When we compare layers of the deposit at a considerable vertical distance, the difference is very distinct, but not so with those in immediate proximity. Hence it seems certain that drift and modified drift are the result of the same general causes, acting under modified conditions of the surface.

Some statements as to the means of distinguishing genuine drift from modified drift, oceanic from fluviatile action, and that of ice from that of water, will be important, preliminary to a description of the several forms of modified drift.

1. *Drift proper* is the lowest part of the alluvial formation. 2. The fragments are coarser and less rounded than in modified drift. 3. The fragments are frequently striated in one direction, as if held firmly, say by being frozen into ice, and pushed over a rocky surface. 4. The materials are not generally sorted, though there is evidence often that water, as well as ice, was acting upon drift, during its production; so that in the same mass we find one portion mixed confusedly together, and another portion more or less stratified and laminated.

1. *In modified drift* the fragments are rounded, smoothed, and more or less destitute of striæ. 2. They are sorted and arranged in layers; the coarser and finer alternating. 3. In the most recent of these layers, which are superimposed upon the others, though usually lying at a lower level, the finer do the materials become, until the almost impalpable powder of alluvial meadows is met. 4. The most recent portions are deposited in a more nearly horizontal position; the surface becomes more and more level topped, and the terraces more regular, as we descend the side of the valley.

1. The deposits formed by the ocean are generally more irregular on their

surface than those from lakes and rivers, and less perfectly stratified. 2. These deposits occur sometimes in positions (as when they fringe the side of a mountain, where there is no corresponding elevation opposite), where no rivers can ever have existed.

1. Deposits by lakes and rivers are found on the sides of valleys, or wide basins, or at the *debouchure* of smaller into larger valleys. 2. These deposits usually slope downward in the direction in which the river runs, and at the same or a more rapid rate than the river. 3. Fluviatile deposits are generally made up of more perfectly comminuted and finer materials than oceanic deposits; as if the former were made in more quiet waters.

Fig. 106.

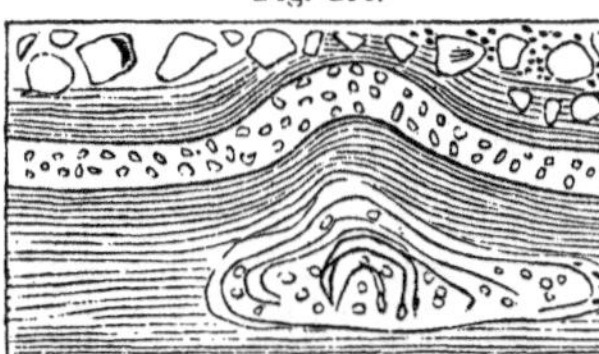

If masses of ice are moved along over the surfaces of stratified sand and gravel, it is obvious they will plough furrows or pile up a ridge in front, and in various ways disarrange the layers. Or masses of ice might be mixed among alluvial deposits, and produce irregularities in the strata by its melting. The curvature in Fig. 106 may have been produced in this way.

Fig. 107, which is the section of a terrace in Newfane, Vt., shows how very coarse modified drift may succeed unconformably to fine clay.

Fig. 108 shows an interesting case in Palmer, Mass. The cliff is mostly gravel, sand, and coarse bowlders, yet in the midst of it are deposits of fine blue clay.

Fig. 107.

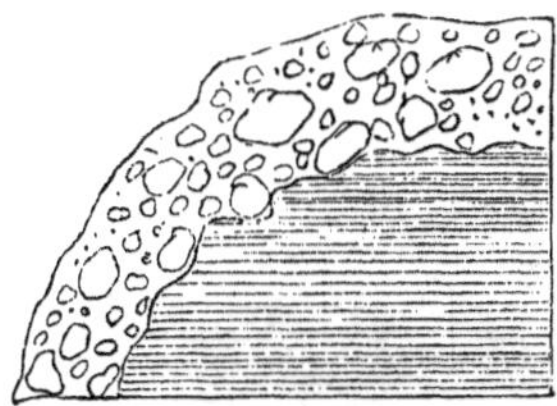

Section in Newfane, Vt.

Fig. 108.

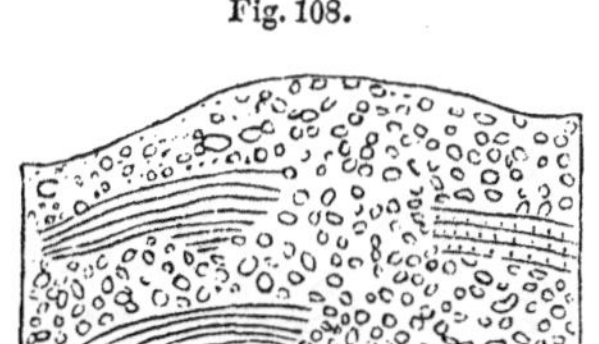

Deposits of loose materials from water alone are distinguished by two circumstances. 1. The materials are, as a general fact, arranged in horizontal layers; although in some places of limited extent they may be urged down a slope, and present a lamination considerably inclined. 2. The materials are sorted into finer and coarser, and arranged into layers one above another; often passing into each other by the most delicate gradation. Hence, wherever we find a deposit possessing both these characters, we may be sure that it is the result of the action of water.

Forms of Modified Drift.—Modified drift occurs in the form of moraine terraces, osars, escars, ancient subaqueous ridges, ancient sea beaches and sea bottoms, and terraces. Stratigraphically they all lie above the unmodified drift.

Moraine Terraces.—These are generally accumulations of modified drift, and are often arranged in heaps and hollows, or conical and irregular elevations, with corresponding depressions. Some of them greatly resemble the moraines of glaciers. But they differ from moraines by their structure, being often more or less stratified, and by their position. Generally they are not in localities favorable to the existence of glaciers, though they commonly occur along the foot of hills and mountains. As they are often associated with or change into terraces, we call them *Moraine Terraces*, thus indicating their affinities.

In New England these accumulations are very common, and sometimes they are so crowded together as to exhibit a picturesque appearance, being made up of tortuous and conical elevations with deep intervening cavities, as if scooped out by the hands of a Titan. There are remarkable examples in the vicinity of Plymouth, in Massachusetts, and near the extremity of Cape Cod, in Truro, where they are sometimes 200 or 300 feet high. In Truro they are composed wholly of sand, and they give a singular aspect to the landscape. Fig. 109 represents a small portion of the surface near what is called the Harbor in Truro.

Fig. 109.

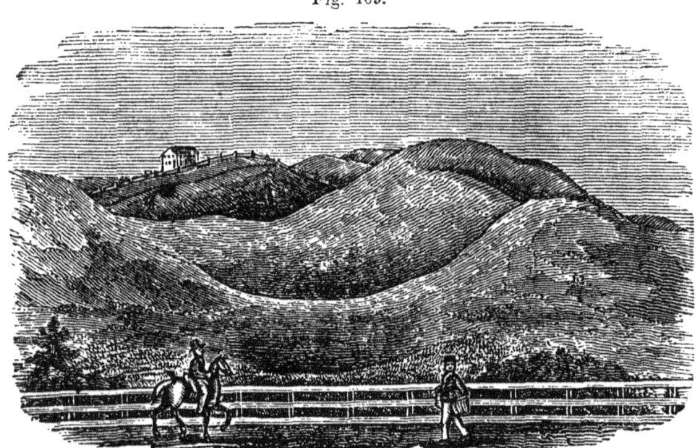

Sketch in Truro.

Fig. 110 shows a row of tumuli, some of them 100 feet high, a little south of the village of North Adams, in Massachusetts, at the foot of Hoosic mountain. The large ridges in the background are made of the same materials as the tumuli.

Moraine terraces are found in other parts of North America, more or less abundant, wherever the drift is found.

In northern Europe, also, and probably in all countries where the drift agency has operated, similar accumulations occur.

It is an interesting fact that these picturesque mounds and depressions have been chosen as the sites of cemeteries. This is the case at Mount Auburn, in Cambridge; Mount Hope, in Rochester; at Plymouth, Massachusetts, the oldest burying ground in New England—at Newburyport, North Adams, etc.

Fig. 110.

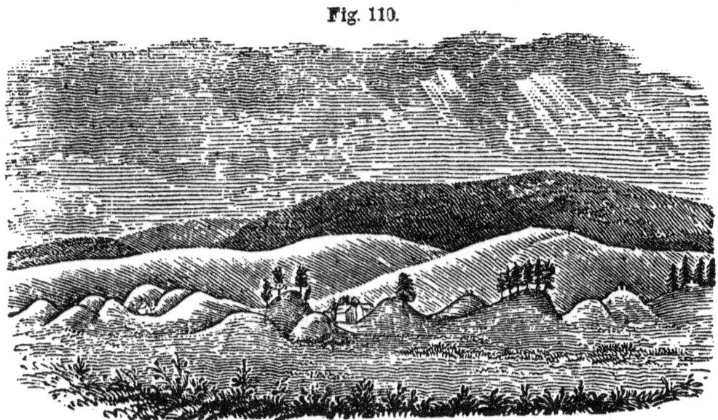

Moraine Terraces in North Adams.

Osars or Œsars.—*Os* in Swedish, signifies a pile of gravel, osar is its plural, though in English it is customary to use it as the singular. They are ridges of sand, gravel, and bowlders, sometimes only a few rods, and rarely a mile long, lying in the same direction as the striæ on the rocks in a given region, having a somewhat rounded back, and not unfrequently proceeding in a train from the *lee* side of a rock or hill. They seem to have been formed by a powerful current, which accumulated the detritus behind the obstruction in a tapering train, resembling in form an inverted canoe. In Sweden and Russia they embrace coarse bowlders, and become, in fact, mere trains of blocks. Sometimes they appear to have accumulated behind stranded icebergs, which subsequently disappeared, as is shown in Fig. 111, which represents the manner in which a remarkable Osar, near Upsale, in Sweden, was probably formed. In this case the lower part is

Fig 111.

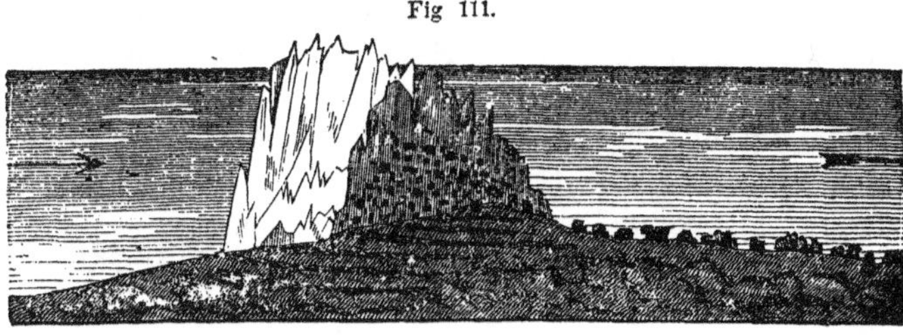

Osar forming behind a stranded Iceberg.

sand and gravel, and the upper part a train of blocks, which probably were derived from the melted floe.

Not many Osars have been pointed out in this country. Mons. Desor, however, who is familiar with such phenomena in Europe, speaks of Osars on the shores of Lake Superior. We incline, as stated on a previous page, to bring under the same designation those trains of angular blocks which we have described in Berkshire Co. and Vermont. (See Fig. 104).

Escars or *Scœurs*.—Those in Ireland consist chiefly of pebbles of carboniferous limestone heaped into narrow ridges forty to eighty feet high, and from one mile to twenty miles long, probably formed in the eddies along the margins of opposing and conflicting currents which piled up the materials from each side. There are ridges of this character in this country, though the pebbles are of all sorts of rock, yet we incline to regard them as Escars. They occur in many parts of the country.

Fig. 112 shows several of these ridges as they occur near the Shawsheen river, in Andover, Massachusetts. One is called the Indian Ridge, and is a mile and a half long. The west ridge is still longer. They are narrow, usually not more than four or five rods wide, and from fifteen to thirty feet high. Some of them are composed of sand and fine gravel, others of coarse gravel with large bowlders intermixed.

Fig. 112.

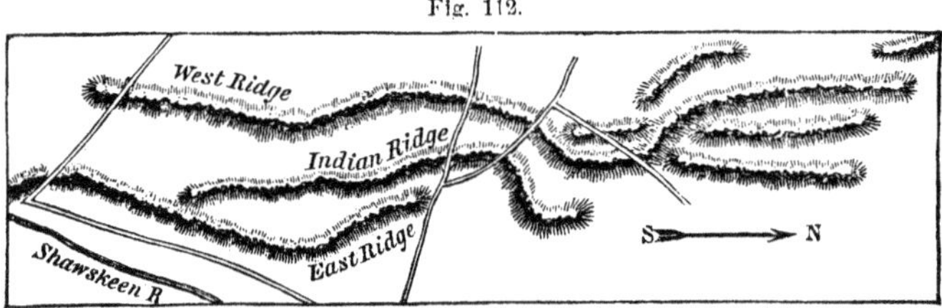

Submarine Ridges, Andover.

These escars are finely developed in Aroostook county, in Maine. They are called "Horsebacks;" and one of them, between Weston and Houlton, is thirty miles long, and nearly straight, running north and south.

Subaqueous Ridges.—These ridges are composed of sand and gravel, which differ from beaches and terraces, by having a double slope, which is usually gentle. They are found around lakes more especially, as lakes Erie and Ontario, and are there called "Ridge Roads." In a longitudinal direction they vary considerable in height, although their general elevation is the same. They form fringes around the lakes.

There are four of them on the south shore of Lake Erie, the lowest 100 and the highest 200 feet high. There are eight upon the north shore of Lake Ontario, from 108 to 762 feet in height. These ridges, however, were not necessarily *submarine*, as a large body of fresh water would produce ridges not at all different from the submarine ridges described by European Geologists. There is one of these ridges on the coast of Massachusetts, between Newburyport and Ipswich, the highest part of the city of Newburyport being situated upon its summit.

Sea Beaches.—Along our coasts these are now in process of formation. They consist of sand and gravel, which are acted upon, rounded and comminuted by the waves, and thrown up into the form of low ridges with more or less the appearance of stratification. With them shells and fragments of shells are usually associated, but not invariably. In passing into the interior from the coast, we occasionally see analogous ridges. A few of them are within 100 feet of the present ocean level, and no one will doubt their marine origin. But as we rise into the higher parts of the country deposits occur which can not be distinguished from these recent beaches, except that they are sometimes much mutilated by erosion. Fossil shells have been observed in these beaches about 540 feet above the ocean level in this country, in the deposits having the provincial name of Champlain clays. No fossils have yet been discovered in the highest beaches.

The most distinct beaches occur below 1,200 feet above the ocean level. A very fine beach, however, is found on the west side of the Green Mountains, in West Hancock, Vt., 2,196 feet high. Others still higher are in Peru, Mass., 2,022 feet; at the Franconia Notch of the White Mountains, 2,665 feet; and at the Notch of the White Mountains, (Gibb's Hotel,) 2,020 feet. Upon comparing together the heights of beaches in different parts of New England, we find a number of them having essentially the same elevation; thus showing that they were formed contemporaneously. For example, there are beaches in Ashfield and Shutesbury, Mass.; in Norwich, Corinth, Elmore, Hardwick, and Brownington, Vt., each 1,200 feet above the ocean, and the most remote are nearly 200 miles apart. Other sets might be named at different elevations than this. On Mt. Snowden, in Wales, the highest beaches are elevated 2,547 feet; in Switzerland, on the west shore of Lake Zurich, 2,105 feet; at Scupsheim, 2,274 feet; and near Berne, 2,640 feet. There is an interesting coast line in Scotland, parallel to its present shore, and continuous around the whole island. It is from thirty to fifty feet above the present ocean level.

Stratigraphically, the beaches lie directly upon the unmodified drift and are formed from its ruins. The striated and angular fragments of rock lose their markings and angles; they are reduced in size, and stratified in successive layers of coarse and fine materials.

Sea Bottoms.—Extensive deposits are accumulating upon the bottoms of present seas and lakes, both of chemical and mechanical origin. These are forming at the same time with the present beaches upon the coast. If, then, we have found ancient sea beaches more than 2,000 feet above the present ocean level, may there not be ancient sea bottoms to correspond with them? There are deposits of great extent in our country, apparently more or less connected with the beaches, which are referable to this class of accumulations. This will not confuse the practical geologist, for he reflects that as the country gradually arose from the ocean, the original sea bottoms would be brought to the surface, and have beaches deposited upon them manufactured from their own ruins. They occupy much more of the surface than all the other forms of modified drift combined. Many of the deposits called *Pleistocene* by geologists are ancient sea bottoms or beaches.

Under this head we embrace all those deposits which contain remains of pelagic animals; as, for example, the lower parts of the Champlain clays in Canada and Vermont. The same kind of deposits at higher elevations may not contain fossils. On the shore of Lake Erie, by rising about 240 feet, the well-marked terraces disappear; and from that level to 650 feet the surface of northern Ohio presents the characters of these ancient sea bottoms. A rise of water 250 feet above Erie, or 850 feet above the ocean, would submerge northern Indiana, Illinois, Michigan, much of New York and Canada West, with much of Wisconsin and Iowa, all which exhibit more or less of these sea bottoms. The same is true of the country near the coast in New England, especially in Rhode Island and Massachusetts. The Pampas of South America and the Steppes of Siberia are also of this class.

The superficial character of sea bottoms is a broad expanse of level or undulating surface, composed entirely of water-worn materials. Many of the Western prairies, especially those confined between ranges of mountains, may be taken for the type. Fine clay and sand, or loam, may compose most of the materials; but bowlders and coarse gravel may have been dropped by melting icebergs, and thus be intermingled with the finer materials.

We introduce here the description of a series of deposits combining both the sea beaches and sea bottoms.

Champlain Clays.—From the mouth of the River St. Lawrence to Lake Ontario, and in the Champlain valley from Montreal to Whitehall, N. Y., and thence to New York City, there are numerous deposits of clay, silt, sand and fine gravel, more or less abounding in marine fossils—molluscs and mammalia. Along the sea-coast from Maine to the Gulf of Mexico, similar deposits occur. These are called *Champlain clays* or *Lawrentian deposits*, from the localities where they are best developed. They extend as high as 540 feet above the ocean, at Montreal, and to 400 feet in the valley of Lake Champlain. The lowest member is a tough, blue clay, containing fossil shells, which must have inhabited very deep water. Those inhabiting the deepest waters were *Foraminifera;* such remains as have been brought up by sounding from the bottom of the Atlantic ocean. These are in the very lowest strata, immediately overlying the bowlder clay. Some of the species of shells observed are extinct; as the *Nucula Portlandica* and *N. Jacksoni*, etc. Thus the character of this lower member is clearly an ancient sea bottom.

Overlying the clay is a mixture of clays, sand, silt and gravel, containing numerous species of littoral shells, such as are now found upon the sea-shore. The most common are *Sanguinolaria fusca*, and *Mya arenaria*, the long clam. Remains of cetacea have been found in Vermont, and of other mammalia in the Southern States. Most clearly, then, all the banks containing these fossils are ancient sea beaches, and the ocean level during this period has been sinking, and the land rising.

Terraces.—The term *terrace* applies to any level-topped surface with a steep escarpment, whether it be solid rock or loose materials. We limit it now to those banks of loose materials, generally unconsolidated, which skirt the sides of the valleys about rivers, ponds and lakes, and rise above each other like the seats of an amphitheatre.

Fig. 113.

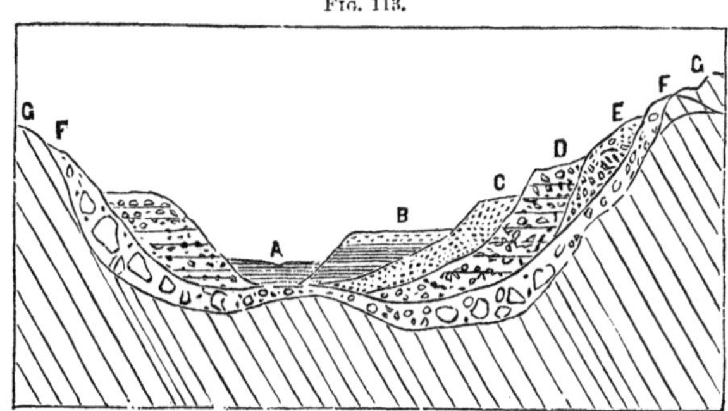

Fig. 113 represents an ideal section of a terraced valley. As we rise from the river, its immediate bank, or meadow, forms the lowest and latest terrace, A, which may be increasing from year to year by alluvial deposits. On the margin of the meadow we come to a steep slope, or talus, whose top, B, forms a second terrace. Very frequently the lower part of this second terrace is composed of clay and the upper part of sand, or small gravel. Another steep slope carries us to a third terrace, C, which is more usually of coarser materials, but thoroughly rounded and mostly sorted. A fourth terrace, D, is still coarser, and the top less level. Indeed it is here, usually, that we find those irregular mounds and ridges already described as *moraine terraces;* that is, they occur upon the highest terraces, and sometimes where no terraces exist; but it is always along the base of mountains or hills. Rising above this we frequently find deposits, E, it may be of sand, gravel, or coarser but water-worn materials, not having a level top, but more or less rounded and reaching a certain level along the side of the hill. These are generally at a great distance from any existing streams, and could not have been produced by them, though they were at a higher level than at present. In fine, these accumulations resemble *beaches*, such as now are forming on the coast. Still higher, as at F, we find the unmodified drift, which lies immediately upon the solid rocks, as at G.

Two facts respecting the occurrence of terraces are illustrated in the last figure: 1. The drift underlies all the beaches and terraces, although it appears upon the surface at a higher level. All the striæ made by the drift underlie deposits of modified drift, and are therefore older than the water-worn accumulations. The beaches underlie the terraces, and each higher terrace underlies each lower terrace. 2. On the opposite side of the valley we may or may not find terraces and beaches. If we do, it is not often that they correspond entirely in number and height on the two sides.

The number of terraces on a river varies with its size, the largest rivers having the smallest number. Thus, on the Connecticut river, the number rarely exceeds three or four; but on some of its tributaries, and those not the largest, as the Ashuelot, at Hinsdale, New Hampshire, and Whetstone Brook, in Brattleboro, Vermont, they rise as high as ten.

The height above the streams which the river terraces attain generally varies directly with the size of the river. The following are some of the highest terraces that have been measured: on Connecticut river, at Vernon, Vt., 237 feet above the river, and 450 feet above the ocean; at White River Junction, Vt., 209 feet above the river, and 529 feet above the ocean; on Black River, at Proctorsville, Vt., 150 feet above the river, and 1,028 feet above the ocean; on Lamoille River, at Hardwick, Vt., 380 feet above the river, and 1,100 feet above the ocean; on Genessee River, at Mount Morris, N. Y., 348 feet above the river. In Peru, Mass., there is a terrace 1,851 feet above the ocean. The highest of the famous Parallel Roads in Glen Roy, in Scotland, is 1,495 feet above the ocean. Robert Chambers has measured the heights of twenty-five successive terraces in this district. A terrace at Rhinefelden, on the Rhine, is 306 feet above the river. In Switzerland the highest terraces are from 1,300 to 4,350 feet above the ocean, but their great elevation may be due to the existence of former barriers of ice, producing basins, in which the terraces were formed without the aid of the ocean.

Terraces occur in basins. There is a series of them from the mouth to the source of a river. For example, there are twenty basins upon the Connecticut river between its mouth and source; and five basins upon Winooski river, in Vermont. Upon lakes and ponds there is but one basin. These basins may be connected with each other directly, or be separated by rocky barriers. About such gorges and obstructions, terraces are usually either higher, or of greater breadth than in other parts of the basin.

River terraces usually slope toward the mouth of the stream, at the same angle with the descent of the river, or even more.

There are four kinds of river terraces: 1. *The Lateral Terrace*, which is the ordinary terrace, parallel with the course of the valley, and continuing for miles along the banks; 2. *The Delta Terrace*, which includes not only the deltas of large streams emptying into the ocean, as the Mississippi, but the former deltas of tributary streams, now cut through by the lowering of the bed of the stream; 3. *The Gorge Terrace*, which includes the deposits about the ends of gorges, intermediate in character between the first and second kinds; 4. *The Glacis Terrace*, which is a ridge sloping rapidly upon the side facing the stream, but gradually upon the opposite side. They are most common in alluvial meadows. It will be seen that lake terraces and maritime terraces are lateral terraces.

Terraces of modified drift occur along rivers in all parts of the world. In South America, Mr. Darwin has described several along the coast; in one part there were seven of them in the distance of 150 miles, rising at length to 1,200 feet. The great chain of lakes in North America have them. Prof. Agassiz speaks of "six, ten, and even fifteen in one spot, forming as it were the steps of a gigantic amphitheatre," on the north shore of Lake Superior. Around the great Salt Lake in Utah, there are not less than thirteen terraces, the highest 200 feet above the plain. In the valley of the Mississippi the *Bottom Prairie* and the *Bluff* are deposits of the terrace period. The latter is somewhat consolidated, and contains fresh water fossils in abundance. It has been worn down by the river in many places, leaving perpendicular banks called *bluffs*, whence the name. It is probably contemporaneous with the *Loess* of the Rhine, which is a silt or fine calcareous clay, without lamination, containing fresh water fossils, and interstratified with beds of volcanic ashes thrown out at intervals by the Eifel volcanos, now extinct.

Changes in the Beds of Rivers.—There are two kinds of deserted ancient river beds. The first and most obvious are depressions in alluvial meadows, connecting at the extremities with curves in the stream. Many of them were occupied by the river since the memory of man. The second kind show a deserted rocky gorge, where once the stream flowed at a higher level than at present. The proof of such a change is found in the existence of pot holes in the rock, situated in a valley connecting with different parts of the principal stream.

In Orange, New Hampshire, on the summit level between the Connecticut and Merrimack rivers, there are pot holes 682 feet above the Connecticut, in the lowest place between the two rivers. They are so situated as to indicate that the current flowed from the Connecticut to the Merrimack. A barrier probably existed at Bellows Falls, so high as to force the Connecticut, or a part of it, into the valley of the Merrimack.

There is proof of the existence of rivers in different channels from the present upon a former continent. On the west bank of the gorge, three miles below Niagara Falls, for instance, at the Whirlpool, the continuity of the bank is interrupted by a deep ravine, filled with gravel and sand. This ravine can be traced to Lake Ontario, four miles west of the present mouth of the gorge, and must have been the bed of the river formerly; for the water must have flowed in the lowest channel. When the continent was under water, this ravine became filled with drift materials so much that the river was forced to seek a new route, and since then has worn away the gorge between Queenstown and the Falls.

In Stratton, Vermont, there is a large pot hole upon the summit level between the waters of the Deerfield and Connecticut rivers, say 1,600 feet above the latter. It is so situated as to make it necessary to suppose the existence of a current to the north, and there is no stream in the neighborhood sufficiently large to have excavated it. As there is a valley extending from Stratton to Canada, with a general northerly descent, it is not improbable that there may have been, in Mesozoic or Palæozoic periods, a river from southern Vermont east of the Green Mountains to the St. Lawrence, although several streams now cross this valley transversely.

Frozen Deposits of Modified Drift, (Frozen Wells,) and Ice Caverns.—In Brandon, Vermont, in November, 1858, a well was dug through layers of gravel and marly clay, to the depth of thirty-five feet. After reaching a depth of about fourteen feet, a frozen mass of the same materials was passed through, from twelve to fifteen feet thick; then a few feet of unfrozen gravel, when water was reached. During the winter the water was frozen over quite hard, and for most of the summer ice lined the stones of the well several inches thick, and the temperature of the water never rose more than 2° or 3° above the freezing point. In the winter of 1859–60, the ice which in September had disappeared, returned.

In Owego, New York, a similar well was dug many years ago, in loose soil, seventy-seven feet deep, which for four or five months in the year was so frozen as to be useless. Another was dug in Ware, Massachusetts, in 1858, in gravel, thirty-five feet deep, which froze over the following winter. Another is described in Lyman, New Hampshire.

On the eastern continent, in the Alps, the Jura, and the Ural Mountains, are numerous caverns in the rocks, where ice forms in the summer, especially, often in such quantity as to be an article of commerce. In all these cases, the caverns have two openings, one at the top the other at the bottom, laterally. This causes a current of air downward in the summer, and upward in the winter. This current evaporates the water upon the sides and floor of the cavern, and thus produces the cold; since evaporation takes up into a latent state nearly 1000° of heat. In the winter the evaporation is less, and the congelation less. On this principle, at Monte Testaceo, in Rome, (which is a hill 300 feet high, made up of broken pottery), excavations are made laterally, connected with chimneys, and thus fine ice houses are formed.

Now, in the case of frozen wells, it seems as if there must be some such circulation of air as in these ice caverns; and why must there not be through the beds of quite clean gravel that occur in the wells, and which sometimes, as at Brandon, we can see cropping out at the surface? The interstices must be filled with air, and at different temperatures this must have motion, even though slow. This would carry off the heat that rises from the earth's interior, while the beds of clay near the surface would prevent the external heat from penetrating far. Thus masses of gravel, frozen during the drift period, may have been preserved to our day, and form a nucleus to which more frost might be added at certain seasons of the years. Such an hypothesis is not without difficulties; but the case of the ice caverns gives it some plausibility.

THEORIES OF SURFACE GEOLOGY.

The origin of drift has long been discussed by geologists. It was formerly thought to have been the result of the deluge of Noah. But this view is now wholly abandoned by geologists, because the remains of man and associated animals living before the flood are not found in it, and because the agency of water, and the brevity of the time involved, are inadequate to explain it. There are three theories proposed to explain these phenomena, which we will state briefly, and endeavor to combine them into a fourth.

The Iceberg Theory.—This theory imputes most of the phenomena of drift to icebergs carried southerly by the currents of the ocean, while the continents where drift occurs were yet beneath the ocean. As they were gradually raised from the deep, the mountains, which would form islands, would send down glaciers to their shores, and thus masses of ice would be broken off to be floated away, loaded with detritus. As the icebergs melted, the detritus would fall to the bottom, and under various circumstances would form all the deposits of unmodified drift. By the stranding of icebergs, the moraine terraces, ridges, escars and osars would be formed. After the ocean had retired, large bodies of water would remain in many places, and by gradual drainage produce the beaches, terraces, sea bottoms, etc.

Theory of Elevations and Earthquake Waves—This theory supposes the phenomena of drift to have resulted from the rise of large areas beneath the Arctic and Antarctic oceans, whereby their waters have been driven southward over a considerable part of Europe and America, bearing along masses of ice loaded with detritus. And further, that there may have been a succession of vertical movements, which produced successive waves; so that the waters may have repeatedly fallen and risen again, and while at their ebb they may have been frozen to the surface, so that as they subsequently rose, vast masses of ice may have been driven along, loaded with detritus, which may have been forced up declivities considerably steep, and thus the surface have been powerfully and rapidly abraded, and the rocks scoured and furrowed. This theory, somewhat modified, has been sustained with great ability by Professors H. D. and W. B. Rogers.

The Glacier Theory.—This theory supposes that at the close of the tertiary period there was a sudden reduction of the temperature of the surface of the earth, whereby all organic life was destroyed; and in high latitudes, at least, glaciers were formed on mountains of moderate altitude; indeed, that vast sheets of ice were spread over almost the entire surface, extending south as far as the phenomena of drift have been observed. The northern regions, especially around the poles, are supposed to have formed one vast Mer de Glace, which sent out its enormous glaciers in a southerly direction by the force of expansion; and the advance and retreat of these glaciers accumulated the moraines and produced the striæ and embossed appearance (*roches moutonnees*) upon the rocks. In Europe the centre of origin was in Scandinavia, whence the glaciers proceeded outward in all directions. In North America this sheet must have been 5,000 feet thick; and by vicissitudes of climate, irregular retreats and advances of the glacial sheet would produce the markings not coinciding with the first set. When the temperature was raised, the melting of the immense sheet of ice produced vast currents of water, which would lift up and bear along huge icebergs loaded with detritus, and thus scatter bowlders over wide surfaces.

Some advocates of this theory suppose that the continents were elevated several thousand feet higher than at present, thus reducing the temperature, and that all the phenomena of drift may be explained by glaciers radiating from the summits, like those now existing in the Alps.

Modified drift, by this theory, is produced by the blocking up of gorges by moraines, thus forming lakes and ponds, in which clay and sand might have been deposited, and afterwards the barriers of these lakes, consisting of loose matter, may have been cut through, and the waters gradually drained off, forming beaches and terraces.

And it is also held that, subsequently to the glacial period, the ocean rose upon the land 500 feet, when the Champlain clays were deposited. Thus this theory supposes an elevation of the continent, then a depression below its present level, and subsequent return to its present height. The elevations are supposed to have been paroxysmal.

This theory was first suggested by Venetz, a Swiss engineer; then advocated by Charpentier; and more recently brought out in its full proportions by Agassiz, in his *Etudes sur les Glaciers.*

General objection.—Against all the preceding theories of drift there lies one general objection. While each one explains some of the phenomena satisfactorily, it leaves others unexplained. They are true causes, but they are not singly sufficient. By combining all these theories, as far as possible, we may find a satisfactory theory, both for drift and modified drift, from the close of the tertiary period to the present moment.

FOURTH THEORY OF SURFACE GEOLOGY.

General Statement.—Since the tertiary period, those countries where drift and terraces exist have been depressed in a great measure beneath the ocean; the United States from 2,000 to 3,000 feet, England 2,300 feet, Scotland from 1,000 to 1,200 feet, and Switzerland from 2,500 to 3,000 feet. Subsequently they have been slowly elevated to their present levels, and drainage has gone on from their entire surfaces.

Drift is mainly the result of these four agencies—glaciers, icebergs, waves of translation, and landslips—acting upon the surface while it was sinking beneath, and rising above the ocean. The forms of modified drift were produced by the same agencies with the addition of rivers. From the close of the tertiary period to the present time, these operations have formed an uninterrupted series. We will now present a particular statement of the condition of this continent at the several divisions of this period.

The Drift Period.—Near the close of the tertiary period there commenced, we suppose, a reduction of the general temperature from the sinking of the land. When sufficiently depressed it would bring oceanic currents from the polar towards the tropical regions. If North America was now submerged, east of the Rocky Mountains, a current from the northwest would flow over it; and if South America was submerged, east of the Cordilleras, a current would flow over it from the southwest.

In connection with this gradual submergence, taking North America and the west part of Europe as an example, two causes would operate to reduce the temperature: 1. The Gulf Stream (the present cause of the higher temperature of Europe than the United States, and of the Atlantic coast above the interior) would be diverted from its present course, and pass along the eastern base of the Rocky Mountains into the northern ocean, and thence perhaps along the coast of Asia. 2. The current from the Arctic regions would be loaded with icebergs, which would be stranded along the shores, and so reduce the temperature that probably the summer could not melt away the ice; and the sea, like that around the poles, might be choked with ice as far south as we now find drift.

As a consequence of this access of cold, while the land was sinking glaciers would form on mountains comparatively low, and where they do not now exist. These would reach to the sea, as they now do, in Arctic regions.

The enormous icebergs that would be moved southerly in such circumstances would grate powerfully upon the bottom of the sea, smoothing and striating the rocks, and especially projecting ledges, upon their northern sides, producing effects which could be distinguished afterwards only with difficulty from those of glaciers, except in the vast extent of country acted upon.

The most reasonable theory of the transport of materials from lower to higher levels is, that as the land sunk, the stranded ice would be lifted higher and higher along the shores, and finally be urged upon and over hills and mountains, carrying detritus along with it. Much of the work of smoothing and scouring down the ledges and accumulating the coarse drift was performed while the continents were sinking.

When the land had sunk 5,000 feet, all the mountains east of the Rocky mountains were submerged, except Mt. Washington, and a few peaks in North Carolina. The glaciers now would be covered up, and the icebergs be the only agency at work. Scarcely any form of life could exist among these icebergs, and only the hardier species when a greater extent of land had risen above the waters.

The land at length began slowly to emerge, and it seems to have been raised as a whole; that is, the whole mass was lifted together, so as not to disturb the relative levels of the surface, just as we know the continent of South America has been raised some 1,400 feet, without disturbing the strata horizontally, or producing the smallest fault or curvature.

As the land rose the water would, to some extent, and in particular places, sort and deposit the detritus worn off. And hence we can account for that mixture of mere mechanical accumulations and aqueous deposits, of which the drift is composed. Especially does it explain why, as we approach the outer (mostly southern) limits of the drift, we find the deposit more and more stratified, and the evidence of glacial action gradually disappearing.

By this submergence and emergence, every foot of surface must have been exposed to the long-continued action of waves, tides, and currents laden with ice; and, consequently, a great amount of detritus must have been broken off.

When the continent was partially submerged, at both the periods of its rise and fall, it is conceivable that large valleys deviating from the usual direction of the currents might incidentally become filled up with ice; and though only a part of the whole force could have acted upon those bergs, according to the laws of the resolution of forces, yet it would be sufficient to produce all the effects of ordinary drift in an unusual direction. In this case the drift may be said to have been deflected from its usual course by valleys. But this will not explain the three general directions of drift, to the southwest, to the south, and to the southeast, in New England. These, with all the minor intermediate variations, must have been produced by variations in the direction of the principal current at different altitudes of the continent.

The Beach and Sea Bottom Period.—When the land had risen to the level of the highest ancient sea-beach—about 2,600 feet in North America—the higher mountains would appear as islands. Oceanic agencies would act upon these, especially in working over and rearranging in sheltered spots the angular and crushed fragments of the unmodified drift. These would be sea-beaches: at first very limited, because the surface acted upon was small, and no streams of much size could exist to aid in the work. Every hundred feet of additional elevation would add to the number and perfection of the beaches.

The irregular accumulations described as Moraine Terraces were formed at this period. If masses of ice were stranded against the sides of hills, and deposits of sand and gravel were mixed with or piled upon them, when the ice melted elevations and depressions of this character would result.

The ancient subaqueous ridges might be formed along the shores of the ancient ocean, just as they are now produced in lakes and seas. Osars might also be formed by the currents sweeping detritus into the rear of obstructions, either of rock or ice, and escars along the eddies. Sea bottoms were deposited at the same time with the beaches.

Some writers have objected to the theory, that the drift, beaches, and terraces, were produced in connection with oceanic agencies, because no organic remains are found in them. We reply: 1. In unmodified drift in this country, the climate may have been so severe as to prevent the existence of such animals as would have left behind traces of their being. Undoubtedly they existed at that time in other parts of the world, beyond the limits of the cold. 2. In the unmodified drift of England and Scotland, broken and comminuted marine shells have been found as high as 2,300 feet above the ocean, the upper limit of the deposit. No one doubts the former presence of the ocean there; but this fact has been only recently discovered. In this country broken marine shells have been found 100 feet above the ocean in unmodified drift, and uninjured specimens more than 500 feet above the ocean in modified drift. It may be that these remains will yet be found in the whole of the unmodified drift, when more thorough explorations shall have been made. 3. Pelagic shells, or such as live in very deep water, have been found at the height of 400 feet in Canada. Hence the ocean must have been nearly a thousand feet deep in the latter part of the drift period. 4. This deposit of pelagic shells lies immediately upon the bowlder clay. Now, had this clay been produced by a glacier, and not by the ocean, the country must have sunk at least 3,000 feet between the deposition of the bowlder clay and the bank of shells. We should then expect to find stratified layers between them, because in so great a period of time materials must have accumulated there. Besides, how much simpler it is to suppose one system of rise and fall of the continent, than to suppose two such systems of oscillation, as the objectors must maintain.

5. The beaches and terraces lie upon the unmodified drift, even in many gorges where one would suppose a barrier might have existed. Hence all the lower forms of modified drift must have been formed in estuaries of the ocean, for no ridge existed to dam up the waters. No one doubts that a beach or terrace a few feet above the level of the ocean was originally formed by its waters. Now, from the lowest to the highest beach there is a continuous series, like a succession of steps. If the first is formed by the ocean, the second must be; likewise the third, and so on to the highest.

Let us look at this point in another light. As beaches are stratified, the materials must have been deposited from water. Now, when we find upon the side of a high mountain a stratified bank of sand and gravel, we know that some body of water must have existed there. But the land slopes from this bank to the ocean, therefore the water in which these materials accumulated must have been oceanic. There is no barrier which could have existed, high enough to have separated this body of water from the ocean. With such proof before us, we can not hesitate to believe that all those deposits called ancient sea beaches must once have formed the margin of the ocean, although there are no marine remains in them.

Fig. 114 represents beaches thus situated upon mountains. *a*, represents a beach on the east side of Mount Washington, (A), *b* one at Franconia Notch, the highest yet discovered in New England, *c* represents another beach in Hancock, Vermont, on the west side of the Green Mountains, (B),

Fig. 114.

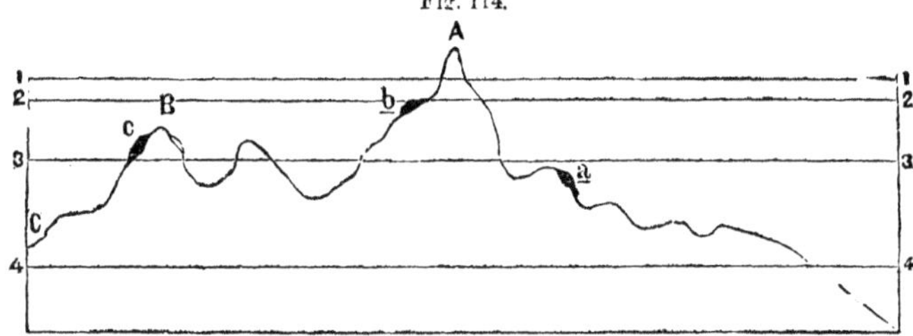

Ideal Section of New England.

C shows the level of Lake Champlain. The line 1, 1, shows the level of the ocean when only the top of Mount Washington peered above the waters; 2, 2, represents the ocean level at the beginning of the Beach Period; and 3, 3, represents the same level at the beginning of the Terrace Period; 4. 4, represents its present level, in the Historic Period, and the base of the figure shows the level of the ocean in the Drift Period, according to the Glacial theory.

THE TERRACE PERIOD.

The country has now risen so much that the great valleys are seen in outline. Rivers of considerable size and length begin to carry into the estuaries a large amount of water worn materials, derived from the washing of the drift and the beaches, and forming small deltas beneath the surface at their mouths. Tides and currents would sweep this along the coast, and after a time the tops of the deposits would be brought above the surface, and no more materials could be deposited upon them by rivers; hence these must push their detritus further into the ocean, and thus a new submarine bank would form outside of the first, and at a lower level. When the second had reached the surface of the water, it would be lower than the first, because the land had been rising during the process of its production. In the same way a third and a fourth bank will form in succession, and thus there is a series of terraces presented to view. These are delta terraces, and it is not essential that they should have been formed under the ocean, but wherever one stream flows into another.

The streams emptying into these estuaries would produce a current toward the ocean, which would spread the detritus along the shores in the same direction, and produce the lateral terraces, having a slope at least as great as that of the current. In order to form successive lateral terraces, it is only necessary to suppose the drainage and erosion to go on till the rivers have sunk to their present beds, which could not take place till the continent had risen above the ocean to its present height, or the water had sunk.

There is another mode in which lateral terraces might have been formed, and are now forming, where a stream must cut its way through alluvial materials. The mere erosion would form terraces of equal height along the stream; or all the detritus on one side might be swept away by the stream, so as to leave a terrace only on the other side. But after a channel has thus been made to some depth, if a freshet occurs, the current will act powerfully upon one or the other of the banks, and sweeping them away will form a meadow when the flood has subsided. In subsequent floods, this meadow will receive fresh accessions of alluvial matter, and of course be somewhat raised up. Meanwhile the river is cutting a deeper and deeper channel, so that at length it can no longer rise high enough in floods to spread over the meadow, which has now become a second terrace, because the sinking of the stream by erosion would prevent the meadow from ever rising as high as the original bank. Being no longer able to overflow the meadow, it begins again, in time of freshet, to wear away the bank, and to form a second and lower meadow, which ultimately becomes, as above described, a third terrace, and thus may the work go on and the number of terraces be increased, as long as the river can deepen its channel.

Gorge terraces connect different basins together, being situated about gorges. The current transporting materials toward a gorge would have its compass diminished by the narrowing of the basin, so much as to cause a deposition of the materials near the gorge. However small these accumulations may be at first, in process of time, they might become even greater than an ordinary terrace. But the same current which transported the detritus to the upper part of the gorge may have its velocity greatly increased in passing through the narrow channel. This would remove much sediment, which would be redeposited at the lower end of the gorge, where the velocity of the current is diminished by contact with the placid waters of the lower basin. This process in many cases would go on only in times of freshets.

The Glacis terraces may have been formed by the unequal deposition of detritus over the surface. Sometimes they are mere modifications of lateral terraces, or undulations in large meadows.

We see then that by the simple drainage of a country, including its rivers, terraces might be formed along the shores of the ocean, lakes and the banks of rivers, supposing only a general slow and perfectly uniform rise of the land or depression of the ocean. Almost all writers, however, suppose these vertical movements to have been by starts, with intervening pauses. At an earlier date, the prevailing theory was, that the terraces were produced by the bursting away of the barriers of lakes, and the sudden sinking of the waters. These are quite natural suppositions to explain the stair-like aspect of terraces. But in respect to river terraces, we have the following decided proof that no such paroxysmal rising or sinking has produced them. 1. By such theories the terraces ought to correspond in number and height on opposite sides of the river, which is very rarely the case, although to the eye it may frequently seem so. Neither do they correspond in number or height in different parts of large lakes. 2. Where tributary streams have cut through the lateral terraces of the principal river, as they have often done near their mouths, the number and height of the terraces on both streams ought to agree. But the reverse is true. Thus, on Connecticut river the number of terraces is usually three or four; but on some of its tributaries, as on the Ashuelot river, at Hinsdale, and Whetstone Brook, in Brattleboro, the number rises as high as ten, and yet the uppermost is no higher than the highest on the main river.

We can, then, explain the formation of terraces without supposing the continent to have risen by a series of paroxysmal movements. They might have been produced by mere drainage, with a slow and equable movement. Yet we would not deny the phenomena of the bursting of barriers, or of sudden elevation at particular localities. For example, the sudden rushing of the waters of Runaway Pond to Lake Memphremagog, by the bursting of the barrier, left behind two lateral terraces. And some have explained the Parallel Roads of Lochaber, in Scotland, by pauses in the rise of the country. Doubtless, also, there are other cases where terraces have been formed by sudden elevation. But if river terraces have generally been formed without paroxysmal movements, as they must have been, and if terraces on lakes and the ocean may have been produced in some cases by the drainage of the country (as was the case upon Lake Lungern, in Switzerland, by artificial drainage), it is reasonable to suppose that such may have been their usual origin.

THE HISTORIC PERIOD.

We are now brought to the period when the country had attained essentially its present altitude. All the agencies that produced drift, viz., icebergs, glaciers, land-slips and waves of translation, are still in operation in some parts of the world, and therefore drift is still being produced. Ever since the tertiary period these causes have been acting, but their intensity has varied in different ages.

The same is true of the agencies that have produced beaches, osars, escars, subaqueous ridges and terraces, viz., the action of rivers and the ocean, combined with the secular elevation of continents. In other words, the agencies producing drift and modified drift have run parallel to each other from the very first. Hence they both are varieties of the same formation, extending from the close of the tertiary period to the present.

The sections describing aqueous, igneous and organic agencies contain the history of this period in detail. The Flora and Fauna are those now existing.

Man has existed on the earth a comparatively short part of the alluvial period. We have a few records of the commencement of this period. There are many examples of river beds on a former continent, which became so filled by drift and modified drift, while the continent was beneath the ocean, that when it emerged, the rivers were compelled to abandon the old beds and seek new channels. And the amount of erosion effected by them since that time is before our eyes. The gorge through which the present Niagara river runs, between the Falls and Lake Ontario, seven miles long, is one of these cases. Another case of similar erosion is the Genessee river between Portage and Mount Morris; where it has cut a channel deeper, in most places, than that of the Niagara, some fourteen miles long. There are other examples in

New England, where the erosions are not as long and deep, but the rock is much harder, and may have required a longer time for their excavation. And what shall we say of the cañons on Red River and the Colorado, a mile deep! Such facts indicate great antiquity to the latter part of the alluvial period only. What then must have been the duration of the unmodified drift period, and all the great systems of earlier date!

During the historic period numerous organic agencies have been producing geological changes, which we will now consider.

AGENCY OF MAN IN PRODUCING GEOLOGICAL CHANGES.

The human race produce geological changes in several modes: 1. By the destruction of vast numbers of animals and plants to make room for themselves. 2. By aiding in the wide distribution of many animals and plants that accompany man in his migrations. 3. By destroying the equilibrium between conflicting species of animals and plants; and thus enabling some species to predominate at the expense of others. 4. By altering the climate of large countries by means of cultivation. 5. By resisting the encroachments of rivers and the ocean. 6. By helping to degrade the higher parts of the earth's surface. 7. By contributing peculiar fossil relics to the alluvial depositions now going on, on the land and in the sea; such as the skeletons of his own frame, the various productions of his art, numerous gold and silver coins, jewelry, cannon balls, etc., that sink to the bottom of the ocean in shipwrecks, or become otherwise entombed.

The best known examples of the entire extinction of the larger animals coëval with man, and probably through his agency, are the following: 1. The *dodo*, a bird larger than the turkey, which existed in Mauritius and the adjacent islands when they were colonized by the Dutch, 200 years ago; but it is no longer to be found; and even all the stuffed specimens that were brought to Europe are lost; so that a head and a foot of one individual in the Ashmolean museum, at Oxford, and the leg of another in the British museum, are all that remains of it, except some fossil bones lately found in the Isle of France. 2. The *Notornis* and *Apteryx australis*, of New Zealand, appear to be on the point of extinction, if not actually extinct. 3. The eleven species of Dinornis formerly inhabiting New Zealand. 4. The *Æpiornis maximus*, a still larger bird, whose bones are found in Madagascar. 5. The *Great Auk*, (*Alca impennis*), of northern regions, "existed in the last century; no specimen has been obtained within the present." (*Owen.*) 6. The large Sirenian animal, like the Manatee, called Stelleria, which formerly inhabited the shores of Siberia is now believed to be extinct.

In particular countries it is a more common occurrence for species to become extinct, as the beaver, wolf, and bear in England. In this country the animals of the forest are disappearing or moving westward as the forests are clearing up. Since the discovery of the island of South Georgia, 1771, one million two hundred thousand seal skins have been annually taken from thence; and nearly as many more from the Island of Desolation. The animal is becoming extinct at these islands. Some have maintained that the climate of Europe is very much warmer than in the times of the Roman Emperors, and have supposed that the extinction of animals is caused by this change.

FORMATION OF CORAL REEFS OR ISLANDS.

Coral reefs are ridges of calcareous rock, whose basis is coral, (chiefly of the genera Porites, Astræa, Madrepora, Meandrina, and Caryophyllia), and whose interstices and surface are covered by broken fragments of the same, with broken shells and echini, and sand, all cemented together by calcareous matter. They are built up by the polypi, apparently on the tops of submarine ridges, and sometimes perhaps, though not generally, on the margins of ancient volcanic craters, beneath the ocean, not generally from a depth greater than twenty-five or thirty feet, yet sometimes 120 or 130 feet. The polypi continue to build until the ridge gets to the surface of the sea at low water; after which the sea washes upon it fragments of coral, drift wood, etc., and a soil gradually accumulates, which is at length occupied by animals with man at their head. The reefs are sometimes arranged in a circular manner, with a lagoon in the centre, where, in water, a few fathoms deep, grow an abundance of delicate species of corals, and other marine animals, whose beautiful forms and colors rival the richest flower garden. Volcanic agency often lifts the reef

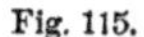

Fig. 115.

Whitsunday; a Coral Island.

far above the waters and sometimes covers one reef with lava, which in its turn is covered with another formation of coral. The growth of coral structures is so extremely slow that centuries are required to produce any important progress. The rate of increase is about half an inch per annum.

The diameter of the circular reefs has been found to vary from less than one to thirty miles. On the outside, the reef is usually very precipitous, and the water often of unfathomable depth. Fig. 115 is a view of one of these circular islands in the South Seas, called Whitsunday Isle; so far reclaimed from the waters as to be covered with cocoanut trees and with some human dwellings. Fig. 116 represents another of the coral islands in the Pacific Ocean.

Fig. 116.

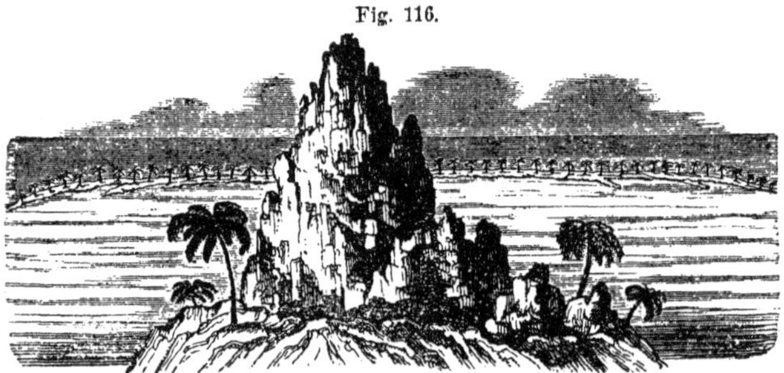

View of the Island of Bolabola.

These islets occur abundantly in the Pacific Ocean, between the thirtieth parallels of latitude. They abound also in the Indian Ocean, in the Arabian and Persian Gulfs, in the West Indies, etc. Usually they are scattered in a linear manner over a great extent. Thus, on the eastern coast of New Holland, is a reef 350 miles long. Disappointment Islands and Duff's Group are connected by 500 miles of coral reefs, over which the natives can travel from one island to another. Between New Holland and New Guinea is a line of reefs 700 miles long, interrupted in no place by channels more than thirty miles wide. A chain of coral islets, 480 geographical miles long, has long been known by the name of the Maldivas. Some groups in the Pacific, as the Dangerous Archipelago, are from 1,100 to 1,200 miles long, and from 300 to 400 miles broad.

Deposits of the Skeletons of Infusoria, and Microscopic Plants.—It is surprising that skeletons of animals and plants, made of silica and iron, requiring thousands of millions to form a single cubic inch, should yet form deposits of considerable extent. At Egea, in Bohemia, there is a stratum two miles long and twenty-eight feet in thickness, mostly composed of shells of infusoria.

At Bilin, in Bohemia, is a similar stratum, fourteen feet thick, every cubic inch of which contains 41,000,000,000 skeletons of *Gaillconella distans*, now generally regarded as a microscopic plant. The city of Richmond, Virginia, stands upon a stratum of infusorial earth twenty feet thick, as described by Professor W. B. Rogers. There is scarcely a town in New England which does not contain extensive deposits of analogous character.

Formation of Soils.—Animal and vegetable substances, when buried in the earth, or the waters, sometimes undergo an almost entire decomposition; at other times, this is very partial; and sometimes the change is so slow that for years scarcely no apparent progress is made. Different substances will be the result of these different degrees of decomposition.

Berzelius embraces all the organic matter of soils in the generic term *humus*. In some places, as on the western prairies, these organic matters of soils increase so as to form a layer several feet thick; but in general they are so much used in the nourishment of plants, that they rarely become more than a few inches thick.

PEAT.

Peat usually consists of soluble and insoluble humus, with a mixture of undecomposed vegetable matter and some earths. Most of it results from the decomposition of certain mosses, especially of the genus *Sphagnum*, which decay at their lower extremity, while the top continues to flourish with vigor. Trees and whatever other organic matter happen to get into these peat bogs, soon become enveloped and assist to swell the amount. In some instances the beds have acquired a thickness of more than forty feet.

In tropical climates, except on high lands, the decomposition of vegetable matter is so rapid that it is resolved into its ultimate elements before peat can be produced. Hence peat is limited chiefly to the colder parts of the globe. In Ireland, the peat bogs are said to occupy one-tenth of the surface, and one of them, on the Shannon, is fifty miles long, and two or three broad. In Massachusetts, exclusive of the four western counties, the amount of peat has been estimated at not less than 120 millions of cords; and probably this falls far short of the actual amount.

By the long-continued action of water and other agents, the humus of peat is changed into bitumen and carbon, which constitute lignite and bituminous coal. In a few instances the process of bituminization has been found considerably advanced in the beds of peat.

Peat bogs are remarkable for their antiseptic power, or the power of preserving animal substances from putrefaction; some remarkable cases of which are on record.

Peat bogs sometimes burst their barriers in consequence of heavy rains, and produce extensive inundations of black mud.

The increase of peat varies so much under different circumstances, that it is of no use to attempt to ascertain its rate of growth. On the continent of Europe, it is stated to have gained seven feet in thirty years.

Where peat is formed in, or transported into estuaries, it is sometimes covered with a deposit of mud; over this another layer of peat forms, and in this way several alternations may occur.

In some peat bogs large trees have been found standing where they originally grew, yet immersed to the depth of twenty feet, as in the Isle of Man.

DRIFT WOOD.

Large rivers, which pass through vast forests, carry down immense quantities of timber. When these rivers overflow their banks, this timber is in part deposited upon the low grounds. But much of it also collects in the eddies along the shores, or is carried into the ocean. After a time it becomes *water-logged*, that is, saturated with water, and sinks to the bottom. Thus a deposit of entangled wood is often formed over large areas. This is subsequently covered by mud; and then another layer of wood is brought over the mud; so that, in the course of ages, several alternations of wood and soil are accumulated. The wood becomes slowly changed into what Dr. Macculloch terms *forest peat;* that is, peat which retains its woody fiber.

The Mississippi furnishes the most remarkable example known of these accumulations. In consequence of some obstruction in the arm of the river called the Atchafalaya, supposed to have been formerly the bed of the Red river, a raft had accumulated in thirty-five years, which in 1816 was ten miles long, 220 yards wide, and eight feet thick. Although floating, it is covered with living plants, and of course with soil. Similar rafts occur on the Red river; and one on the Washita concealed the surface for seventeen leagues. At the mouth of the Mississippi, also, numerous alternations of drift wood and mud exist, extending over hundreds of square leagues.

Similar deposits of wood and mud are found in the river Mackenzie, which empties into the North Sea, and in the lakes through which it passes. At the mouth of the river, which is almost beyond the region of vegetation, are extensive deposits brought from the more southern region through which the river passes.

A part of the drift wood which is brought down the Mississippi and other rivers, along the coast of America, is carried northward by the Gulf Stream and thrown upon the coasts of Greenland. The same thing happens in the bays of Spitzbergen and on the coasts of Siberia.

In the history of common peat and drift wood, we see the origin of the beds of coal which exist in the older strata; for it needs only that the layers of peat (in which term we include submerged drift wood) should be bituminized, and the intervening layers of sand and mud be consolidated, in order to produce a genuine coal formation. Common marsh peat alone can have originated but a small part of the beds of coal.

CONSOLIDATION OF LOOSE MATERIALS.

Having described a variety of natural processes by which just such materials as form the fossiliferous rocks are produced, it remains to inquire whether any agents are now in operation to effect their consolidation.

A considerable degree of solidity is sometimes produced by mere desiccation.

When clay is exposed for a long time to the sun, it becomes as hard as some rocks:—*ex. gr.*, the marly clay dug from the bottom of Lake Superior. Some rocks, when dug from a considerable depth in the earth, in so soft a state as to be readily cut with a knife, become very hard on exposure to the atmosphere.

Carbonate of lime, conveyed in a state of solution among the loose particles of gravel, sand, clay, or mud, and there precipitated, becomes a very efficient agent of consolidation.

EXAMPLES.—1. On the shores of the Bermuda and West India Islands, extensive accumulations of broken shells, corals, and sand, are formed upon the shores by the waves; and these are subsequently consolidated, frequently into very hard rock, by the infiltration of the water which contains carbonate of lime in solution. The famous Guadaloupe rock, in which human skeletons, along with pottery, stone arrow heads, and wooden ornaments, are found, is of the same kind. 2. The Mediterranean delta of the Rhone is ascertained to be, in a good measure, solid rock, produced by the numerous springs that empty into it, that contain carbonate of lime in solution. The same is true of other rivers on the Mediterranean, especially on the east coast, where the ancient Sidon, formerly on the coast, is now two miles inland. 3. In Pownal, Vt., coarse gravel is cemented by carbonate of lime. 4. The fragments of marble accumulating at the quarries, are sometimes, in the lapse of a few years, cemented together as firmly as marble, by streams of water passing over them, saturated with carbonate of lime. An example is in West Stockbridge, Mass.

Another agent of consolidation is the red or peroxide of iron, or rather the carbonate of iron, since the peroxide is not soluble in water without carbonic acid.

EXAMPLES.—1. On the northern coast of Cornwall, England, large masses of drifted sand have been cemented by iron into rocks, solid enough sometimes to be employed for building stones. 2. A similar case occurs on the coast of Karamania, and other parts of Asia Minor. 3. In the United States it is common to find the sand and gravel of the drift and tertiary strata more or less consolidated by the hydrated peroxide of iron.

Silica dissolved in water appears to have been, in former times, an important agent in consolidating rocks; but at the present day it seems to be limited chiefly to deposits from thermal waters, since it is only water in this condition that will dissolve silica in much quantity.

Heat is an important agent in the consolidation of rocks, the most so when it produces complete fusion; yet this is not necessary to the production of a good degree of solidification.

In many of the cases that have been described, great pressure assists in the work of consolidation. Indeed, it is sometimes sufficient of itself to bring the particles within the sphere of cohesive attraction.

GENERAL INFERENCE.

From the facts detailed. in this section, it appears that all the stratified fossiliferous rocks of any importance may have resulted from causes now in operation.

Proof and Examples.—1. Beds of clay need only to be consolidated to become clay slate, or shale. 2. The same is true of fine mud. 3. Sand, consolidated by carbonate of lime, will produce calcareous sandstone; by iron, ferruginous sandstone. 4. Drift, in like manner, will form conglomerates of every age, according to variations in the agents of consolidation. 5. Marls need only to be consolidated to form argillaceous limestones; and if sand be mixed with marl, the limestone will be silicious. 6. Coral reefs and deposits of travertin, subjected to strong heat under pressure, will produce those secondary limestones that are more or less crystalline—but more of this under the sixth section. 7. We have already seen how beds of lignite and coal may be produced from peat and drift wood. 8. The formation of such extensive beds of rock salt and gypsum as occur in the secondary and tertiary rocks is more difficult to explain by any cause now in operation. And yet, in respect to the former, it is said that the lake of Indersk, twenty leagues in circumference, on the steppes of Siberia, has a crust of salt on its bottom more than six inches thick, hard as stone, and perfectly white. The lake of Penon Blanco, in Mexico, yearly dries up, and leaves a deposit of salt sufficient to supply the country. We have also described a somewhat similar case at the lake of Ooroomiah, in Persia. According to Dr. Daubeny, thick beds of rock salt exist at the bottom of Lake Elton, and of several other lakes adjoining the Caspian Sea.

SECTION V.

OPERATION OF IGNEOUS AGENCIES IN PRODUCING GEOLOGICAL CHANGES.

Volcanic action, in its widest sense, is the influence exerted by the heated interior of the earth upon its crust. Igneous agency has a still more extensive signification; embracing all the action exerted by heat upon the globe, whether the source be internal or external. The history of the former will prepare us better to appreciate the influence of the latter.

Volcanic agency has been at work from the earliest periods of the world's history; producing all the forms and phenomena of the unstratified rocks, from granite to the most recent lava. Modern volcanoes will first come under consideration.

These are of two kinds, *extinct* and *active*. The former have not been in operation within the historic period; the latter are constantly or intermittingly in action.

A *volcano* is an opening in the earth from whence matter has been ejected by heat, in the form of lava, scoria, or ashes. Usually the opening called the *crater* is an inverted cone; and around it there rises a mountain in the form of a cone, with its apex truncated, produced by the elevation of the earth's crust and the ejection of lava. The volcanic cones vary in height from 90 feet, as in the volcano of the Island of Reguain, near Sumatra, to 23,900 feet in Aconcagua, in Chile. The lower volcanoes are usually the most active.

When nothing but aqueous and corrosive vapors have been emitted from a volcanic elevation for centuries, such elevation is called a *solfatara*, or *fumerole*.

When volcanos exist beneath the sea, they are called *submarine;* when upon the land, *subaërial.*

As a general fact, volcanic vents are arranged in extensive lines or zones; often reaching half around the globe.

Examples.—1. Perhaps the most remarkable line of vents is the long chain of islands commencing with Alaska on the coast of Russian America, which passes over the Aleutian Isles, Kamtschatka, the Kurilian. Japanese, Philippine and Moluccan Isles, and then turning, includes Sumbawa, Java and Sumatra, and terminates at Barren Island in the Bay of Bengal. 2. Another almost equally extensive line commences at the southern extremity of South

America, and following the chain of the Andes, passes along the Cordilleras of Mexico, thence into California, and thence northward as far at least as Columbia river; which it crosses between the Pacific Ocean and the Rocky Mountains. 3. A volcanic region, ten degrees of latitude in breadth and 1,000 miles long, extending from the Azore Islands to the Caspian Sea, abounds in volcanoes, though very much scattered. The region around the Mediterranean is perhaps better known for volcanic agency than any other on the globe; because no eruption occurs there unnoticed.

Volcanoes not arranged in lines or zones are called central volcanoes, and are more or less insulated. Examples will be found in Iceland, the Sandwich Islands, Society Islands, Island of Bourbon, and a region in Central Asia of 2,500 square geographical miles, from 800 to 1,200 miles from the ocean.

The number of active volcanoes and solfataras on the globe, is estimated at 407, and the number of eruptions about twenty in a year, or 2,000 in a century; though on both these points there is room for considerable uncertainty.

The following table will show how the active volcanoes and solfataras are distributed on the globe.

	On Continents.	*On Islands.*	*Total.*
CENTRAL VOLCANOES.			
In the Mediterranean Sea	1	3	4
In the Atlantic Ocean		24	24
In the Indian Ocean		3	3
In the Pacific Ocean		8	8
Asiatic Continent	3		3
VOLCANOES ARRANGED IN LINEAR SERIES.			
Parts of Europe and Asia	3	4	7
Australia	13		13
Oceanica		188	188
North America	17	35	52
Central America	38	10	48
South America	54		54
Antarctic Continent		3	3
	129	278	407

278 of these volcanoes, or more than two-thirds, are situated upon the islands of the sea; and of the remainder, the greater part are situated upon the borders of the sea, or a little distance from the coast. Hence it is inferred that water acts an important part in volcanic phenomena; indeed, it seems generally admitted that the immediate cause of an eruption is the expansive force of steam and gases. It ought not to be forgotten, however, that some volcanoes are far inland, as Jorullo, in Mexico, and the volcanoes in central Asia.

Intermittent Volcanoes.—Only a few volcanoes are constantly active; in most cases their action is paroxysmal, and is succeeded by longer or shorter intervals of repose. This interval varies from a few months to seventeen centuries. In the island of Ischia the latter period has been known to intervene between two eruptions.

Hence some of the volcanoes of America, generally regarded as extinct, (as Chimborazo, and Carguairazo in Quito, Tacoza, in Peru, and Nevado de Toluca, in Mexico), may yet break forth and show themselves to belong to the class of active volcanoes.

PHENOMENA OF AN ERUPTION.

A volcanic eruption is commonly preceeded by rumbling sounds in the earth, or earthquakes, in the vicinity; stillness of the air, with a sense of oppression; noises in the mountain; and the drying up of fountains. The eruption commences with a sudden explosion, followed by vast clouds of smoke and vapor, with flashes of lightning, jets of acids and mud, and showers of stones; and at length by streams of red hot lava, which break out in irregular intermitting springs of molten earthy matter, and spread over the surrounding country. The eruption is terminated by showers of ashes.

Volcanoes whose summits are far above the snow line, present many peculiar appearances; a sudden melting of the snow indicates the approach of an eruption, even before smoke appears; and this rapid thawing of the accumulated snows occasions destructive floods and violent torrents, in which heaps of smoking ashes are floated away on thick blocks of ice.

Probably the most remarkable eruption of modern times took place in 1815, in the island of Sumbawa, one of the Molucca group. It commenced on the 5th of April, and did not entirely cease till July. The explosions were heard in Sumatra, 970 geographical miles distant, in one direction, and at Ternate in the opposite direction, 720 miles distant. So heavy was the fall of ashes at the distance of forty miles, that houses were crushed and destroyed beneath them. Toward Celebes, they were carried to the distance of 217 miles; and toward Java, 300 miles, so as to occasion a darkness greater than that of the darkest night. On the 12th of April, the floating cinders to the westward of Sumatra were two feet thick; and ships were forced through them with difficulty. Large tracts of country were covered by the lava; and out of 12,000 inhabitants on the island only twenty-six survived.

During the great eruption of the volcano of Cosiguina, in Guatemala, on the shores of the Pacific, in 1835, ashes fell upon the island of Jamaica, 800 miles eastward; and upon the deck of a vessel 1,200 miles westward.

The situation of Vesuvius and Etna has made their history better known than that of most volcanoes. More than eighty eruptions of the latter are on record, since the days of Thucydides; and more than forty of the former, since the first century of the Christian era. That which occurred in Vesuvius, A.D. 79, is best known, from the fact that it buried three cities, Herculaneum, Pompeii, and Stabiæ, which were flourishing at its base. Not much lava appears to have been thrown out at the eruption, but other volcanic products, such as sand, ashes, cinders, and stones. Not only were the cities buried in this loose material, but the buildings, cellars, and vaults, were filled by currents of mud produced by copious showers, resulting from the condensation of aqueous vapors ejected from the volcanoes, mixed with ashes and fine sand. In Herculaneum these deposits are from 70 to 112 feet thick.

Hence it is, that when these cities were first excavated, more than a hundred years ago, every thing enveloped was in a most perfect state of preservation—the pavements of lava, with deep ruts worn by the carriage wheels; the names of their owners over the doors of the houses; the frescoed paintings as bright as though put on but yesterday; fabrics in the shops still showing their texture; vessels of fruit so well preserved as to be easily recognized; bread retaining the stamp of the baker, and medicine yet remaining on the apothecary's counter. The whole constitute perfect examples of fossil cities!

In 1759, in the elevated plain of Malpais, in Mexico, which is from 2,000 to 3,000 feet above the ocean, and at the distance of 125 miles from the sea, a volcanic eruption took place, producing six volcanic cones; now varying in height from 200 to 1,600 feet. Around these cones, and covering several square miles, are a multitude of small cones, from two to six feet high, called *hornitos*, which continually give off hot aqueous vapor and sulphuric acid.

Sometimes during a violent eruption the whole mountain, or cone, is either blown to pieces or falls into the gulf beneath, and its place is afterwards occupied as a lake.

EXAMPLES.—1. In 1772, the Papandayang, a large volcano in the island of Java, after a short and severe eruption, fell in and disappeared over an extent of fifteen miles long and six broad; burying forty villages, and 2,957 inhabitants. 2. In 1638, the Pic, a volcano in the island of Timor, so high as to be visible 300 miles, disappeared, and its place is now occupied by a lake. 3. Many lakes in the south of Italy are supposed to have been thus formed. 4. A volcano occupying the same spot as the present Vesuvius, is supposed thus to have been destroyed in 1779, and its remains to constitute the circular ridge, called Somma, which is several miles in diameter.

VOLCANOES IN HAWAII.

In Hawaii, one of the Sandwich Islands, are the most remarkable volcanoes, perhaps, in the whole world. There are three of them; the first, Mauna Kea, in the northern part of Hawaii, 13,950 feet high, now extinct; the second, Mauna Loa, in the southern part of the island, 13,760 feet high; the third Kilauea, upon a table land at the base of Mauna Loa, 3,970 feet high. Kilauea is the most interesting, as it is constantly active. In approaching the crater it is necessary to descend two steep terraces, each from 100 to 200 feet high, and extending entirely around the volcano. The outer one is 20, and the inner one 15 miles in circumference; and they obviously form the margin of vast craters, formerly existing. Arrived at the margin of the present crater, the observer has before him a crescent shaped gulf 1,500 feet deep, at whose bottom, which is from five to seven miles in circumference, the top being from eight to ten miles, is a vast lake of lava, in some places molten, in others covered with a crust; while in numerous places (some have noticed as many as fifty at once), are small cones with smoke and lava issuing out of them from time to time. Sometimes, and especially at night, such masses of lava are forced up, that a lake of liquid fire, not less than two miles in circumference, is seen dashing up its angry billows, and forming one of the grandest and most thrilling objects that the imagination can conceive. Fig. 117 is a view of this volcano taken by Rev. Mr. Ellis, an English missionary.

Eruptions from Kilauea are repeated every few years. There was a powerful eruption in May and June 1840. For several years the great gulf had been gradually filling up, until it was not more than 900 feet deep, and this molten mass was raging like the ocean when lashed into fury by a tempest. At length the lava found a subterranean passage, and flowed eight miles under ground, when it reached the surface, and sweeping forest, hamlet, plantation, and everything before it, rolled down with resistless energy to the sea, a distance of thirty-two miles, where, leaping a precipice of forty or fifty feet, for three weeks, the stream of half a mile in width and twenty feet in thickness, poured in one vast cataract of fire into the deep below, with fearful hissings and loud detonations. The atmosphere in all directions was filled with ashes, spray, and gases; while the burning lava, as it fell into the water, was shivered into millions of minute particles, and being thrown back into the air, fell in showers of sand on all the surrounding country.

Fig. 117.

Volcano of Kilauea, Sandwich Islands.

Generally Kilauea is quiescent while Mauna Loa is active; but in 1849, during a partial eruption of the latter, Kilauea was unusually active. If there is any connection between the two volcanoes it must be very deeply seated, otherwise the products of Mauna Loa would empty themselves through Kilauea, the lower opening of a great syphon.

In 1843 Mauna Loa sent forth two great streams of lava, one of them being twenty-five miles long and a mile and a half wide. During the eruption a rent twenty-five miles long was produced in the mountain. In 1852 there was another eruption of great power. Persons who visited the crater say that in the midst of the roaring, upheaving ocean of fire, there was a *fountain of lava* of dazzling brilliancy, now shooting up to the height of 700 feet, and now dwindling down to 200 feet, but varied on the top and sides by points and jets, like the ornaments of Gothic architecture; thus producing a fountain constantly varying in form, dimensions, color and intensity, and far surpassing all the possible beauties of any artificial water fountain. In 1855 another eruption commenced, which caused great anxiety to the inhabitants of Hilo. A rushing torrent of lava, from three to five miles wide, flowed from the crater in a direct course for the city, for seven or eight months. But after flowing for a distance of twenty miles the supply was exhausted, and the stream was stayed.

These eruptions, although so vast, commenced with no earthquake, no internal thunderings, or any premonitions discernible at the base of the mountains. The eruptions themselves were comparatively quiet and noiseless; the mountains opened, the lavas flowed out. This stands out in distinct contrast with the bellowing explosive eruptions of Vesuvius and Etna. Hence there are two types of volcanic action,—the one exemplified by Mauna Loa and the other by Vesuvius.

Barren Island.—Fig. 118 is a view of Barren Island, in the Bay of Bengal, which is volcanic.

Fig. 118.

Barren Island, Bay of Bengal.

Fig. 119 shows the summit of Cotopaxi, in South America, emitting smoke. It is nearly 19,000 feet high.

DYNAMICS OF VOLCANIC AGENCY.

We can form an estimate of the power exerted by volcanic agency from three circumstances: first, the amount of lava protruded; secondly, from the distance to which masses of rock have been projected; and thirdly, by calculating the force requisite to raise lava to the tops of existing craters from their base.

Vesuvius, more than 3,000 feet high, has launched scoria 4,000 feet above the summit. Cotopaxi, nearly 19,000 feet high, has projected matter 6,000 feet above its summit; and once it threw a stone of 109 cubic yards in volume, to the distance of nine miles.

Fig. 119.

Cotopaxi.

Taking the specific gravity of lava at 2.8, the following table will show the force requisite to cause it to flow over the tops of the several volcanoes whose names are given, with their height above the sea. The initial velocity which such a force would produce, is also given in the last column.

Name	Height in feet.	Force exerted upon the Lava.	Initial velocity per second.
Stromboli, (Chicciola) . . .	2947	231 Atmospheres.	371 feet.
Vesuvius	3948	320	496
Etna	10874	884	832
Teneriffe	12182	990	896
Mauna Kea, Sandwich Islands	136[illegible]5	1109	966
Cotopaxi, Quito	18875	1493	1104
Aconcagua, Chile	23910	1943	

There can be but little doubt but the chimney of a volcano extends generally as much below the level of the sea as it does above; and often probably fifty times as deep; so that the actual force pressing upon the lava in its reservoir, may be far greater than the second column of the preceding table represents; and the initial velocity much greater than in the third column.

The amount of melted matter ejected from Vesuvius in the eruption of 1737, was estimated at 11,839,168 cubic yards; and in that in 1794, at 22,435,520 cubic yards. But these quantities are small compared with those which Etna has sometimes disgorged. In 1660, the amount of lava was twenty times greater than the whole mass of the mountain; and in 1669, when 77,000 persons were destroyed, the lava covered eighty-four square miles.

According to Professor Dana, 15,400,000,000 cubic feet of matter flowed from Kilauea in the eruption of 1840—a mass equal to a triangular ridge 800 feet high, two miles long, and a mile wide at the base.

New Islands formed by Volcanic Agency.—History abounds with examples of new islands rising out of the sea by volcanic action. Such were Delos, Rhodes, and the Cyclades, situated in the Grecian Archipelago, and described by Pliny, the naturalist, and other ancient writers. In more modern times, small islands have risen in the Azore group; such as Sabrina, in 1811, which was 300 feet high, and a mile in circumference; but after some time it disappeared; another in 1720, was six miles in circumference. In 1707, the island called Isola Nuova, was thrown up near Santorini, and continues to this day. Just before the great eruption of Skaptar Jokul in Iceland, in 1783, a new island appeared off the coast; which, however, subsequently disappeared. In 1796, a new island rose to the height of 350 feet, having two miles of circumference, in the Aleutian group, east of Kamtschatka, which is permanent. In 1806 another permanent island rose in the same vicinity, four geographical miles in circumference. In the same archipelago, in 1814, another peak arose, which was 3,000 feet high; and which remained standing a year afterwards. In those where the cone does not sink back beneath the sea, it is probably composed of the more solid lavas, such as trachyte, or basalt.

On Fig. 120 is exhibited the eruption by which Sabrina, mentioned above, was produced.

The rise of these islands is sometimes connected with submarine volcanoes. In July, 1831, a volcanic island rose up through the sea off the coast of Sicily, and was called Graham's Island. In August it was 180 feet high, and one and a third miles in circumference; but the part above water being composed of loose materials, disappeared in two or three years, leaving a rocky shoal.

These islands are not always raised to their full height by a single paroxysm of the volcanic force; but by a succession of efforts for months and even years.

Very many large islands appear to be wholly, or almost en-

Fig. 120.

Submarine Volcano (Sabrina).

tirely, the result of volcanic action; and to be composed chiefly of lava and rocks upheaved by this agency, such as sandstone and limestone. Examples may be found in the Sandwich Islands, of which Hawaii, the largest, contains 4,000 square miles of surface, and rises 18,000 feet above the ocean; in Teneriffe, 13,000 feet high; in Iceland, Sicily, Bourbon, St. Helena, the Madeira, and Faroe Islands; and a great part of Java, Sumatra, Celebes, Japan, etc.

Character of Molten Lava.—Lava in general is not very thoroughly melted; so that when it moves in a current over the country, its sides form walls of considerable height, and a crust soon forms over its surface, which serves still more to prevent its spreading out laterally. It is kept in a semi-fluid state by the water which it incloses, and which is prevented from escaping by the hard crust.

Hence a lava current may be deflected from its course by breaking away its crust on one side; and in this way it has sometimes been turned away from towns that were threatened by it. In one instance, the inhabitants of Catania attacked a lava current and turned it towards Paterno, whose inhabitants took up arms and arrested the operation.

The crust forming upon lava soon becomes a good non-conductor of heat; and hence the mass requires a long time to cool; *ex. gr.*, the case of Jorullo, in Mexico, 1,600 feet high, which was ejected a hundred years ago, but is not yet cool.

This explains a curious fact. In 1828, a mass of ice was found on Etna, lying beneath a current of lava. Probably before this flowed over it, the ice might have been covered by a shower of volcanic ashes, which are a good non-conductor of heat, and might have prevented the immediate melting of it, while the superimposed lava has preserved it from the period of its eruption to the present.

When lava is thrown out upon the dry land, with only the pressure of the atmosphere upon it, it is apt to become vesicular and scoriaceous; but when cooled slowly and under great pressure, it becomes compact and may be even crystalline. The porous varieties result from the cooling of the lava while expanded with the contained gases. Scoria and pumice may often be regarded as the froth or foam of the volcano.

Volcanoes constantly Active.—A few volcanic vents have been constantly active since they were first discovered. They always contain lava in a state of ebullition; and vapors and gases are constantly escaping.

EXAMPLES.—1. Stromboli, one of the Lipari Islands, has been observed longer probably than any volcano of this class; and for at least 2,000 years it has been unremittingly active. The lava here never flows over the top of the crater; though it is sometimes discharged through a fissure into the sea, killing the fish, which are thrown upon the shore ready cooked. It is said to be more active in stormy than in fair weather; likewise more so in winter than in summer: a fact explained by the different degrees of pressure exerted by the air upon the lava at different times. When the air is light, the internal force predominates; but when heavy, it restrains the energy of the volcano.

2. In Lake Nicaragua is a volcano which is contantly burning. Villarica, in Chile, so high as to be seen 150 miles, is never quiet. The same is said to be the case with Popocatepetl, in Mexico. Ever since the Spanish conquest of Mexico it has been pouring forth smoke. Kilauea is the most remarkable active volcano on the globe, and has already been described.

Seat of Volcanic Power.—Volcanic power must be deeply seated beneath the earth's crust.

Proof.—1. The melted lava is forced out from beneath the oldest rocks, as gneiss and granite; for masses of these rocks are frequently broken off and thrown out. 2. Lines or trains of volcanoes indicate some connection between the vents; and the great length of these lines, several thousand miles in some instances, can be explained only by supposing that the fissure or cavity by which the connection is made must extend to a great depth. 3. When, in 1783, a submarine volcano on the coast of Iceland, ceased to eject matter, immediately another broke out 200 miles distant, in the interior of the island. 4. Were not the power deep-seated, volcanoes would become exhausted; as they sometimes throw out more matter at a single eruption, than the whole mountain melted down could supply.

EXTINCT VOLCANOES.

Many writers maintain that there is a marked difference between the matters ejected from active and extinct volcanoes. It is said that the more modern lavas have a harsher feel, are more cellular, and more vitreous in their appearance, and also less feldspathic than the ancient. But it is doubtful whether any character will satisfactorily distinguish them, except the period of their eruption.

The extinct volcanoes are of very different ages. Some of them were active during the tertiary period, some during the drift period; and some since that time. In some instances, as a mountain called the Puy de Chopine, in Auvergne, which stands in an ancient crater, and rises 2,000 feet above an elevated granitic plain, itself about 2,800 feet above the sea, there is a mixture of trachyte and unaltered granite.

The extinct volcanoes of Auvergne, and the south of France, have long excited deep interest; and have been fully illustrated by Scrope, Bakewell, and others. Near Clermont, the landscape has as decidedly a volcanic aspect as in any part of the world; of which Fig. 121 will convey some idea.

Fig. 121.

Extinct Volcanoes; Auvergne.

Extinct volcanoes exist also in Spain, in Portugal, in Germany, along the Rhine, in Hungary, Styria, Transylvania, Asia Minor, Syria and Palestine. To the east of Smyrna in Asia Minor, is a region called the *Burnt District* (Katakekaumena of the Greeks), because it shows such striking marks of extinct volcanoes. In the valley of the Jordan, especially around Lake Tiberias, extending as far northwest as Safed, volcanic rocks abound, with warm springs and occasional earthquakes.

The region about the Dead Sea is decidedly volcanic; but appears more like a region of extinct than active volcanoes. Yet the destruction of the cities of the plain, Sodom and Gomorrah, as represented in the Bible, seems to have been caused by volcanic agency. Some suggest that Sodom and Gomorrah were built upon a mine of bitumen, that lightning kindled the combustible mass, and that the cities sunk in the subterranean conflagration. The principal difficulties in the way of this hypothesis are, first, to see how the bitumen, buried beneath a considerable thickness of soil, could have burnt rapidly enough suddenly to destroy the cities and their inhabitants; and secondly, to conceive of a bed of bitumen so thick, as by its combustion to sink the surface from the present high-water mark to the bottom of the sea. Dr. Robinson describes the high-water mark as seen by him "a great distance," south of the margin of the sea at that time. The surface, therefore, must have suffered a great depression. Would it not somewhat relieve these difficulties to suppose volcanic action combined with the combustion of the bitumen? No geologist will doubt the correctness of Von Buch's opinion, that a fault extends from the Red Sea through the valley of Arabah and the Jordan to Mount Lebanon; and along that fissure we might expect volcanic agency to be active. But it might have produced very striking effects without the ejection of lava. Earthquakes sometimes cause the surface to sink down many feet, and flames have been seen to issue through the fissures which they produce. Thus might the *slime pits* (literally *wells of asphaltum*) have been set on fire, immense volumes of steam, smoke and suffocating vapors have been set at liberty, perhaps, too, the remarkable ridge of rock salt called Usdam have been protruded, and finally, by the subsidence of the surface after the destruction of the cities, might the waters of the lake have flowed over the spot. In a similar manner was the city of Euphemia, in Calabria, destroyed in 1638. "After some time," says Kircher, who was near the spot, "the violent paroxysms (of the earthquake) ceasing, I stood up, and turning my eyes to look for Euphemia, saw only a frightful black cloud. We waited till it had passed away, when nothing but a dismal and putrid lake was to be seen, where once the city stood."

Mt. Ararat in Asia, is an extinct volcano. A large proportion of the lofty peaks of the Andes and the mountains of Mexico belong to the class of extinct volcanoes, as well as large districts of the region between the Rocky Mountains and the Pacific Ocean.

The size of ancient volcanic cones and craters was often very large.

In the middle and southern parts of France, extinct volcanoes cover several thousand square miles. Between Naples and Cumea, in the space of 200 square miles, according to Brieslak, are sixty craters; some of them larger than Vesuvius. The city of Cumea has stood three thousand years in a crater of one of these volcanoes. Vesuvius stands in the midst of a vast crater, whose remains are still visible, called Somma. The volcanic peak of Teneriffe stands in the centre of a plain, covering 108 square miles, which is surrounded by perpendicular precipices and mountains, which were probably the border of the ancient crater. According to Humboldt, all the mountainous parts of Quito, embracing an area of 6,300 square miles, may be considered as an immense volcano, which now gets vent sometimes through one, and sometimes through another of its elevated peaks; but which must have been more active in former times to have produced the results now witnessed. Of the two ancient craters of Kilauea, one is fifteen and the other twenty miles in circumference. Two other ancient craters exist in Maui, one of the Sandwich Islands, the one twenty-four and the other twenty-seven miles in circuit.

From such facts many geologists have inferred that volcanic agency in early times was more powerful than at present, and that it is gradually diminishing. Lyell and others, however, are of a different opinion, and quote as equalling any of the ancient eruptions, the outbursts from Skapter Jokul, in 1783, and from Mauna Loa and Kilauea in the nineteenth century.

EARTHQUAKES.

Earthquakes almost always precede a volcanic eruption; and cease when the lava gets vent.

Hence, the proximate cause of earthquakes is obvious; viz., the expansive efforts of volcanic matter, confined beneath the earth's surface.

Hence, too, the ultimate cause of volcanoes and earthquakes is the same, whatever that cause may be.

During the paroxysm of the earthquake, heavy rumbling noises are heard: the ground trembles and rocks; fissures open on the surface, and again close, swallowing up whatever may have fallen into them; fountains are dried up; rivers are turned out of their courses; portions of the surface are elevated, and portions depressed; and the sea is agitated and thrown into vast billows.

The *concussions* of earthquakes, or the violent commotions of the surface, are of three kinds: the first being distinguished by a series of perpendicular, the second by horizontal or undulatory, and the third by rotatory motions, following each other in rapid succession. The perpendicular motions act from below upwards; as during the destruction of Riobamba, in 1797, when dead bodies were thrown upon a hill several hundred feet high. The horizontal motions act in an undulating manner, causing an alternate rising and sinking of the earth. The rotatory or circular motions are the most rare, but are the most destructive. They consist of whirling movements of the earth, whereby buildings without being overturned are twisted, parallel rows of trees deflected, and fields when covered with grain made to change their relative positions.

The *progression* of earthquakes is generally in a linear direction, undulating with a velocity of from twenty to thirty geographical miles in a minute. Sometimes the progression is in concussion circles or great ellipses, in which, as from a center, the vibrations extend with decreasing force to the circumference.

Several thousand cases of earthquakes have been recorded. During many of them tracts of land have been elevated or depressed. The following are a few of them. In 1692, a part of Port Poyal, in the West Indies was sunk;

in 1755, a part of Lisbon; in 1812, a part of Caraccas. About the same time numerous earthquakes agitated the valley of the Mississippi, for an extent of 300 miles, from the mouth of the Ohio, to that of the St. Francis, whereby numerous tracts were depressed, and others elevated, lakes and islands were formed, and the bed of the Mississippi was exceedingly altered. There is a remarkable subsidence, twenty miles in length, and a mile in width just above the Falls, in Columbia river, in Oregon. Through the whole distance the trees are standing in the bottom of the stream, at an average depth of twenty feet. The region appears to be one of extinct volcanoes.

The most extensive elevation of land on record by means of earthquakes, took place on the western coast of South America, in 1822. The shock was felt 1,200 miles along the coast; and for more than 100 miles the coast was elevated from three to four feet; and it is conjectured that an area of 100,000 square miles was thus raised up.

In 1783, a large part of Calabria was terribly convulsed by earthquakes, over an area of 500 square miles. The shocks lasted for four years; in 1783, there were 949, and in 1784, 151. A vast number of fissures of every form were made in the earth, and of course a great many local elevations and subsidences; which, however, do not appear to have exceeded a few feet. In some sandy plains, singular circular hollows a few feet in diameter, and in the form of an inverted cone were produced by the water which was forced up through the soil. Some of these are exhibited on Fig. 122.

Fig. 122

Holes formed by an Earthquake.

The ocean is almost always agitated during earthquakes, thus producing waves of translation, often of great size and power. Their effects have been alluded too in the previous section.

The number of earthquakes is about twenty annually, corresponding to the number of volcanic eruptions.

The effects of earthquakes in changing levels may not be permanent, because a depression of a tract may be counterbalanced by a subsequent elevation.

THERMAL SPRINGS.

Hot springs are very common in the vicinity of volcanoes; such as the well-known geysers in Iceland. Some of these are intermittent, probably in consequence of the agency of steam within subterranean cavities. The great geyser consists of a basin fifty-six by forty-six feet in diameter; at the bottom of which is a well ten feet in diameter and seventy-eight feet deep. Usually the basin is filled with water in a state of ebullition; but occasionally an eruption takes place, by which the water is thrown up from 100 to 200 feet, until it is all expelled from the well, and there follows a column of steam with amazing force and a deafening explosion, by which the eruption is terminated. These waters hold silica in solution; as do those of the Azore Islands; and extensive deposits are the result. The coating over of vegetables by this silicious matter, has given rise to the common opinion that certain rivers and lakes possess the power of rapid petrifaction.

Fig. 123 represents the great geyser of Iceland in action.

Thermal springs are not confined to the vicinity of volcanoes. They occur in every part of the globe; and rise out of almost every kind of rock. They frequently contain enough of mineral substances to constitute them mineral waters. But one of their most striking properties is the evolution of gas; such as carbonic acid, nitrogen, oxygen, sulphuretted hydrogen, etc., in a free state.

Theory of Thermal Springs.—When these springs occur in volcanic districts, their origin is very obvious. The water which percolates into the crevices of the strata becomes heated by the volcanic furnace below, and impregnated with salts and gases by the sublimation of matter from the same focus. The thermal springs not in volcanic districts, in a large majority of cases rise either from the vicinity of some uplifted chain of mountains, or from clefts and fissures caused by the disruption of the strata; and therefore, in all such cases are probably the result of deep-seated volcanic agency, which may have been long in a quiescent state.

Fig. 123.

Great Geyser of Iceland.

TEMPERATURE OF THE GLOBE.

The principal circumstances that determine the temperature of the globe and its atmosphere are the following: 1. Influence of the sun. 2. Nature of the surface. 3. Height above the ocean. 4. Oceanic currents. 5. Temperature of the celestial spaces around the earth. 6. Temperature of the interior of the earth, independent of external agencies.

1. *Solar Heat.*—The solar rays exert no influence, as a general fact, at a greater depth than about 100 feet. (Baron Fourier mentions 130 feet as the maximum depth; Poisson fixes it at seventy-six feet.) A thermometer placed at that depth remains stationary all the year. The diurnal effect does not extend more than three or four feet. In receding from the tropics, the amount of solar heat diminishes. During six months it continues to increase, and to diminish the remaining six months. The decrease of the mean temperature from the equator towards the poles is nearly in proportion to the cosines of latitude. Prof. Forbes has made some observations near Edinburgh, from which it appears that the oscillations of annual temperature would cease at the depth of forty-nine feet in trap tufa, sixty-two feet in incoherent sand, and ninety-one feet in compact sandstone.

Solar heat is the fundamental element on which depends the surface temperature of the globe and the character of the climate.

2. *Nature of the Surface.*—The radiating and absorbing power of land is quite different from that of water. Ice and snow are still different; and the nature of the soil affects sensibly its power to imbibe or give off heat. Hence low islands have a higher temperature than large continents in the same latitude; and the ocean possesses greater uniformity of climate than the land.

On these facts Sir Charles Lyell has founded an hypothesis for explaining the high temperature of the surface of the globe in northern latitudes in early times. He supposes that but little land then existed in the northern parts of the globe, and that this produced so great an elevation of temperature above what it is at present, that tropical animals and plants might then have inhabited regions now subjected to almost perpetual winter. That the quantity of dry land in the northern hemisphere, during the deposition of the older fossiliferous rocks, was much less than at present is very probable; and this might affect the climate somewhat. But if the action of currents from tropical regions extending to the frigid zone, at the present day, is not sufficient to render the climate temperate, we can not think a greater depression in ancient times would be adequate to the production of a climate in which tropical plants and animals might flourish.

3. *Height above the Ocean.*—The temperature of the air diminishes one degree Fahrenheit for 300 feet of altitude; two degrees for 595 feet; three degrees for 872 feet; four degrees for 1,124 feet; five degrees for 1,347 feet; and six degrees for 1,539 feet.

Hence, at the equator perpetual frost exists at the height of 15,000 feet, diminishing to 13,000 feet at either tropic. Between latitudes 40° and 59° it varies from 9,000 to 4,000 feet. In almost every part of the frigid zone this line descends to the surface. These results, however, are greatly modified by several circumstances; so that, in fact, the line of perpetual congelation is not a regular curve, but rather an irregular line descending and ascending.

4. *Oceanic Currents.*—The surface of all oceans is occupied by currents. Some flow from the poles toward the tropics, carrying with them cold water, thus lowering materially the temperature of the warmer regions. Others flow from the tropics to the colder regions, carrying warm water and the products of warm climates, with effects to correspond. Most of the irregularities in the isothermal curves, where they cross oceans, are produced in this way. A familiar illustration may be seen along our coast. A cold current from Baffin's Bay passes near the eastern shore of North America, and makes the isotherm bend to the south along the whole distance; while the Gulf Stream, passing in the opposite direction, outside of the cold current, renders the climate of northern Europe much warmer than our shores at the same degree of latitude.

5. *Temperature of the Celestial Spaces around the Earth.*—This can not be much less than the temperature around the poles of the earth, where the solar heat has scarcely any influence. Now the lowest temperature hitherto observed near the poles (as recorded by Dr. Kane in North Greenland) is 70° below zero; and this has been assumed as the temperature of the planetary spaces. Hence it follows that there must be a constant radiation of heat from the earth into space.

6. TEMPERATURE OF THE INTERIOR OF THE EARTH.

In descending into the earth, beneath the point where it is affected by solar heat, we find that the temperature regularly and rapidly increases. If this rate is continuous, all the interior of the earth, below a crust of 100 miles thick, is at present in a state of fusion. Fig. 124 represents the proportion of melted and unmelted matter in the earth, on the supposition that the crust, which is represented by the black line, is 100 miles thick.

The evidences of an increase in the temperature of the interior are these: the higher temperature of the earth in all artificial excavations of considerable depth; the existence of thermal springs; and the existence and distribution of volcanoes.

Proof 1.—There are three sources of information from artificial excavations. 1. The temperature of springs which issue from the rocks in mines. 2. The temperature of the rock itself in mines. 3. The temperature of the water from Artesian wells. The following table gives many of the particulars:

TEMPERATURE OF SPRINGS IN MINES.

Countries.	Mines.	Depth in feet.	Temperature.	Mean temperature at surface.	Depth in feet for one degree Fahrenheit.
Saxony	Lead and Silver Mine of Junghohe Birk	256	48.9°	46.9°	102.4
	do of Beschertgluck	712	54.5	46.4	87.
Brittany	do of Poullauen	123	58.4	52.7	182.
	do of Huelgoet	197	54.	51.8	89.5
Cornwall	Dolcoath Mine	1440	82.	50.	45.
Mexico	Guanaxato, Silver Mine	1713	98.2	68.8	45.8
TEMPERATURE OF THE ROCK IN MINES.					
1. *In loose matter near the face of the rock.*					
Cornwall	United Copper Mines	1142	87.4	50	30.5
		1201	88.		31.1
Carmeaux, Fr.	Coal Pit of Ravin	597	62.8	52	55.8
	do of Castellan	630	67.1	52	40.8
2. *In the rock near its surface.*					
Saxony	Mine of Beschertgluck	591	52.2		101.
	do do	813	59.	46.4	67.
	do do	1246	65.7		64.4
3. *Three feet three inches within the rock.*					
Cornwall	Dolcoath Mine. Register kept 18 months	1381	75.6	50.	54.
Saxony	Lead and Silver Mine of Kurpinz	413	59.6		81.3
	do do do	686	62.5		42.6
	do do do	1063	67.7		49.9
E. Virginia	Coal Mines	780	68.7	56.7	60.
TEMPERATURE OF ARTESIAN WELLS.					
Paris, near the Barrier de Grenelle		1800	83.	51.1	59.
Paris, Fountain de St. Venant		828	57.2		49.
Tours		459	63.5	52.7	42.5
La Rochelle		369	64.6	58.4	83.
Near Berlin, in Prussia		675	67.6	49.1	36.3
Louisville, Ky		2086	76.5		
Charleston, S. C.		910	69.	82.3	64.
New Brunswick, N. J.		394	54.		72.

Artesian wells have lately been applied with success in Wurtemburg, to prevent frost from stopping machinery which was moved by running water, and also for warming a paper manufactory. Who knows but this application may prove of immense benefit to some regions of the globe?

The increase of temperature from the surface of the earth downwards does

Fig. 121.

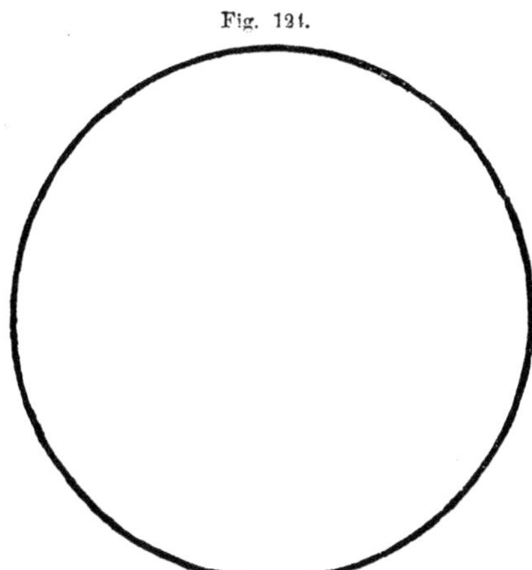

not appear to be at the same rate in all countries. The mean of all the observations which have been made in England, gives 44 feet for a change of one degree. In some mines in France the increase is much slower, and in a few it is faster. The mean is reckoned at about 45 feet for each degree. In Mexico according to the only observation given above, it is 45.8 feet. In Saxony it is considerably greater, not far from 65 feet to a degree. The few observations in this country, given in the preceding table, indicate an increase of 57 feet to a degree.

The average increase for all the countries where observations have been made is stated by the British Association to be at the rate of 45 feet to each degree, and this may be used for the present.

At this rate, assuming the temperature of the surface to be 50°, a heat sufficient to boil water would be reached at the depth of 7,290 feet, or more than a mile; a heat of 6,400°, sufficient to melt all known rocks, would be reached at 59.23 miles; and if the temperature continued to increase uniformly, at the center of the earth it would amount to 475,000°. But it is probable that the degree of heat is uniform after reaching a certain point.

Another method of calculating the thickness of the crust has been proposed from the Precession of the Equinoxes. This change of the earth's position is caused by the attraction of the sun and moon upon the protuberant ring of matter around the equator. It is claimed that the amount of this attraction will vary in proportion to the amount of fluid matter in the earth. If the earth were entirely fluid or entirely solid, the amount of precession would vary to the one or other side of its present rate. Thus the present rate is a sort of medium between two extremes; and hence it is calculated that the solid crust of the earth must be at least 800 miles thick to be consistent with the present amount of precession. This view would relieve many of the difficulties urged against the doctrine of internal heat; but the solution of the problem depends upon so many niceties, that it would be well to suspend our judgment upon the results of the calculation until all the preliminaries are satisfactorily established.

This argument for the internal heat of the earth receives strong corroboration from the fact that *not one exception to this increase of internal temperature has ever occurred, where the experiment has been made in deep excavations.*

It appears from the experiments and profound mathematical reasoning of Baron Fourier, that even admitting all the internal parts of the earth to be in a fused state, except a crust of thirty or forty miles in thickness, the effect of that internal heat might be insensible at the surface, on account of the extreme slowness with which heat passes through the oxidized crust. He has shown that the excess of temperature at the surface of the earth, in consequence of this internal heat, is not more than 1-17th of a degree (Fahr.), nor can it ever be reduced more than that amount by this cause. This amount of heat would not melt a coat of ice 10 feet thick in less than 100 years; or about one inch per annum. The temperature of the surface has not diminished on this account, during the last 2,000 years, more than the 167th part of a degree; and it would take 200,000 years for the present rate of increase in the temperature, as we descend into the earth, to increase the temperature at the surface one degree; that is, supposing the internal heat to be 500 times greater than that of boiling water. From all which it follows, that if internal heat exist, it has long since ceased to have any effect practically upon the climate of the globe.

Proof 2.—Until some fact can be adduced showing that the heat of the earth ceases to increase beyond a certain depth, nothing but hypothesis can be adduced to prove that it does not go on increasing, until at least the rocks are all melted; for when they are brought into a fluid state, it is not difficult to see how the temperature may become more equalized through the mass, in consequence of the motion of the fluid matter; so that the temperature of the whole may not be greatly above that of fused rock. Now, if the hypothesis of internal fluidity have other arguments (which follow below) in its favor, while no facts of importance sustain its opposite, the former should be adopted.

Proof 3.—This is derived from the existence of thermal springs.

Vast numbers of these occur in regions far removed from any modern volcanic action; generally upon lofty mountain ranges; as upon the Alps, the Pyrenees, the Caucasus, the Ozark mountains in this country, where are nearly seventy, in California, etc. Their temperature varies from about summer heat to that of boiling water. Nor can their origin be explained without supposing a deep-seated source of heat in the earth.

Proof 4.—*The existence of* 400 *active volcanoes, and many extinct ones, whose origin is deep seated, and which are connected over extensive areas.* If these were confined to one part of the globe, or if after one eruption the volcano were to remain forever quiet, we might regard the cause as local and the effect of particular chemical changes at those places, aided perhaps by electro-magnetic agencies. But if the internal parts of the earth are in a melted state, that is, in the state of lava; and if this mass be slowly cooling, occasional eruptions of the matter ought to be expected to take place by existing volcanoes. Assuming the thickness of the earth's crust to be sixty miles, the contraction of this envelope one 13,000th of an inch, would force out matter enough to form one of the greatest volcanic eruptions on record. More probably, however, the percolations of water to the heated nucleus, or other causes of disturbance, more frequently produce an eruption than simple contraction.

Some geologists have proposed chemical theories to account for the phenomena of volcanoes. We will consider the two most important ones.

Hypothesis of the Metalloids.—This hypothesis, originally proposed, though subsequently abandoned, by Sir Humphrey Davy, supposes the internal parts of the earth, whether hot or cold, fluid or solid, to be composed in part of the metallic bases of the alkalies and earths, which combine energetically with oxygen whenever they are brought into contact with water, with the evolution of light and heat. To these metalloids water occasionally percolates in large quantities through fissures in the strata, and its sudden decomposition produces an eruption. Dr. Daubeny, the most strenuous advocate of this theory, has brought forward a great number of considerations which render it quite probable that this cause may often be concerned in producing volcanic phenomena, even if we do not admit that it is the sole cause.

Many of the phenomena of volcanoes may be explained upon this view, as the formation of vapor, the extrication of gases, and the sublimation of sulphur, salts, etc., and the connection of volcanoes with water. But it does not satisfactorily account for the constantly active vents. Moreover, silica, the most common chemical combination in lava, can not be produced by the union of silicium and oxygen under any heat known to chemists. Silicium is unaltered before the blowpipe, and is incombustible in oxygen gas. Aluminum also unites with oxygen very slowly, even under powerful heat.

Modified Chemical Theory.—Some geologists, as Lyell, have called in the aid of electricity to assist in the decompositions and recompositions that result from volcanic agency. By this means the temperature of the uncombined metals is raised, so as to cause them to become oxidized more readily.

Against the universality, at least, of both these theories, may be brought to bear the type of volcanic action in Mauna Loa and Kilauea. It is a quiet, gradual overflow, not a sudden decomposition and recomposition of elements, evolving great heat with explosive accompaniments. The chemical theories suppose violent action.

OBJECTIONS TO THE DOCTRINE OF INTERNAL HEAT.

Objection 1. *It has been maintained that the high temperature of deep excavations may be explained by chemical changes going on in the rocks*; such as the decomposition of iron pyrites by mineral waters, the lights employed by the workmen, the heat of their bodies, and especially by the condensation of air at great depths.

Answer. In the experiments that have been made upon the temperature of mines, care has been taken to avoid all these sources of error except the last (which are indeed sometimes very considerable), and yet the general result is as has been stated; nor is there a single example on the other side to invalidate that result. As to the condensation of air in mines, Mr. Fox has shown that the air which ascends from their bottom is much warmer than when there; so that it carries away instead of producing heat.

Objection 2. *The temperature of the ocean.* The temperature of the ocean diminishes as the depth increases; at first rapidly, then very slowly. The deepest measurements do not indicate any increase of heat, but only a uniform coldness. From observations made by Lieut. Maury and others, it is found that the change of temperature in the ocean is as follows: for the first 2,500 feet the temperature diminishes 40°; from 2,500 to 14,000 feet the reduction is about 3°. The temperature at any lower depth is unknown.

Answer. There are many local exceptions to this decrease of temperature, especially in northern latitudes; yet, on the whole, we should expect that the temperature of the sea would decrease downwards, until it had reached a temperature below which it would rarely descend; after which we should expect a uniform temperature to the greatest depths. Moreover, it is a fact that the warmest particles of a liquid body always rise to the top, as is the case in a vessel of water that is heated over a fire; that is, the strata of water arrange themselves according to their specific gravities. Hence we should expect to find the warmer portions upon the surface. The temperature of sea-water, at great depths, can never be less than 25.4°, because that is the point of its greatest density. If it was colder than this, it would rise, in consequence of being specifically lighter by expansion.

Objection 3. *Circulation of the internal heat.* If the central heat were as intense as is represented, there must be a circulation of currents, tending to equalize the temperature of the resulting fluid, and the solid crust itself would be melted. For example, if the whole planet were composed of water, the exterior crust of fifty miles thickness being condensed to ice, and the interior ocean having a central heat about 200 times that of the melting point of ice, then the ice, instead of being strengthened annually by new internal layers, would be melted, and the whole spheroid assume an equable temperature throughout. This is the objection of Sir Charles Lyell.

Answer 1. It is not essential to the doctrine of central heat that a temperature very much exceeding that requisite to melt rocks (6,400° F.) should exist in any part of the molten nucleus. It may even be admitted that the whole globe was cooled down very nearly to that point before a crust began to form over it. For still, according to the conclusions of Fourier, it would require an immense period to cool the internal parts, so that they should lose their fluid incandescent state after a crust of some twenty miles thick had been formed over them.

2. There is the case of currents of lava, which cool at their surface, so as to permit men to walk over them, while for years, and even decades of years, the lava beneath is in a molten state, and sometimes even in motion. And if a crust can thus readily be formed over lava, why might not one be formed over the whole globe, while its interior was in a melted state; and if a crust only a few feet in thickness can so long preserve the internal mass of lava at an incandescent heat, why may not a crust upon the earth, many miles in thickness, preserve for thousands of years the nucleus of the earth in the same state? True, if we immerse a solid piece of metal in a melted mass of the same, the fragment will be melted; *because it can not radiate the heat which passes into it;* but keep one side of the fragment exposed to a cold medium, as the crust of the earth is, and it will require very much stronger heat to melt the other side. If the crust of the globe were to be broken into fragments, and these plunged into fluid matter beneath, probably the whole would soon be melted, if the internal heat be strong enough. But so long as its outer surface is surrounded by a medium, whose temperature is at least —70°, nothing but a heat inconceivably powerful, can make much impression on its interior surface.

3. A globe of water intensely heated at its center, and covered by a crust of ice, is not a just illustration of a globe of earth in a similar condition, covered by a crust of rocks and soils. For between the ice and water there is no intermediate or semi-fluid condition. As soon as the ice melts, there exists a perfect mobility among the particles; so that the hottest, because the lightest, would always be kept in contact with the surrounding crust of ice, and melt it continually more and more; especially as ice, being a perfect non-conductor of heat, would not permit any of it to pass through, and by radiation prevent the melting. On the other hand, between solid rock and perfectly fluid lava, there is every conceivable degree of spissitude; and of course every degree of mobility among the particles. Hence, they could not in that semi-fluid stratum, arrange themselves in the order of their specific gravities; and therefore, the layer of greatest heat would not be in contact with the unmelted solid rock. True, the heat would be diffused outwards, but so long as the hardened crust could radiate the excess of temperature, the melting would not advance in that direction. This would take place only when the heat was so excessive, that the envelope could not throw it off into space.

FORMER IGNEOUS FLUIDITY OF THE EARTH.

We have already shown that probably the interior of the earth is in a state of igneous fluidity. We now advance a step farther, and say, that previous to the formation of the lowest solid rocks, the whole globe was in a state of igneous fusion, and that its present crust has been formed by the cooling of the surface by radiation. Many of the proofs of this position are also the strongest arguments for the present internal heat of the earth.

Proof (1). *The Spheroidal Figure of the Earth.*—This is the strongest of all the proofs. The form of the earth is precisely that which it would assume, if while in a fluid state, it began to revolve on its axis with its present velocity; and hence the probability is strong that this was the origin of its oblateness. But if originally fluid, it must have been igneous fluidity; for since the solid matter of the globe is at present 50,000 times heavier than the water, the idea of aqueous fluidity is entirely out of the question. If the rate of revolution had been greater than it is now, the poles would have been flattened more, and if the rotation had been less rapid, the poles would have been flattened less than at present. To have formed a perfect sphere, there must have been no rotation at all.

Proof (2). *All the crust of the globe has been in a melted state.*—As to the older unstratified rocks nearly all admit that they are of igneous or aqueo-igneous origin. As to the aqueous and metamorphic rocks, also, it will be admitted that they were originally made up of fragments derived from the unstratified rocks; and consequently that they have been melted. Hence if the entire crust of the globe has been melted, it is a fair presumption that it was the result of the fusion of the whole globe.

Proof (3). *The universality of a tropical or ultra-tropical climate, in the earlier geological ages, and in high latitudes.*—If the earth has passed through the process of refrigeration, there must have been a time, when the whole surface had such a high temperature, as is denoted by its organic remains. A climate, also, chiefly dependent on subterranean agency, would be more uniform over the whole globe, than one dependent on solar influence: and the first appears to have been the agency in those remote ages.

Other Suppositions.—1. Lyell has proposed an hypothesis, dependent upon the relative height of land in high latitudes at different periods, to explain the tropical character of organic remains, without the aid of a secular refrigeration. But this has already been treated of. 2. Another hypothesis has been advanced with much confidence by certain writers, not, however, practical geologists, to the same effect. It supposes these organic remains to have been drifted after death from the torrid zone. But their great distance in general from the torrid zone, the perfect preservation, in many cases, of their most delicate parts, with other evidences of quiet inhumation near the spot where they lived, such as the preservation in several cases of the softer parts of the animals, render such a supposition wholly untenable.

Proof (4).—*This theory furnishes us with the only known adequate cause for the elevation of mountain chains and continents.*

OTHER SUPPOSED CAUSES OF ELEVATION.

1. *Earthquakes.*—Examples have been given in another place of small and limited elevations of land, produced by earthquakes. And it has been maintained that an indefinite repetition of such events might elevate the highest mountains, if they took place on no larger scale than at present. But it seems to be satisfactorily proved, that some elevations at least, such as those producing the enormous dislocations in the north of England, have occurred to an extent of several thousand feet, by a single paroxysmal effort; whereas the mightiest effects of a modern earthquake have produced elevations only a few feet, and in most cases the uplifted surface has again subsided. Again, there is little probability that a succession of earthquakes should take place along the same extended line through so many ages, as would be necessary to raise some existing mountain chains. Earthquakes may explain some slight vertical movements of limited districts; but the cause seems altogether inadequate to the effect, when applied to the elevation of continents.

2. *Expansion of Rocks by Heat.*—A block of granite, five feet long, if its temperature be raised 96°, will expand 0.027792 inch; a block of crystalline marble, 0.03264 inch; sandstone 0.054914 inch. By these data it appears, that were the temperature of a portion of the earth's crust ten miles thick to be raised 600°, it would cause the surface to rise 200 feet. A greater heat would produce greater results, such as have taken place in the earth's history.

This cause, therefore, though it may perhaps explain such vertical movements of particular regions as are taking place in Scandinavia, Greenland, Italy, England, etc., seems inadequate to account for the permanent elevation of large continents. If they had been raised in this manner, and the same remark applies to some extent to earthquakes, we should hardly expect to find several distinct systems of elevation on the same continent, nor so many examples of vertical strata.

3. *Unequal contraction and expansion of land and water by cold and heat.*—Assuming the mean depth of the ocean to be ten miles, and that it had cooled from boiling heat to 40° F., its volume would contract about 0.042; while the contraction of the land would be only 0.00417. This would produce a sinking of the ocean of 697 feet. An increase of temperature would produce an opposite effect; viz., the partial submersion of the land; though it would be less than the desiccation, because of the greater area over which the water would flow. Admitting these changes of temperature to have taken place, and the theory of central heat supposes the former, that is, the refrigeration, they could not account for the desiccation of the globe, because the tilted condition of the strata shows that the land has been raised up; whereas this theory implies a mere draining of the waters.

4. *A change in the position of the poles of the globe.*—This hypothesis—not long since so much in vogue—would explain how continents once beneath the ocean are now above it, if we admit the form of the earth before the change, to have been the same as at present: viz., an oblate spheroid. But it would not explain the tilted condition of the strata, nor is it sustained by any analogous phenomena which astronomy describes.

EFFECTS OF THE EARTH'S REFRIGERATION.

The consequences of the earth's cooling are these: 1. Solidification of the surface, so as to form a crust. 2. Contraction, involving both subsidence and elevation of parts of the crust. 3.

Fissures of the crust, or faults, producing displacements in the strata. 4. Escape of heat and eruptions of melted matter from the igneous nucleus through these fissures. 5. Earthquakes. 6. Configuration of the earth's surface, or the courses of mountains and coast lines, and the general forms of the continents.

1. *Solidification of the Surface.*—This process must have been extremely gradual at the first, and still slower after the formation of a crust. These conditions would be favorable to crystallization; and there may have been a general uniformity in the crystalline structure, so that there should be two directions of easiest fracture in the crust, a north-east and a north-west course. It is probable that large circular or elliptical areas continued open as centers of volcanic action, which have been growing smaller to the present time, or may have become extinct.

2. *Contraction.*—As the globe continued to cool, its size would diminish. After a crust had been formed, the interior portion would gradually lose its heat and contract, perhaps leaving a vacant space between itself and the crust. Where the tension was too great to be sustained, ***the consolidated crust would collapse upon the contracted interior nucleus***, and thus gradually produce the present ridged and furrowed condition of the surface.

Fig. 125.

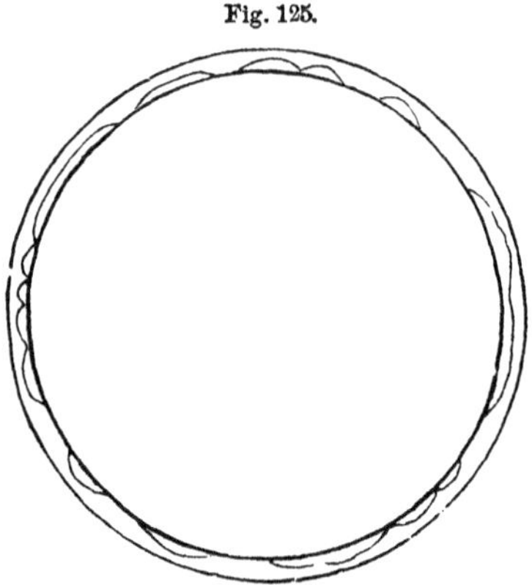

Fig. 125 will illustrate this point. The outer circle represents the crust of the earth, after it had become consolidated above the liquid mass within. This heated nucleus would go on contracting as it cooled, while the crust would remain nearly of the same size. At length, when it became necessary for the crust to accommodate itself to the nucleus, contracted say to the inner circle, it could do this only by falling down in some places and rising in others; as is represented by the irregular line between the two circles. Thus would the surface of the earth become plicated by the sinking down of some parts by their gravity, and the elevation of correspondent ridges by the lateral pressure.

It has been objected that such a shortening of the earth's diameter as this hypothesis supposes would increase the rapidity of its rotary motion, and shorten the length of the day; whereas astronomy shows that for 2,000 years no such change has taken place.

But that period is too short fairly to test the point; since it requires a long time for the tension upon the crust of the globe to become so great as to produce a fracture; and this may not have occurred since that time. If there be any flexibility, however, in the earth's crust, gravity must produce some depression of it in some places, and elevation in others, before the tension is great enough to produce a fracture. And possibly this may be the origin of some cases of slight subsidence or elevation on record.

Thus a contraction of the nucleus beneath the crust causes a subsidence of the surface in one place and an elevation in another, by lateral pressure. This leads us to speak more particularly of the *depression of the beds of the oceans*, and the *vertical movements of continents.*

DEPRESSION OF THE BEDS OF OCEANS.

The subsidence of the surface would be greatest where the crust was thinnest, and least where the crust was thickest. When the crust was sufficiently cool to allow the presence of water, large lakes would collect in the lowest places, where the subsidence had been the greatest. The parts elevated would continue to increase in thickness more than the depressed portions, because the heat would radiate less rapidly through the former. We must suppose that this process of the subsidence of the basins and the elevation of the shores to have continued until oceans and continents were formed. Some maintain, as Professor Dana, whose excellent views we mostly adopt upon the whole subject of refrigeration, that our present oceans and continents have never changed places from the earliest times, but that the oceans have been constantly growing deeper, and the continents higher, though subject to frequent minor variations.

These principles may be more clearly understood by an explanation of Fig. 126. The dotted line *a a* represents the outline of the globe before contraction; the line *c c*, its present surface, having a large ocean, *d d*, in a depression of the surface. At first there may have been numerous small depressions in the district now occupied by the ocean, too shallow to contain all the water. As the depth increased the water would leave the higher lands and occupy the oceanic depression, while the continental shores E E are enlarging and rising. The depression may not be uniform, but may be studded with islands, *f f*, often volcanic, as the internal fires show themselves where the crust is weakest. From this figure we see that the amount of lateral action claimed is very great, as the heights E E on the scale adopted correspond to elevations more than twelve miles above the sea level.

Fig. 126.

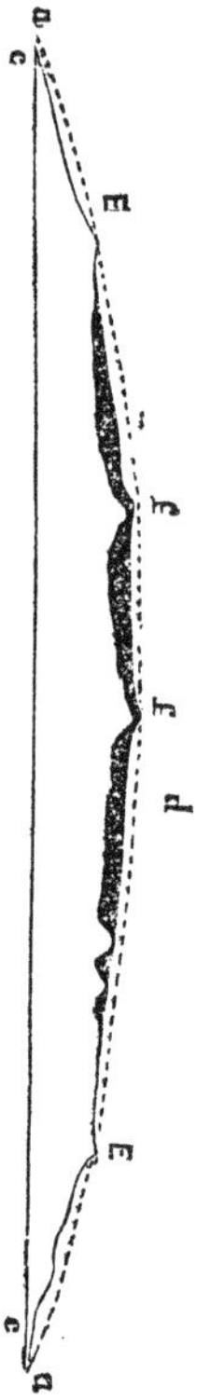

Fig. 1, of our first section, illustrates these principles in the Atlantic Ocean. The West Indies Islands at the west end of the section may be regarded as a part of the western continent, as their position shows them to belong to the continental area. Fig. 126 corresponds more nearly with the arrangement of land and water in and about the Pacific Ocean.

VERTICAL MOVEMENTS OF CONTINENTS.

It is a well established fact, that large tracts of land, and even continents, are now undergoing vertical movements, both of elevation, depression, and as the result, sometimes a see-saw movement. These changes of level can not have been produced by earthquakes.

Examples of Elevation.—The most certain example of elevation of an extensive tract of country, in comparatively recent times, is that of the northern shores of the Baltic, investigated with great ability by Von Buck and Lyell. Some parts of the coast appear to have experienced no vertical movement. But from Gothenburgh to Torneo, and from thence to North Cape, a distance of more than 1,000 geographical miles, the country appears to have been raised up from 100 to 700 feet above the sea. The breadth of the region thus elevated is not known, and the rate at which the land rises (in some places towards four feet in a century) is different in different places. The evidence that such a movement is taking place, is principally derived from the shells of the mollusca now living in the Baltic being found at the elevations above named; and some of the barnacles attached to the rocks. They have been discovered inland in one instance 70 miles.

In other countries similar proofs are relied upon to show elevation. In Scotland, England and Wales beaches containing existing sea-shells are found in many places at various altitudes, from a few feet to 2,300. In North America, as already mentioned, similar relics are abundant in the Northern States and along the southern coast as high as 540 feet. But it has been shown in Section IV. that our country has been elevated at least 2,500 feet during the alluvial period.

Mr. Darwin has shown, beyond all question, that the eastern part of South America has been raised in the most quiet manner, without disturbing the horizontality of the strata, from 100 to 1,400 feet, over an extent of 1,180 miles, since the drift period. It is difficult to explain such a movement by common earthquake action.

Example of Depression.—In the southernmost part of Sweden, in the province of Scania, there has been a loss instead of a gain in the land, amounting to several feet.

Examples of the See-saw Movement.—In Finmark, in Scandinavia, the terraces show that at one end of a district, forty geographical miles in extent, the land has sunk fifty-eight feet; while at the other extremity there has been a rise of ninety-six feet. It has been supposed that the whole of Greenland was gradually sinking. But the observations of Dr. Kane show that only about 300 miles of the southern part is sinking, while the northern parts are rising. The axis of oscillation is at the latitude of 77°. The evidences of elevation are in the successive terraces or beaches, often containing marine shells, which line the sides of the fiords. Upon Mary Minturn river there are forty one of these shelves, the highest of which is 480 feet above the ocean. They were compared to the Parallel Roads of Glen Roy in their general aspect. This elevation is of a comparatively recent date, because deserted stone huts were seen, which had been abandoned by the natives in consequence of their elevation.

Darwin and Dana have shown that over a wide area of the Pacific Ocean a part of the islands are rising and a part sinking by this same oscillatory movement.

Submarine Forests—On the shores of Great Britain, France and the United States, usually a few feet beneath low-water mark, there occur trees, stumps and peat, seeming to be ancient swamps which have subsided beneath the waters, sometimes to the depth of ten feet. In many cases the stumps appear to stand in the spots where they originally grew; yet it requires great care to ascertain this fact.

The Origin of these Forests.—It is probable that this phenomenon results from several causes. 1. When the barrier between a peat swamp and the sea is broken through, so that the water may be drained off, a subsidence of several feet may take place in the soft spongy matter of the swamp, sufficient to bring it under water. 2. In a case on Hogg Island, in Casco Bay, it is inferred that some submarine forests may have been produced by the gradual removal of the contents of a peat swamp, by the retiring tide, after the barrier between it and the ocean has been removed so as to form a slight slope into the water. At the spot referred to, the process may be seen partly completed. 3. But probably most submarine forests were produced by earthquakes, or other causes of subsidence, which we find to have operated on the earth's surface.

Temple of Jupiter Serapis. There has been an interesting subsidence and elevation of land at Pozzuoli, near Naples, as exhibited by the ruins of an ancient Roman temple. The temple was originally built at the level of the sea, for the convenience of sea-bathers. Subsequently the ground subsided, and a lake was formed in the interior of the temple, in which incrustations were deposited from a hot spring as high as 4.6 feet. Then the sea brought in ashes and sand to the height of seven feet. Next the area was subjected to a violent incursion of the sea, other materials were brought in, and the subsidence continued to the height of nineteen feet above the pavement. After this third subsidence the sea remained quiet, and lithophagous molluscs attached themselves to those parts of the columns within seven or eight feet of the surface. At length the land gradually rose, until the pavement of the temple is now on the level of the sea. The shells were left in the cavities excavated by their inhabitants, and thus indicate to us the former subsidence.

Fig. 127 shows the present aspect of the columns. Those parts of them which are covered by the markings of the Modiolæ are indicated by transverse lines in the figure.

Fig. 127.

Temple of Jupiter Serapis.

FOLDING OF STRATA.

The energy exerted in raising vast tracts of land has produced the foldings in the strata described in Section I. These flexures are of three kinds. When both sides of the curve, or anticlinal axis, are inclined at the same angle, the flexure is said to be *symmetrical.* Fig. 128 represents the *normal* and most common of the curves. It is the direct result of a lateral force crowding the strata. If the lateral force continues to act after the formation of the normal flexure, it will make both sides steeper, until the side most remote from the lateral force is bent under the other, as in Fig. 129. This is the *folded* flexure.

Fig. 128. Fig. 1

It is a curious fact that in the continental elevations adjoining the oceans, there is a succession of these three kinds of flexures.

Nearest the oceans the inverted anticlinals alone are found ; but gradually becoming less inclined, till the normal, and finally the symmetrical flexures are reached. Fig. 19 shows a natural section of the rocks across the Alleghany range in the eastern part of the United States, exhibiting this succession of flexures. In Pennsylvania the folded and normal curves appear ; west of which are the symmetrical flexures of the Upper Palæozoic rocks of Ohio. For the discovery of this beautiful series of curves the world is indebted to Professors H. D. and W. B. Rogers.

As the strata are elevated to form these curves, igneous matters would fill the vacancies beneath the arches, and thus assist in the process both by sustaining the strata and by increasing their pliability through the transmission of heat. The Professors Rogers, in accounting for these flexures, admit a degree of lateral action, but argue that this action proceeded from the propelling force or thrust of moving waves of igneous matter, or the natural undulations of the liquid interior.

3. FISSURES AND DISPLACEMENTS OF THE CRUST.

Fissures in the crust of the earth are produced by the unequal contraction of its different parts—the weight of one portion being too great to be sustained by other portions of the crust; hence there will be a forcible rupture of the strata, and the layers, once continuous, may be displaced, sometimes hundreds of feet. The direction of the fissures may coincide with the tendency of the rocks to cleave in a general northeasterly or northwesterly direction, or be modified by the size and relative positions of large areas contracting unequally.

Many of the fissures thus produced may be arranged in systems of uninterrupted or parallel lines instead of single lines of great length. Sometimes these lines, or systems of lines, are curved. Fissures often occur along anticlinal or synclinal lines, as in Fig. 40, because the rocks are weakest along these axes.

4. ESCAPE OF HEAT AND MELTED MATTER THROUGH FISSURES.

Dykes are eruptions of melted matter filling up fissures; their injection is an *effect*, not a cause of displacement.

Like the fissures, dykes are arranged in systems, either linear or curved. It is an interesting fact that these systems correspond in their mode of distribution with volcanoes. For a fine illustration we refer to Percival's remarkably accurate map of the dykes of Connecticut. (*See Geol. Report of Connecticut.*)

The continental thermal springs, dykes of igneous rocks and volcanoes are found chiefly along the mountain ranges near oceans; that is, along those portions of the crust which have been elevated and fractured by the lateral forces produced by contraction. The larger these ranges are, the greater has been the action of heat. For example, along the ranges of the Pacific coast of North America, some of them 18,000 feet above the ocean, there are immense active and extinct volcanoes, besides vast basaltic overflows; while along the Atlantic coast, where only a few peaks exceed 6,000 feet, there are no volcanoes, but a few thermal springs, metamorphic rocks, and fewer dykes.

The subjects of eruptions and earthquakes have already been treated of. They are the legitimate results of the former igneous fluidity of the earth, but they are also the proofs of the doctrine, and must, therefore, be described before stating the theory.

6. CONFIGURATION OF THE EARTH'S SURFACE.

The earth is a flattened spheroid, marked with elevations and depressions—the former constituting continents and islands, and the latter forming the beds of the oceans. The average height of the land above the level of the ocean is 1,008 feet; and the average depth of the ocean beneath the same level, is from two to three miles. The proportion of the surface covered by land to that covered by water, is as three to eight; or fifty-three millions of square miles of land, to 144 millions of square miles of water.

There are evidences of systematic structure in the relative arrangements of land and water, but especially in the configuration of the continents themselves. The northern hemisphere contains more than three-fourths of all the land on the globe; and if the north pole be shifted to the south part of England, nine-tenths of all the land would be in the northern hemisphere, while the water would be mostly in the southern hemisphere. The land is in two great areas, the Eastern and Western Hemispheres. They nearly unite about the north pole, but towards the south pole divide into three great peninsulas, diverging in different directions; viz., South America, Africa, and Australia, which last, with the adjacent islands, may be viewed as a southeast prolongation of Asia. There is a tendency to a triangular form in the subdivisions of the land, as in Africa and the two Americas. Encircling the earth in the tropics, there is a nearly continuous, though irregular belt of water.

An order of arrangement may be seen more clearly in the *trends* or directions of islands, coast lines, and mountains. The islands in the Pacific Ocean, which are properly the peaks of submerged mountain ranges, have a prevalent northwesterly trend, and there are several systems of islands nearly parallel to one another. The coast lines have mostly a northwesterly or northeasterly direction; and it is the constancy of these two directions which has given to so many of the continental subdivisions a triangular form. All the shores of the Americas and Africa, the western of Europe, and the southeastern of Asia, have one of these courses. And as in all these continents there are great ranges of mountains parallel to these shores, it will be observed that these two general directions belong to them also.

A careful examination of all the trends of island groups, mountains and coast-lines, results in the following laws:

(*a*.) The ranges are made up of shorter consecutive and sometimes parallel lines, instead of being uninterrupted for long distances. All the parts of Fig. 130 illustrate this law.

Fig. 130.

(*b*.) The ranges are more commonly curved, than straight or corresponding to a great circle of the earth, as in *b*, *c*, *d*, and *f*.

(*c*.) The straight ranges may have either straight or curved constituent lines, as *a* and *e*.

(*d*.) Curved ranges may arise from a general curvature in the whole, as is represented by the dotted lines in *b* and *d*, or from the positions of the constituent parts.

(*e*.) The same range may vary greatly in its course in different portions of the whole.

(*f*.) When two courses intersect each other, they meet nearly at right angles, but they may directly unite by a curve.

Fig. 130 illustrates these different laws. The straight or curved lines represent the parts of the ranges; that is, successive peaks of mountains, or islands in the ocean; and the general direction of the smaller lines shows the course of the ranges. If one examines the latest and most accurate maps, he will find all these ranges illustrated in the great mountain ranges of the globe, and especially in the islands and coasts of the Pacific Ocean. For example, the system of curves in *e* may be seen by observing the curves along the Asiatic coast from Alaska, in Russian America, to Siam. *f* may be illustrated by the Azores or Western Islands.

We may conclude from these laws that there is a system in the grand outlines of the earth, although we may not as yet be familiar with all its details; that all over the globe, northwest and northeast lines prevail, corresponding to the general cleavage structure of the crust; and that these lines or ranges do not conform to the great circles of the earth, but are often quite irregular, and even intersect one another.

Typical Form of Continents.—The simplest continental feature is that of a great basin, bordered by mountain ranges, and having a general triangular shape. Sections across all the continents illustrate this feature. Fig. 131 shows a section across North America, having the Appalachian ranges upon the eastern border, the Rocky Mountain ranges upon the western border, and the great plains of the western and southern United States for the interior and depressed portion of the basin. Figs. 132, 133, 134, 135, show the same features in South America, Africa, (as ascertained by Dr. Livingstone), Europe, and Asia.

We will state a few particulars respecting the continental features of North America, which will apply in general to all continents.

The general outline of North America is triangular, and in this respect it is a type of all continents. All the shore lines correspond with the northeast or northwest trend. They may be observed upon both shores of Greenland, the northeast coast of Labrador, and the Atlantic coast, from Labrador to Panama. The western coast has the northwesterly trend throughout.

The prominent mountain ranges are situated upon the borders of the continent, parallel to the coast lines. The Laurentian mountains in Labrador and Canada have the same trend with the Green Mountains in New England, and the Appalachian ranges of the Middle and Southern States. The great ranges of the Pacific coast are continuous from Russian America through the Rocky

Mountain ranges to Central America, there uniting with the prolongations of the Andes ranges of South America.

It will be observed that the smallest of these ranges face the smallest oceans; and that the greatest elevations are opposite to the largest oceans—the low Appallachians facing the *small* Atlantic, and the lofty Rocky Mountains, a double line of heights, facing the *broad* Pacific. By referring to Figs. 131 to 135, it will be seen that everywhere the highest mountains stand fronting the largest and deepest oceans.

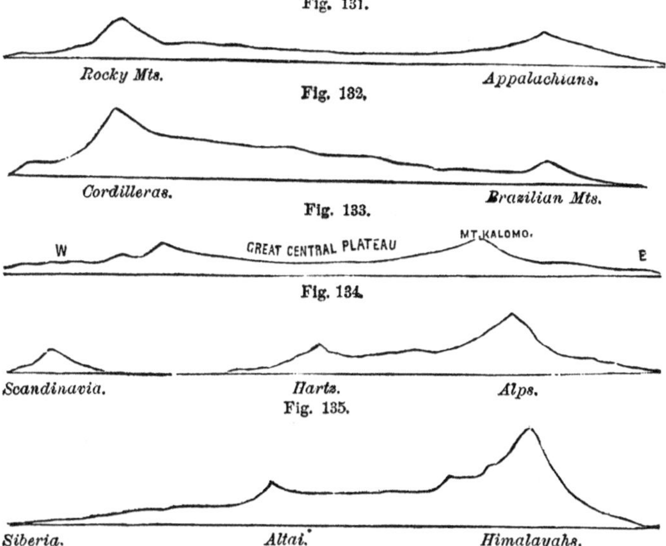

Fig. 131.

Fig. 132.

Fig. 133.

Fig. 134.

Fig. 135.

The coasts of North America in general are so turned as to face the widest range of ocean. The Appalachian coast does not face Europe, but southeast, towards the great opening of the Atlantic ocean, between America and Africa. So the western coast does not face Asia, but the broadest range of the Pacific ocean. This is a principle of universal application.

In North America the larger ranges show greater action of heat. The volcanoes are found only upon the Pacific slope: and the effects of heat are mostly confined to the borders. There are no volcanoes and scarcely any metamorphic rocks in the great interior basin, while the effects of heat are everywhere seen near the oceans. These facts are in accordance with the general principle, that the nearer the water the hotter the fire.

So, too, the strata along the borders are contorted and often overturned, while in the interior they have scarcely been disturbed from their original position. Hence the general principle, that the nearer the water, the vaster the plications of the rocks.

Professor Dana holds that this continent has always had the same shape it now has; that from the earliest times it has gradually been growing, just as a tree continues to increase in size, retaining the same proportions; and that all the continents have always been the more elevated portions of the crust, and the oceanic basins have always been the more depressed portions of the crust.

ELEVATION OF MOUNTAINS AND SYSTEMS OF MOUNTAINS.

The present configuration of the earth's surface has been acquired gradually. The different parts of each continent appear to have been elevated at different epochs. Mountain ranges are not the result of denudation, but of elevation.

Fig. 136.

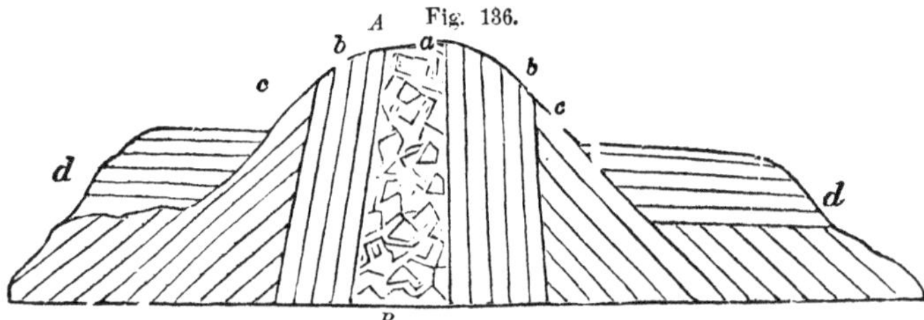

Let A B, Fig. 136, represent a mountain ridge, with an axis, *a*, of unstratified rock. Let the three systems of strata, *b b*, *c c* and *d d*, rest upon the axis *a*, and upon one another, unconformably, and dip at different angles, except *d d*, which suppose horizontal. Now it is obvious that the formations *c c* and *b b* must have been elevated previous to the deposition of *d d*; otherwise the latter would have partaken of the upward movement. And if there be no regular member of the series of rocks wanting between *d* and *c*, it is obvious that we thus ascertain the geological though not the chronological epoch, when *c c* was elevated. *c c*, however, is unconformable to *b b*; and therefore *b b* was partially elevated before the deposition of *c c*; in other words, *b b* has experienced at least two vertical movements. Now this is a just representation of the actual state of things in the earth's crust; and hence, by ascertaining the dip of the formations that are in juxtaposition, we ascertain the different epochs of elevation.

By the application of these principles, it is found that the mountains of Europe have been elevated at no less than twenty-four different epochs: the oldest of which dates as far back as the time when the slates of Westmoreland were tilted up; and the most recent (Etna and Vesuvius) is said to be subsequent to the deposition of the tertiary strata.

The distinguished French geologist, M. Elie de Beaumont, has proposed an elaborate theory for the elevation of mountains and systems of mountains. In his work he endeavors to show that mountain chains have been ridged up by the plication of the earth's crust as it contracted, in the direction of great circles on the surface. He considers those chains parallel which lie upon different great circles, although those circles cut one another in two opposite points. Hence, if we prolong the course of a system of mountains, or of strata, so as to form a great circle on the globe, we shall discover what other mountains were of nearly contemporaneous elevation. How far on different sides of such a circle we may regard the parallel chains as contemporaneous, is not definitely settled. But it is clear that some latitude is allowable in this respect.

Another fact comes in to modify inferences from the preceding statements. It is found that along, or near, the same line of dislocation and elevation, mountains have been raised up at different epochs.

Hence coincidence or parallelism of direction does not prove systems of strata to be contemporaneous. But we must rely on the age of the formations disturbed to prove the epoch of elevation.

Beaumont thinks he can identify, in North America, at least four of the systems of mountains which he has described in Europe, by a prolongation hither of the great circles with which they coincide in Europe. One is the *System of Morbihan*, which is very ancient, and which shows itself in Labrador and Canada, and passes northwest of Lake Superior to the Lake of the Woods. The second is the *System of Ballons*, which embraces a large part of the coal fields of New England, Pennsylvania, Virginia, and Tennessee. The third is the *System of the Thuringerwald*, which he finds in the copper region of Lake Superior, etc. The fourth system is that of the *Pyrennees*, which was plicated between the cretaceous and the tertiary periods.

It is supposed by Beaumont that mountain chains have been, to a great extent, suddenly elevated by paroxysmal movements, not by slow upheaval, and that such sudden emergence of large areas has produced those destructions of life on the globe, which seem for the most part to have been sudden and general.

Most geologists adopt the fundamental principle of Beaumont's theory, but are unwilling to accept his ultimate conclusions. There is too much room for the play of fancy in tracing out contemporaneous mountain chains on distant continents. Most of the ranges may be theoretically accounted for, by the reciprocal influence of oceans and continents upon each other.

THE EARLIEST STATE OF THE EARTH.

The theory of internal heat extends no further back in the world's history than to the time when the globe was in a state of fusion from heat. But the mind naturally inquires what the condition of the world was at its commencement, or at the earliest period of which we can obtain any glimpse. The earliest records are so vague that different answers may appear equally satisfactory to the same question. Hence it is that so many *theories of the earth* have been failures, as well as a cause of ridicule to geologists. It belongs to other sciences than geology to investigate the most remote condition and the causes of the earth.

An hypothesis is advanced by the advocates of original igneous fluidity, which supposes that previous to that time, the matter of the globe had been in a state so intensely heated, as to be entirely dissipated, or converted into vapor and gas. As the heat was gradually radiated into space, condensation would take place; and this process would evolve a vast amount of heat, by which the materials would be kept in a molten state, until at length a solid crust would be formed as already explained.

Analogies in favor of this hypothesis.—1. The nature of comets shows that worlds may be in a gaseous state. They have less solidity of coherence than a cloud of dust or a wreath of smoke, as stars are visible through them, with no perceptible diminution of their brightness. Some of them have more density toward their nucleus, and others appear to become denser throughout, at each successive return. They are self luminous. In these facts there is a striking resemblance between comets and the early condition of our planet, according to this hypothesis.

2. The nebulæ appear to be similar in composition to comets, though not yet actually converted into comets. They prove that a vast amount of the matter of the universe actually exists in the state of vapor.

3. The sun, and probably the fixed stars, appear to be examples of immense globes so far condensed as to be in a fluid state of intense heat.

GEOLOGY OF OTHER WORLDS.

If we assume the history of the earth to be according to this hypothesis, we have a standard by which to judge of the advance of other worlds in the process of refrigeration. The comets and some of the nebulæ appear to be in the earliest stage of the process. They are gaseous, probably from excess of heat, yet are gradually condensing. The sun is apparently in a state of igneous fusion; such a condition as the earth was in during the second stage of refrigeration. In the third stage of the process, worlds become opaque, like the planets; but we may suppose them to be in different degrees of advancement. The planets beyond Mars, (excluding the asteroids), appear to be in a liquid condition, but not from heat, and therefore may be composed of water, or some fluid lighter than water; or at least be covered by such fluid.

Mars, Venus, and Mercury, most nearly resemble our world. Astronomers have fancied that upon Venus the outlines of continents can be traced, and that her poles are annually covered with snow, which in its season is melted and disappears. And it seems to be enveloped by an atmosphere like our earth.

But the state of the moon, as the nearest heavenly body, has been most accurately ascertained; and it exhibits the most astonishing examples of volcanic action, though it is not certain that any volcanoes upon it are now active. But craters, cones, and circular walls, or mountains, exist of extraordinary dimensions. Some of the cones are nearly 25,000 feet high; and some of the craters, 25,000 feet deep, below the general surface; and the latter are of various diameters, even up to 150 miles. The inside of some of these craters presents all the wild and jagged appearance of similar rocks on our earth. Of the mountains and cavities of the moon, about 1,100 have been measured with great accuracy, and we have a more accurate map of the surface of the moon turned towards us, than of our own planet. There appears to be no water or air upon its surface.

Fig. 137 is a sketch of one of the most remarkable volcanic regions in the moon, seen a little obliquely, called Heinsius.

Fig. 137.

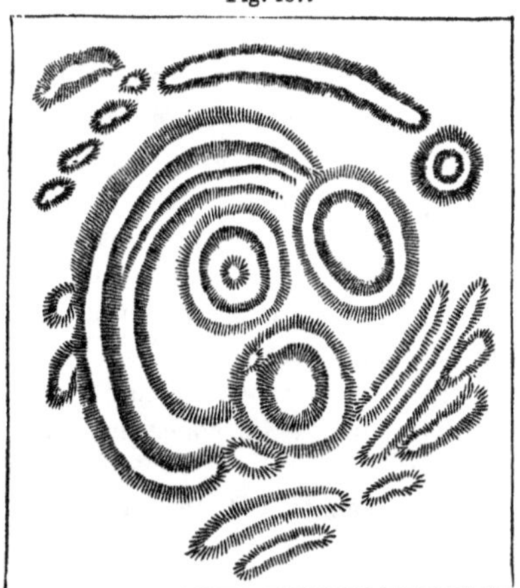

Volcanoes in the Moon.

The question of the habitability of other worlds geology does not answer, further than to suggest that beings of such an organization as man could not exist in the intensely heated celestial bodies, or upon planets whose specific gravity or want of water and air present insuperable obstacles to his abode. Whether such worlds are inhabited by other orders of beings is a matter of conjecture. If the moon was inhabited by beings like men, it must be that their works would be noticed by our powerful telescopes, for objects can be discerned through them having a diameter of 300 feet.

SECTION VI.

METAMORPHISM OF ROCKS.

THE metamorphism of a rock is its transformation from one kind into another. Consequently it takes place after the original formation of the rock.

The term "metamorphic rocks" has been used by Sir Charles Lyell and others in a much more limited sense, to designate a class of rocks (mica schist, talcose schist, gneiss, etc.) that have been so transformed as to have become crystalline, and to have lost, for the most part, their original structure. But this is only one case of metamorphism. Professor John Phillips, also, limits metamorphism to rocks that have been altered by heat; whereas it appears that water and other agents have played quite as important a part in the change as heat.

Agents of Metamorphism.—Heat is a most important agency, and a certain degree of it is probably indispensable; and yet other agencies effect important transformation of rocks at a temperature not above that of the atmosphere generally. Yet the most striking examples of metamorphism were first observed in the vicinity of trap dykes, where chalk was changed into crystalline limestone, clay into clay slate and mica schist, and fossils were obliterated. Hence it was natural to suppose that whenever such effects were seen, dry heat had been the cause, since the trap dykes were regarded as having been once in a melted state. But it has been found that other agencies might be concerned even in the case of dykes.

Water is one of these agents. It acts in two ways: first in connection with heat, secondly by its power of dissolving all rocks, and as the carrier of chemical reagents to aid in the work. There is a third mode in which it sometimes prepares the way for chemical metamorphic action, viz., by freezing in the minute fissures of rocks, and thus opening them to the influence of decomposing agencies.

Professors W. B. and R. E. Rogers subjected forty-eight species of silicious minerals, rocks, glass, porcelain, etc., to the action of pure water and of water charged with carbonic acid. The minerals and rocks were such as feldspar, hornblende, augite, shorl, mica, talc, chlorite, serpentine, epidote, dolomite, chalcedony, obsidian, gneiss, greenstone, lava, etc., and the result was, that all of them were acted upon by the carbonated water, and in a slight degree by pure water. Quartz was not among them. This, in a pure state, is absolutely insoluble by water or by any acid save the fluoric. There is a form of silica which is soluble, and if it be converted into silicates, as in most of the minerals used by Professors Rogers, it is soluble, and is found in most mineral waters. The decomposition of these silicates is accomplished in a variety of ways, and usually leaves an excess of silica in a free state, which forms quartz.

How deep water penetrates into the crust of the earth we know not. But we know that it possesses an astonishing power of working its way into fissures and pores. Especially when converted into steam, and kept in by strong pressure, we can hardly set bounds to its interpenetration. We know that rocks deposited in water are several miles thick, and in some of them water is chemically combined.

We might suppose that the increasing heat as we descend into the earth would expel all the water, or at least drive it near the surface. But the phenomena of volcanoes leads to a different conclusion. The immense quantities of steam that are poured forth from the craters demonstrate the presence of water at a great depth, as do the eruptions of mud, called Moya, in South America and in the Caucacus, and which in one volcano in Java became a river of mud and diluted sulphuric acid. But the most remarkable fact of all is, that ejected molten lava probably owes its liquidity to water. When a stream of it is poured forth, steam escapes from the surface, and a crust is formed in consequence, which prevents the escape of the condensed steam within, except when cracks are formed; and hence the fluid state is preserved within for a long time; nor till that has escaped will it be consolidated; so that in the opinion of some of the ablest writers on volcanoes, such as Scrope, liquid lava is an aqueo-igneous fusion. The heat is found to be not high enough to produce liquidity without water.

Suppose, now, the water in the stratified rocks to be highly heated, and yet essentially imprisoned by impervious strata at the surface; it is easy to conceive that they might reduce the rocks to a fluid or semi-fluid condition without destroying the planes of stratification or producing a complete fusion like that of lava. In that state such chemical changes might occur as would give a crystalline structure, form new simple minerals and produce planes of cleavage, foliation and joints.

But though hot water and steam would produce powerful metamorphic effects, they would be very much increased if we suppose that water to contain in solution chemical agents of great power; for instance, carbonic, sulphuric and muriatic acids, sulphuretted hydrogen and alkaline carbonates. There is no rock, except, perhaps, pure quartz, that could withstand their combined action. They would all be softened and made so plastic, that in the course of centuries all the changes exhibited by metamorphic rocks might be brought about.

We have a very striking example of such agencies in the account given us by Forest Shepherd, Esq., of the "Pluton Geysers, of California." These are hot springs, which throw out intermittingly and spasmodically, powerful jets of steam and scalding water; their temperature varying from 93° to 169° Fahr. Sulphuric acid and sulphuretted hydrogen, at least, according to Mr. Shepherd's account, are present, and probably other energetic ingredients. Says Mr. Shepherd, "you find yourself standing not in a solfatara, nor in one of the salses described by the illustrious Humboldt. The rocks around you are rapidly dissolving under the powerful metamorphic action going on. Porphyry and jasper are transformed into a kind of potter's clay. Pseudo-trappean and magnesian rocks are consumed, much like wood in a slow fire, and go to form sulphate of magnesia and other products. Granite is rendered so soft that you may crush it between your fingers, and crush it as easily as bread unbaked. The feldspar appears to be converted partly into alum. In the meantime the bowlders and angular fragments brought down the ravine and river by floods, are being cemented into a firm conglomerate, so that it is difficult to dislodge even a small pebble, the pebble itself sometimes breaking before the cementation yields." Mr. Shepherd adds: "the metamorphic action going on is at this moment effecting important changes in the structure and conformations of the rocky strata. It is not stationary, but apparently moving slowly eastward in the Pluton Valley." (Am. Journ. Sci., vol. xii., No. 3, pp. 157, 158).

This spot seems to be an opening into the great laboratory of nature, where we get a glimpse of the mighty work she has been carrying on in almost every part of the earth's crust during the past geological ages. We have reason, however, to believe that the action was more powerful in past times than at present, because the earth's crust was thinner, and volcanic agency more common and energetic. Yet at the Pluton Geysers it is energetic enough, and that too at a very moderate temperature, to melt and transform all known rocks, unless it be pure quartz.

But though a very high degree of heat does not seem to be necessary to most cases of metamorphism, yet it is essential that there should be an increase of it in newly formed strata, that they may be changed; and how may we suppose this to have been accomplished.

An eminent mathematician, Professor Babbage, in 1834, proposed a theory to show how the surfaces of equal temperature within the earth's crust might experience changes in a vert cal direction. Thus, suppose A B, Fig. 139, to be the ocean's surface, and A D its bottom, rising into a continent above A. Let G F be a line of equal temperature, say two miles below the ocean's bed, which line would be essentially parallel to the surface. Let now the accumulations of sand, clay, and gravel, which are constantly going on in the ocean, raise its bottom to A C. This coating of non-conducting materials would prevent the escape of the heat, which rises from the heated interior, and cause it to accumulate at a higher level, so that the isothermal G F, (line of equal temperature), would rise to G E; that is, as high as the bottom of the ocean had been filled. The increase of heat might be sufficient to produce the metamorphisms which we find many of the stratified rocks to have undergone.

Fig. 138.

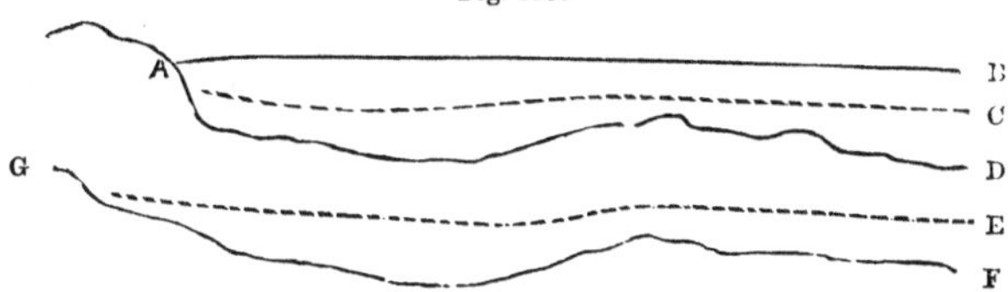

Another consideration deserves to be taken into the account. Different beds of rock require for their fusion, or semi-fusion, very very different degrees of heat. Hence, heat permeating upward through the successive beds A, B, C, D, E, Fig. 139, might almost entirely melt some, (D) partially fuse others, (B) obliterate the

Fig. 139.

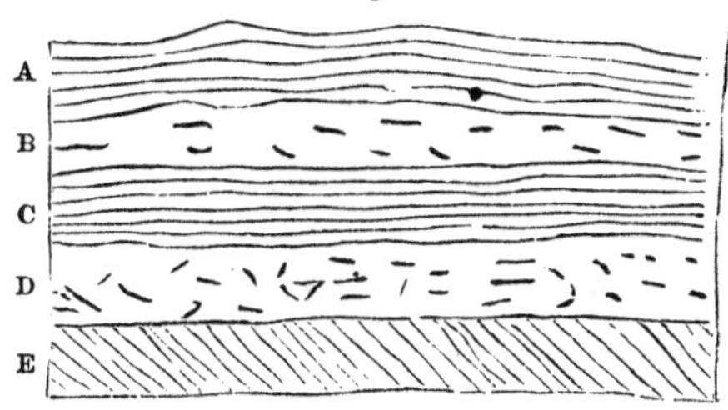

fossils in one, (E) and leave them more or less distinct in others (C and A). This is exactly what we find in the earth, and what we might expect in theory.

Another fact may be explained on the same principles. If we examine a rock formation over its whole horizontal surface, we shall find sometimes that it has undergone very different degrees of metamorphism in different parts. In one portion of the field we find that the original rock has been transformed into gneiss, and in another into mica schist, (as in the Hoosic Mountain range in Massachusetts), in another part (as in Canada), but little altered, and containing organic remains. The statements above made show us how these different degrees of metamorphism might have occurred, either by the different degrees of fusibility in the materials, their different composition, or the greater or less amount of heat introduced into them.

The above facts and reasonings authorize a more sweeping conclusion, viz., that almost every rock is capable by metamorphism of being converted into almost any other. It is usual to suppose that we are to find in the metamorphic rock only the ingredients that exist in that from which it was derived. But if the latter be made plastic by aqueo-igneous agency, why may not the water present contain other ingredients not in the original rock? And who can set limits to the varieties of rocks that might thus be produced.

In view of such facts, also, we can readily assent to Bischoff's conclusion, when he says, "the mineral kingdom, therefore, contains nothing that is unchangeable, unless, perhaps, it be the noble metals, gold and platinum."

A third important agency in metamorphism is the atmosphere. Its four constituents, nitrogen, oxygen, carbonic acid, and aqueous vapor, all act upon the rocks, not merely at the surface, but by means of water they are carried deep into the earth, to furnish probably a large part of the chemical agents that are active in metamorphism. Thus nitrogen and oxygen uniting form nitric acid, and nitrogen combining with the hydrogen resulting from organic changes, forms ammonia; and both these agents, nitric acid and ammonia, carried by water into the crust of the earth, form very energetic agents of change, we know not how deep. Carbonic acid, also, is soluble in water, and is thus introduced among the rocks, which it dissolves by direct action and by uniting with other ingredients to form other reagents. There is enough in the atmosphere to contain 2,800 billion pounds of carbon, and this carbon acts as a carrier of the atmospheric oxygen, first introducing it among the plants and rocks as carbonic acid, and leaving it by other combinations to escape again. These atmospheric agents operate quietly, but the amount of disintegration exhibited almost everywhere by the rocks show that the work is a mighty one. The atmosphere, which, as we breathe it, seems so bland and inefficient, is, in fact, silently crumbling down the solid rocks, we know not how deep, with a power compared with which the effects of the quarryman and the miner are mere infinitesimal blows.

A fourth metamorphic agency at work in the earth is galvanism. All chemical changes do, indeed, imply the presence of this force; but we know of no other agency which, in rocks but partially plastic, could transfer ingredients from one part of the mass to another, as seems to have been done and to be now doing. Thus, a vein of copper ore has been divided by a transverse crack, so that the two ends were separated some inches. But the fissure was subsequently filled with sand, and after some years it was found that the vein was continued across the opening by the introduction of copper ore. Again, how but by galvanism can we explain the production of cleavage, foliation, and joints? These have required a polarizing force, and galvanism is such a force.

Having pointed out the most important agents of metamorphism, we proceed to enumerate their effects as they have been traced out in nature.

1. *Plasticity of the older rocks subsequent to their consolidation.*

This has not hitherto been laid down as an admitted principle: but satisfactory proofs of its truth have fallen under our notice, which its importance leads us briefly to state:

1. It is admitted generally by geologists that the stratified rocks were deposited from water, and consequently with the exception of a very few, perhaps, that crystalized at once from solution, they must have been in a soft state. In fact, they must have been mere accumulations of materials more or less ground down and brought together by mechanical agency.

2. These materials must subsequently have hardened into rock, in order to form shales, sandstones, conglomerates, and fossiliferous, earthy and compact limestones. Though the cementing material of such rocks must have been under the influence of chemical agency, yet the grains and fragments of the body of the rock remain nearly unchanged, bearing decided marks of the mechanical forces by which they were crushed, rounded, and comminuted.

3. But subsequently these rocks must have been brought into a state more or less plastic. This was indispensable in order to produce the following effects, which we find these rocks to have experienced.

1. Their texture has been more or less changed from mechanical into crys-

talline. We find the process indeed in all its stages, and this enables us to prove that it has actually taken place.

2. The organic remains in these rocks have been sometimes elongated, or otherwise distorted, so as it could have been done only while in a plastic state. More often these remains have disappeared entirely, and a crystalline texture has supervened. This could have been done only by chemical agency, while the materials were in a yielding state. For a change of crystallization can take place only where the particles are free to obey the laws of molecular action for bringing them into new positions.

.3. The strata and folia of rocks now highly crystalline, have been subject to remarkable foldings and contortions, such as only a plastic state of the material will explain. Fig. 18, in Section I., represents a block of gneiss, and interstratified hornblende schist, six feet long, obtained from the bed of Deerfield river, at Shelburne Falls, in Massachusetts, and now a part of the geological collection at Amherst College. No geologist will doubt that mechanical pressure must have produced the beautiful curvatures of the layers. It may, indeed, be supposed that the folding took place when the materials were in the form of clay. But it is doubtful whether such great perfection in the curvatures could have been produced in clay, and retained through all the subsequent changes which have resulted in a highly crystalline condition. Moreover, some of the foldings in this rock have an angular sharpness which we have never seen in clay, as in the subjoined sketch in Fig. 140, taken at

FIG. 140.

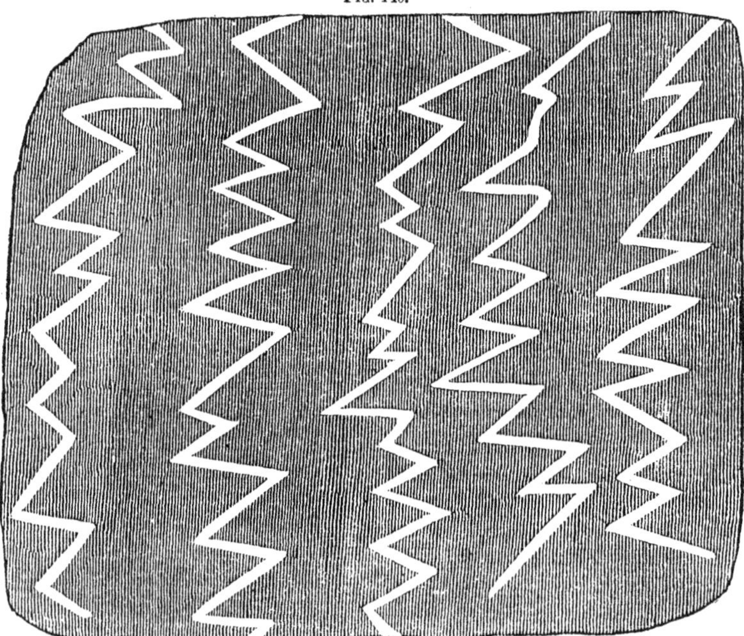

the same locality; and, indeed, the specimen exhibited on Fig. 18 shows to the eye a multitude of such curvatures too minute to be exhibited on the drawing. They seem to be the result of strong pressure on fine folia of rock, and having somewhat of stiffness, so that a lateral force would crumple it up, rather than produce regular curves.

4. Granitic veins and trap dykes in the crystalline rocks have been subject to dislocations and foldings, such as indicate a semi-plastic condition of the rock into which they have been injected. We give only two examples to illustrate this argument. The figure below (Fig. 141) shows a vein of granite (nearly all feldspar) in micaceous limestone, which has neither foliation nor stratification. But that the vein fills a crack in the limestone is obvious from its tapering to a point at one end. Yet it is impossible that a rock could have been split open in such a serpentine course, with pieces of the rock projecting one or two inches on the side, yet often not a quarter of an inch thick. Our theory, therefore, is, that when the crack was made and filled (not, probably, by injection, but by deposition from aqueo-igneous fusion), it was not so crooked as it now is, but was subsequently crumpled up and the folia obliterated by the semi-fusion. The vein is certainly not one of segregation, as that term is usually understood; for how improbable that granite should be segregated from limestone? Neither can melted matter have been injected into it mechanically, without tearing off the projecting laminæ of rock.

Fig. 141.

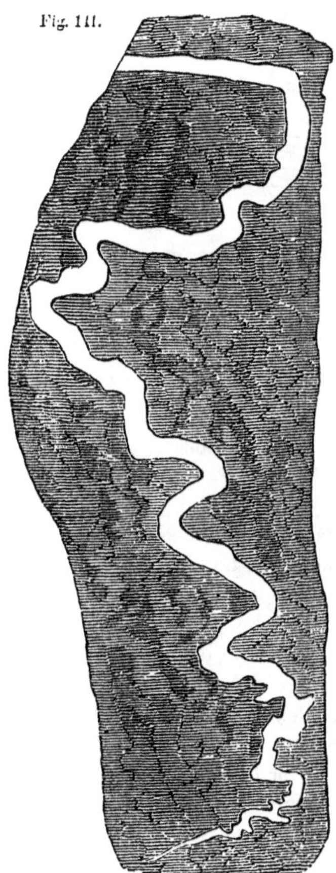

The next case is that of a trap or greenstone, possibly doleritic dyke in gneiss, in a bowlder four feet in diameter, found in Pelham, Massachusetts. This dyke, nowhere more than two inches wide, encircles the whole bowlder; but on one side the two tapering extremities are separated about five inches by intervening gneiss. On the other side the dyke appears to have been fractured and the ends separated, say half an inch, as shown in Fig. 142 Yet the whole rock shows very distinctly the foliated structure of firmly compacted gneiss, whose layers are entirely parallel, save that for a few inches between the extremities of the dyke they are turned aside a few degrees, as shown below.

No one can look at this rock without being satisfied that the gneiss must have been in a plastic state when the dyke was formed, and especially when, by a lateral movement, it was broken off. In this case there is no appearance

Fig. 142.

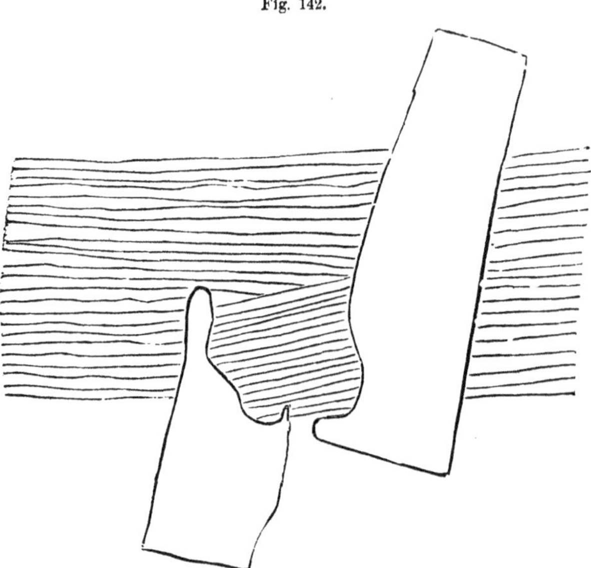

of a cross fracture along which the broken dyke might have slid. Whatever lateral movement there was—and there must have been one—the materials were in such a state as to fill up the fissures entirely with crystalline rock. But this must have been subsequent to the consolidation of the rock, for it must have been more or less solid in order that a fissure should be formed in it for the introduction of the trap.

5. Some conglomerates, with a paste of talcose schist as a cement, and therefore highly metamorphic, have had their pebbles elongated and flattened since their original consolidation. This is the most decisive of all the proofs of the proposition under consideration, and as it has never been adduced, to our knowledge, we must dwell a little upon it.

We have found two very decided localities, and widely separated, to which we can appeal for examples. One is near Newport in Rhode Island, only a little over two miles east of the town, but within the limits of Middletown. Perhaps the phenomena are most strikingly exhibited at the place called Purgatory. But the range of remarkable conglomerate commencing there extends four or five miles northerly, retaining essentially the same character. In looking at the rock one is struck with three peculiarities.

One is, that it seems to consist almost entirely of pebbles and bowlders, with very little cement. Another is, that the pebbles, evidently rounded by attrition, are elongated, and the longer axes are all parallel. The extension is often very great. We have seen some of the pebbles or bowlders four and even six feet long, and not more than a foot in diameter in the middle. They often resemble a mass of candy or wax that has been drawn out when warm—as in the pebble ten inches long shown below, on Fig. 143.

Fig. 143.

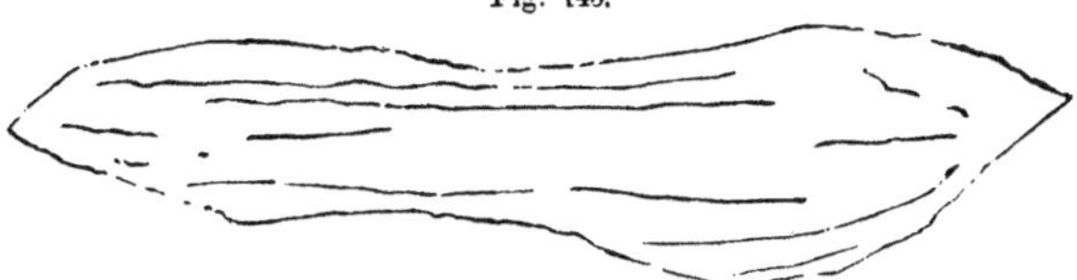

A third circumstance is, that the masses of this rock are crossed at short distances by very distinct joints, which generally cut the pebbles in two with the cleanness and often the smoothness of a knife. They run east and west, or at right angles to the strike of the rock, and are perpendicular, so that when masses of the rock have been removed vertical walls remain, often ten to twenty feet high, showing the cut-off pebbles most distinctly. Acres of these walls may be seen in an hour's walk.

If the pebbles be carefully examined, many of them will be found flattened by a force acting at right angles to their longer axis, so that a cross section will approach a square or parallelogram, as in the sketches below, on Fig. 144, taken from two pebbles from two to fourteen inches across.

Fig. 144.

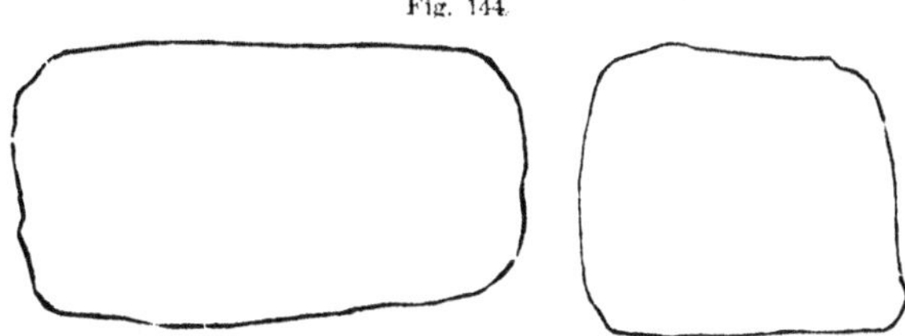

The pebbles are nearly all a gray, rather compact quartz, sometimes white, and approaching the hyaline variety. On breaking the pebbles, however, many of them seem to have undergone some change, verging towards talcose schist, and their surfaces, as well as the talcose cement, are thickly set with small octahedral crystals of magnetic iron ore.

These facts lead inevitably to the conclusion that the rock has been in a somewhat plastic condition since its original consolidation. The case is even stronger than that of the elongation and distortion of organic remains, and these by common consent are thus explained. Whether we can tell exactly how the elongating and compressing force operated, we are sure that these pebbles must have been acted upon by it in a direction perpendicular to the strike; and if they had not been softened, though they might have been crushed, they would neither have been elongated nor flattened.

At the other locality, which is in North Wallingford, Vermont, where the pebbles are cemented by talcose schist, they are not as much elongated as at Newport, perhaps; but some features are shown more distinctly. Though the pebbles are mostly quartz, they are occasionally granite and probably some other rocks. The quartz is white, almost hyaline, and much purer than that at Newport. Yet have its pebbles been so compressed and bent as to prove them to have been in a plastic condition. Fig. 145 shows them as they appear on the surface of a joint crossing the layers at right angles. Fig. 146 shows a single pebble, ten inches long, not only elongated but bent.

Fig. 145.

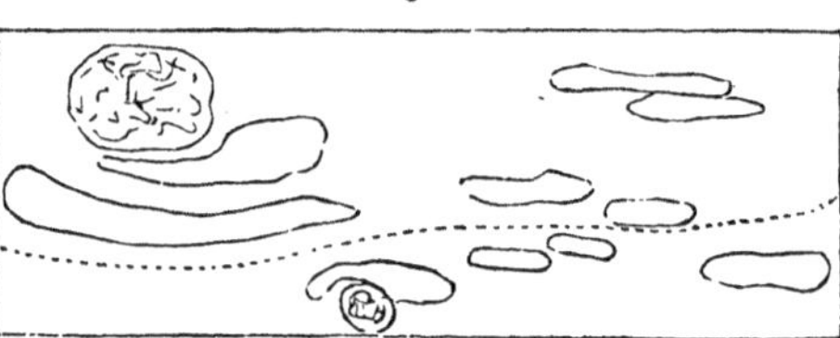

Fig. 146.

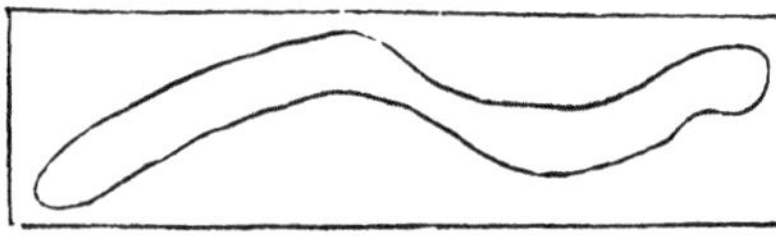

The same bending is manifest on the cross sections at Newport, but we did not notice any examples quite so striking as at Wallingford. The conglomerate at the latter place, which frequently passes into the green talcose schist that forms its cement, has a high westerly dip. The pebbles were derived for the most part from quartz rock such as is abundant along the west side of the Green Mountains, which may be of Devonian age. To prove this semi-plastic subsequent to consolidation, is to make it probable that all rocks exhibiting a similar metamorphism were so also; for quartz is the most difficult of all to bring into that condition, and did not facts compel us to admit it, we should perhaps say it is impossible.

6. The superinduced structures in the crystalline slates and schists, show that they must have been in a semi-fluid state when these were made. We refer to cleavage, foliation and joints. Whatever theory we adopt as to their mode of formation, a yielding state of the ingredients was essential, whether we suppose with Sir John Herschell, that cleavage is a sort of crystallization in plastic materials, or that, as Sharp and Sorby maintain, it has resulted from compression and extension; or, as to foliation, if, as David Forbes supposes, it has resulted from chemical action; or, as to joints, if we regard them as due solely to shrinkage and fracture. In all these cases, however, of cleavage, foliation, and joints, (we add the latter because they seem to us to belong to the same general class of phenomena, and not to be explicable by simple mechanical agency,) we must, with Professor Sedgwick, suppose polarizing forces (*ex. gr.*, heat or galvanism), to have been concerned, and these require the molecular movement among the particles which only plasticity can give. We know that joints are sometimes found in rocks that have not been much softened, and of course chiefly by mechanical agencies; but we do not believe that such as occur in the quartzose conglomerates of Rhode Island and Vermont could have been formed, such as they are, if the whole mass had not been plastic. But here is not the place to go into details on such a point.

7. The insensible passage of schistose into unstratified rocks, (gneiss, for instance, into granite,) affords a presumption that the former have been in a semi-plastic state. For all admit the fluidity of granite, either simply igneous or aqueo-igneous. But if we can hardly tell, often, where the one ends and the other begins, it is fair to conclude that the unstratified have resulted from the more thorough and complete operation of the same agency that produced the stratified crystalline group. This argument, however, would only show that the schistose rocks have been plastic, but gives us no information as to their previous consolidation.

We should not have spent so much time on this subject had it been discussed in other elementary works, and did it not seem to have a most important bearing on the whole subject of metamorphism. Admit the schists to have been in a plastic state subsequent to their consolidation, by the agency of hot water, steam and other agents, and the whole subject of metamorphism is easily explained. But deny this, and the phenomena seem inexplicable.

We resume now a detail of the effects of metamorphism. Several of these, however, have been touched upon in the preceding argument, and will need but little further notice.

2. *A second effect of metamorphism is the abstraction of one or*

more of the ingredients of rocks and simple minerals. If a calcareous sandstone, for instance, should be permeated by water containing some ingredients that would abstract the carbonic acid, a porous quartz rock would remain, with perhaps some silicate of lime. Most of the magnesia of the talcose schists of the Green and Hoosic Mountains has been removed, as the chemists prove by analysis. If one or more proportions of oxygen were abstracted from peroxide of iron or manganese, quite different ores would result. In this way have many of the simple minerals and many of the rocks been essentially changed.

3. Similar results, only more complicated, would result from the introduction of new ingredients, held in solution by the water diffused through the plastic materials. Hence mineralogists reckon a large number of what they call pseudomorphs; that is, minerals which have the crystalline form of other minerals whose cavities they occupy. In this way, too, the characters of rocks may be essentially changed.

4. Though the problem be often quite difficult, yet chemical geologists have been able to point out a great number of these metamorphoses in the rocks with much probability, by comparing the composition of the unchanged with the changed. We give some examples.

Clay slate has been converted into mica schist, talcose schist, gneiss, and granite.

The origin of clay slate from clay is obvious to the most common inspection.

Almost any of the silicious sedimentary rocks can be converted into mica schist. Indeed, hand specimens of micaceous sandstones can hardly be distinguished from mica schist. This rock has also been derived from chlorite schist and from greenstone.

Mica may be produced from feldspar. That in sandstone was not improbably formed by the agency of meteoric water, subsequent to the deposition of the sandstone.

Talc, steatite, and chlorite have been found to result from the decomposition of feldspar, hornblende, augite, garnet, mica, etc. The excess of silica in these minerals may have produced the quartz in talcose and chloritic schists.

Pulverulent carbonate of lime, such as chalk and marl, readily becomes crystalline or saccharine by being brought into a liquid condition, as is sometimes seen in the vicinity of trap dikes.

Bischoff contends that dolomite, which is a double carbonate of lime and magnesia, is produced wherever there is "a formation of carbonate of lime by water containing bicarbonate of magnesia, which is one of the most common constituents of spring water." Hunt, of the Canada Survey, maintains that "dolomites, magnesites, and magnesian marls, have had their origin in sediments of magnesian carbonate, formed by the evaporation of solutions of bicarbonate of magnesia:" which solutions have resulted from the decomposition of sulphate or chloride of magnesium by bicarbonate of soda.—*Am. Jour. Sci., vol. xxviii.*, p. 383.

Serpentine and other varieties of rock that come under the general denomination of Ophiolites, are essentially hydrous silicates of magnesia. Talc, chlorite, and steatite, have so nearly the same chemical constitution, that they may easily pass, and doubtless have often passed, into one another—more often probably from the schist into serpentine than the reverse; and that, too, most likely, in the wet way, although serpentine is usually as unstratified as granite, and sometimes has in it distinct veins of chlorite; as at New Farne, in Vermont. Greenstone and diorite, also, pass into serpentine, which is probably formed out of them. Hornblende, feldspar, and mica, have likewise been converted into serpentine. In the very probable opinion of Sir William E. Logan, the abundant serpentines of the Green Mountain range have resulted from changes in silicious dolomites and magnesites. Other minerals and rocks might be named as capable of producing serpentine by metamorphism; such as garnet, olivine, chondrodite, gabbro, etc. As it is one of the final products of mineral alteration, it is one of the most permanent of rocks.

Quartz rock, being insoluble by water or acid, "appears," in the opinion of Bischoff, "in all cases to be a product of the decomposition of silicates in the wet way." This opinion certainly seems plausible. But when we examine mountains of almost pure compact quartz, certainly 1,000 or 2,000 feet high, it seems difficult to conceive how all the other ingredients could have been separated so entirely, and leave the quartz in such enormous solid masses; and we must think that geologists have yet something to learn as to its origin, and that they will find that in some way or other it has been in a plastic state.

The changes that are found to have taken place in the ores of iron, are a good example of metamorphism. Starting with the carbonate, it is first changed into hematite, both hydrous and anhydrous, next into specular ore, and then into the magnetic protoxide.

Carbonaceous matter affords another good example. Peat, which is partially decomposed vegetable matter, is the first stage of the metamorphosis. This, permeated for ages by water, and covered by aqueous deposits, will become lignite, or brown coal. The next step develops bitumen, even without much increase of heat above the ordinary surface temperature. By still more powerful metamorphic action, probably heat expels the bitumen and leaves anthracite. A further step in the process produces graphite, or black lead, and perhaps the ultimate produce is diamonds.

Change of slate schistose rocks, conglomerates, and breccias into granitic rock is metamorphism. Theory makes such changes quite possible and probable, and observation shows that they have been made. For example, along the west side of Connecticut River, in Vermont and Massachusetts, are numerous hills and mountains of syenite and granite. In several places, as in Granby, Mt. Barnet, Ascutney, and Whately, the granitic rocks are a syenitic conglomerate, that is, the syenite contain rounded pebbles of quartz and schist, and on the margin of the deposit, as on Little Ascutney, we find the original conglomerate from which the syenite is formed. There is reason to suppose that a large part of the granitic rocks of New England are merely transformed slates, schists, and conglomerates. Granite seems to be the most complete form of metamorphosis.

The following statements may be regarded as inferences from the doctrines of metamorphism as above developed.

1. *We see how it is that azoic schists may be interstratified with fossiliferous strata.*—A few examples of this sort have been pointed out, especially in the Alps, where wedged-shaped masses of fossiliferous limestone, of liassic age, have been intercalated among the strata of gneiss. Indeed, strata of eocene tertiary have been converted into crystalline gneiss, mica schist, and even into granitic beds. In our country not many analogous cases have been pointed out. We present one, which fell under our notice in the town of Derby, on the east shore of lake Memphremagog, in Vermont. The section below, in Fig. 147, will give an idea of this case, as we understand it. Here, as we ascend a hill of moderate elevation, the strata succeed one another in the following order; mica schist, granite, fossiliferous limestone, (Devonian), granite, mica schist, granite, mica schist, limestone, schist, etc. Some of these masses, especially the granite, may be somewhat wedge-shaped, especially as we follow on in the direction of the section. The mica schist is highly crystalline, containing that peculiar species of mica denominated Adamsite, by Professor Shepard.

Fig. 147.

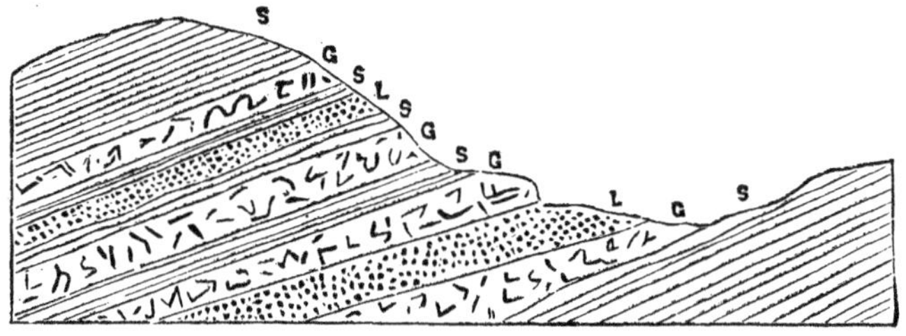

Section in Derby, Vt.

Here we have highly crystalline granite and mica schist lying above limestone of Devonian age, in which we found encrinal stems, and scarcely at all crystalline. But we have shown how this might take place, viz., by the greater fusibility of the superimposed and intercalated beds, or possibly by a lateral permeation of heat and water.

2. *The process of metamorphism is still going on.*—We see it more strikingly at the surface, especially in regions that have not experienced the erosions of the drift agency. There the rocks

are manifestly changed, often to the depth of several feet. But when we open the most solid rocks, or descend into the deepest mines, we shall find minerals undergoing alteration—new ones taking the place of old ones. Wherever water penetrates, even though the temperature be not raised, we may expect metamorphism. Indeed Bischoff, whose great work on Chemical Geology forms almost a new era in geology, regards these changes as universal. "All rocks," says he, "are continually subject to alteration, and their sound appearance is not any indication that alteration has not taken place." (Vol. 3, p. 426). If it be so, it shows us how wide and difficult is the field which lies open for geological research.

3. *Metamorphism shows us that the earliest formed rocks on the globe may have all disappeared.*—If we suppose, what geologists now generally admit, that the globe has cooled from a molten state, the earliest formed crust may have been a granitic rock. Now this crust, as a general fact, has been thickening. But the process in many places, and, perhaps, alternately all over the globe, has been reversed. Suppose, by the slow process of erosion, materials have accumulated in the bottom of the ocean to a great thickness, the effect would be to cause the line of fusion to ascend, it may be so far as to melt off all the rock originally deposited. In other places erosion might have worn off the upper part of this crust, and though this would cause the line of fusion to descend, and thus add new rock, yet between those agencies above and beneath, continued through countless ages, none of the first formed crust may remain. Or if any of it is left, it would be impossible to distinguish it from subsequent formations. So that the idea of a primary granite, or any other rock, in the strict sense of the term, has no foundation in nature.

4. *Metamorphism furnishes the most plausible theory of the origin of the azoic stratified rocks.*

The hypothesis that these rocks were deposited in a crystalline state, in an ocean so hot that the materials would crystallize, is not consistent with what we now know of chemical geology. For water can not hold in solution silicates enough for the purpose, nor does the order in which the materials are arranged correspond with that in which they would crystallize if they were in solution. No possible reason can be given, for instance, for the alternate layers of quartz and mica, or feldspar, or hornblende, or talc, which occur in the foliated rocks.

The theory of metamorphism has fewer difficulties. It supposes these rocks, originally deposited as sand, clay, pebbles, marl, etc., after consolidation, to have been converted again into a plastic state by the permeation of hot water and steam charged with powerful chemical reagents. We know that this agency is sufficient to bring the silicates into a sort of aqueo-igneous plasticity, and that is all that is necessary to produce the imperfect kind of crystallization which the azoic stratified rocks exhibit. It is not that complete crystallization which would result from thorough solution, either aqueous or igneous, but the original mechanical texture sometimes exhibits itself, and many degrees of crystallization are often manifest.

Some may be inclined to impute the hypozoic, and perhaps, in general, more highly crystalline foliated rocks, to some other agency than metamorphism. But we often find rocks of the same kind, and often as highly crystalline, so connected with fossiliferous rocks that we are compelled to regard them as metamorphic, and it seems difficult to conceive that the others have not had the same origin. All the difference between the two classes is the more complete metamorphism of the hypozoic. We seem compelled, therefore, to admit the metamorphic origin of all the azoic foliated rocks, or to deny it to them all; and we can not take the latter ground but in defiance of the plainest facts.

5. *Metamorphism may have obliterated successive systems of life.* We know it to have done this in some of the foliated rocks—in the schists, for instance—that overlie or are interstratified with fossiliferous rocks. It may have done the same with all the hypozoic, in all of which no certain examples of fossils have yet been found, though some bodies of doubtful nature have been described in them.

If this conclusion be admitted, it follows that we can not tell when life first appeared on the globe, because we know not but an indefinite number of organic systems may have been obliterated. This inference, which some eminent geologists have adopted, would be fair, were it not for certain other facts, which we will state in the words of Sir Roderick I. Murchison. "In Bohemia," says he, "as in Great Britain and North America, the lowest zone containing organic remains is underlaid by very thick buttresses of earlier sedimentary accumulations, whether sandstone, schist, or slate, which, though occasionally not more crystalline than the fossiliferous beds above them, have yet afforded no sign of former beings." "The hypothesis that all the earliest sediments have been so altered as to have obliterated the traces of any relics of former life which may have been entombed in them, is therefore opposed by examples of enormously thick and varied deposits beneath the lowest fossiliferous rocks, and in which, if animal remains had ever existed, some traces of them would certainly be detected." (*Siluria, pp.* 20, 21.) These facts ought, at least, to make us cautious how we assert the destruction of other life economies earlier than the Silurian.

6. *Metamorphism throws light upon the origin of the granitic rocks*, such as granite, syenite, and perhaps some varieties of porphyry. The prevailing opinion has been that they consist of melted volcanic matter, thrust into every crack in the overlying strata, and cooled and crystallized under great pressure and with extreme slowness. It is found also, that other rocks adjacent to the granitic have suffered mechanical displacement, and such chemical changes as heat only could produce.

Now all these statements are to some extent true, and they show the presence of a considerable amount of heat, and some mechanical action by the granitic rocks. But more careful examination shows that granite does not generally form the axis of mountains, nor do the stratified rocks dip away from them on opposite sides, but often the granite lies between the strata, and instead of having been the agent by which they have been lifted up, it has only partaken of the general movement which has resulted from some other and more general cause. Moreover, the heat requisite to keep granite in a melted state must be higher than it seems to have possessed; for Bischoff says he could not melt it perfectly in the most powerful blast furnace. Again, if it crystallized from such fusion, the quartz would be first consolidated, because least fusible; whereas it is found to have been the last. Granite, also, contains not a few hydrated minerals, or such as must have been produced in the wet way, and its own ingredients can hardly have had any other origin. If now we admit the foliated rocks to have been brought into a plastic state by the joint action of heat and water, why not admit the same as to the granitic rocks? for often we can not draw the line between them—between gneiss and granite, for instance. Their composition is the same, and they differ only in the schistose or foliated structure, which often is so nearly obliterated in gneiss that we are in doubt whether it be present. What can granite be, then, but an example of metamorphism carried to its utmost limit; carried far enough to obliterate all traces of stratification, lamination and foliation? If water be admitted as a principal agent, heated by caloric from the earth's interior, and prevented from escaping by thousands of feet of superincumbent rock, complete plasticity would result at a temperature far below that required to melt granite in a dry state.

By this view a large proportion of granitic rocks may be only metamorphosed schists. If so, it explains why they have disturbed or changed the adjacent strata so little—the chemical influence rarely being traceable more than a quarter or half of a mile. In some instances, they may have been thrown up from the melted interior of the earth, and possibly in a state of fusion, without water. If only five or ten per cent. of water be present, it is calculated that the heat need not be as high as redness to produce the requisite plasticity.

If it be doubted whether water penetrates so deep into the earth's crust as we know granite to extend, it should be recollected that the stratified rocks, all of which were originally deposited from water, and so far as we can judge, retain more or less of it still, are from ten to twenty miles thick. But if even lava owes its fluidity in a measure to water, it may be supposed to be present in liquid granite with equal reason. In short, whoever admits the aqueo-igneous origin of the crystalline foliated rocks, will feel compelled to admit the granitic rocks to have resulted from essentially the same causes. Nor is the theory very different, after all, from that which usually prevailed. It admits fluidity from heat in the materials, and only introduces water as an important auxiliary in the work. It is by no means the old Wernerian theory revived, for that made granite a deposit from an ocean.

7. *Metamorphism throws light upon the formation of dykes and veins, whether they belong to the granitic, trappean, or volcanic groups of rock.*—It does this by introducing water along with

heat as an essential agent; for this agency will explain some facts in the history of veins and dikes, which, on the common theory of fusion from dry heat, were inexplicable. Thus, when we find veins not thicker than writing paper (and those of granite, epidote, etc., are sometimes as thin, and some of trap are less than half an inch), it is difficult to see how they could have been filled by injection of melted rock, especially if the walls were not very hot; but by means of water the materials could be introduced wherever that substance would penetrate. Again, in the Silurian rocks of Vermont, on the shores of Lake Champlain, we find numerous dikes, both of greenstone and feldspathic rock, either trachytic or feldstone porphyry. These dikes are in some cases partially filled with a conglomerate, or breccia, composed of limestones, sandstones, gneiss, quartz, and granite; of the rocks, in fact, that occur in the region. Now the limestone fragments have lost none of their carbonic acid. But this would have been driven off, as in a lime kiln, if the dike had ever been heated to redness, or to 1000° of Fahrenheit; for carbonates are decomposed below that temperature. This is all consistent if the partial plasticity of the dike was aqueo-igneous; but inexplicable if dry heat alone were concerned. Moreover, such dikes must have been filled mechanically from above, and this might have been done by the currents of an ocean, sweeping into the crevices on its bottom the rounded pebbles accumulated there.

8. *The facts of metamorphism teach us that most rocks have undergone several entire changes since their original production.*—Take the unstratified rocks. These have all been cooled and most of them crystallized from a state of fusion, either entirely igneous, or aqueo-igneous. Here is one change; but from the vertical movement of the isothermal line in the earth's crust, and the erosions at the surface, probably all the original igneous rock has been remelted and recooled, much of it perhaps several times. If any mass has escaped this second fusion, we know not where it is to be found.

Take the stratified rocks. These being derived by abrasion from the unstratified, have, of course, passed through the same changes. But abrasion has brought them under another, a mechanical change, and water has collected the fragments together at the bottoms of lakes and oceans. Subsequently, by consolidation and some chemical agent transfused by water through the mass, it has become converted into detrital fossiliferous rock. Buried afterwards beneath vast accumulations of other rocks, the heat has increased and hot water has permeated the strata, reducing them to a state more or less plastic, causing a crystalline to take the place of a mechanical structure, obliterating the fossils, and substituting cleavage or foliation for lamination. In some cases the heat might be so great that all traces of stratification are blotted out, and granitic or trappean rocks are the result. It may be, after all this, that erosion has again converted these rocks into detritus, and the process of deposition and of metamorphism begins again; nor can we tell how many times these changes may have been repeated. When they have passed through one cycle of change, they are as fresh as ever to commence another.

9. *The final conclusion is, that the entire crust of the globe has undergone metamorphism, and is not now in the condition in which it was created.*—We are sure that every part of it has been in a molten state; and we have every reason to suppose that every part of it has gone through other changes; nor is there any evidence that a portion of the first consolidated crust remains.

Men are accustomed to look upon the solid rocks as emblems of permanency. But in fact science teaches us that they are in a constant state of flux. They may be permanent when measured by the life of an individual, but when we compare their condition in the different and vast geological periods, change is the most impressive lesson they teach; and all those changes most wisely and beneficently ordered.

To give an idea of the extent to which rocks have been metamorphized, we subjoin the following section of the stratified rocks, with the names on the right of the azoic rocks into which we know from reliable observation the fossiliferous have been transformed. It must not be understood that the two kinds are generally interstratified, though they are sometimes so; but usually the azoic are proved to be identical with the fossiliferous by following the line of their strike and finding a gradual change from one into another. Or when a part of a formation is found to be azoic, it is the lower part; and even though it be as high in the series as the tertiary, none but azoic rocks will be found beneath it. This shows that the metamorphic action is deep seated, and it may be that the granitic and trappean veins and dykes are connected with the molten interior of the earth. It is possible, indeed, to conceive that a bed of stratified rock may be converted into one unstratified by heat and water permeating upward through a subjacent bed which is not so changed; in which case we should have granite and trap independent of the molten interior. But the records of geology give us few examples of this kind, and the presumption, therefore, is, that the unstratified rocks require for their production a more powerful metamorphic action than could be communicated through any other rock, without producing a correspondent change in that also.

Alluvium.	
Tertiary.	Eocene Flysch changed into Mica Schist, Gneiss and Protogine.
Chalk.	Into crystalline Limestone.
Oolite.	Lias into Mica and Talcose Schists, and Gneiss.
Trias.	
Permian.	
Carboniferous.	Into Talcose Schist and Gneiss.
Devonian.	Into Clay Slate, Talcose Schist, Calciferous Mica Schist, Gneiss and Granite.
Upper Silurian.	
Lower Silurian.	Into Mica Schist, Talcose Schist, Gneiss and Azoic Limestone.
Cambrian.	Into Mica and Chlorite Schist, and Gneiss
Hypozoic.	
Granitic.	

PART II.

PALÆONTOLOGY.

SECTION I.

PRELIMINARY DEFINITIONS AND PRINCIPLES.

In all the stratified rocks above the Azoic, we find more or less of the remains of animals and plants. These are called *Organic Remains.* When changed into stone they are sometimes called Petrifactions.

Palæontology is the science which describes these organic remains; the word means a history of ancient beings. Some limit it to animals; but we prefer to use it in its more extended sense.

By some able writers the history of fossil animals is called Palæozoology, and that of plants Palæophytology.

A *Fossil* is the body, or any known part or trace of an animal or plant, buried by natural causes in the earth. Hence a mould or mere footmark is a fossil.

This is a difficult term to define, and the above definition may include some organic substances which come not within the province of geology to describe. It might perhaps embrace the frogs that are found alive deep in gravel or enclosed in rock. But may not these properly be regarded as fossils?

Some able writers have thought it necessary to introduce into their definition of a fossil the time and circumstances of its burial. But we prefer the phrase with no other limitation than that given above.

Among the ancients there were some (Strabo, the geographer, for instance,) who noticed and had correct notions about fossil shells. In modern times geological facts first began to excite attention in Italy, in the early part of the sixteenth century. Two questions were argued respecting fossils; first whether they ever belonged to living animals and plants; and secondly, if they did, whether their petrifaction and situation can be explained by the deluge of Noah.

These questions occupied the learned world nearly 300 years. At the commencement of the controversy in Italy, in 1517, Fracastoro maintained, in the true spirit of the geology of the present day, that fossil shells all once belonged to living animals, and that the Noachian deluge was too transient an event to explain the phenomena of their fossilization. But Mattioli regarded them as the result of the operation of a certain *materia pinguis*, or "fatty matter," fermented by heat. Fallopio, Professor of Anatomy, supposes that they acquire their forms in some cases, by "the tumultuous movements of terrestrial exhalations;" and that the tusks of elephants were mere earthy concretions. Mercati conceived that their peculiar configuration was derived from the influence of the heavenly bodies; while Olivi regarded them as mere "sports of nature." Felix Plater, Professor of Anatomy at Basil, in 1517, referred the bones of an elephant, found at Lucerne, to a giant at least nineteen feet high; and in England similar bones were regarded as those of the fallen angels!

At the beginning of the 18th century, numerous theologians in England, France, Germany, and Italy, engaged eagerly in the controversy respecting organic remains. The point which they discussed with the greatest zeal, was the connection of fossils with the deluge of Noah. That these were all deposited by that event, was for more than a century the prevailing doctrine, which was maintained with great assurance; and a denial of it regarded as nearly equivalent to a denial of the whole Bible.

The questions also, whether fossils ever had an animated existence, was discussed in England till near the close of the 17th century. In 1677, Dr. Plot attributed their origin to "a plastic virtue latent in the earth." Scheuchzer in Italy, however, in ridicule of this opinion, published a work entitled, *Querulæ Piscium* or the *Complaints of the Fishes;* in which those animals are made to remonstrate with great earnestness that they are denied an animated existence.

Such discussions led to the accumulation of facts; and these at length led to just views on the subject, and the great works on Comparative Anatomy and Palæontology now extant, by such men as Cuvier, Owen, Agassiz, D'Orbigny, Pictet, Bronn, Brongniart, Lindley, Hutton, and a multitude of others, are the result.

Character of Fossils.—In a few instances animals have been preserved entire in the more recent rocks.

About the beginning of the present century, the entire carcass of an elephant was found encased in frozen mud and sand in Siberia. It was covered with hair and fur, as some elephants now are in the Himalayah mountains. The drift along the shores of the Northern Ocean, abounds with bones of the same kind of animals; but the flesh is rarely preserved. In 1771, the entire carcass of a rhinoceros was dug out of the frozen gravel of the same country.

Many well-authenticated instances are on record, in which toads, snakes, and lizards, have been found alive in the solid parts of living trees, and in solid rocks, as well as in gravel, deep beneath the surface. But in these instances the animals undoubtedly crept into such places while young, and after being grown could not get out. Being very tenacious of life, and probably obtaining some nourishment occasionally by seizing upon insects that might crawl into their nidus, they might sometimes continue alive even many years.

Frequently the harder parts of the animal are preserved in the soil or solid rock, scarcely altered.

Sometimes the harder parts of the animal are partially impregnated with mineral matter; yet the animal matter is still obvious to inspection.

More frequently, especially in the older secondary rocks, the animal or vegetable matter appears to be almost entirely replaced by mineral matter, so as to form a genuine *petrifaction.*

Sometimes after the rock had become hardened, the animal or plant decayed and escaped through the pores of the stone, so as to leave nothing but a perfect *mould.*

After this mould had been formed, foreign matter has sometimes been infiltrated through the pores of the rock, so as to form a *cast* of the animal or plant when the rock is broken open. Or the cast might have been formed before the decay of the animal or plant.

Frequently the animal or plant, especially the latter, is so flattened down that a mere film of mineral matter alone remains to mark out its form.

All that remains of an animal sometimes is its track impressed upon the rock.

The mineralizer is most frequently carbonate of lime; frequently silica, or clay, or oxide or sulphuret of iron, and sometimes the ores of copper, lead, etc.

2. *Nature and Process of Petrifaction.*

Petrifaction consists in the substitution, more or less complete, by chemical means, of mineral for animal or vegetable matter.

The process of petrifaction goes on at the present day to some extent, whenever an animal or vegetable substance is buried for a long time in a deposit containing a soluble mineral substance that may become a mineralizer.

EXAMPLE 1. Clay containing sulphate of iron, will in a few years, or even months, produce a very perceptible change toward petrifaction in a bone buried in it. Some springs also hold iron in solution; and vegetable matters are in the process of time thoroughly changed into oxide of iron. This is seen often where bog iron ore is yearly depositing.

EXAMPLE 2. M. Goppert placed fern leaves carefully in clay, and exposed the clay for some time to a red heat, when the leaves were made to resemble petrified plants found in the rocks.

Theory of Petrifaction.—In all cases of petrifaction, chemistry acts a part. In many instances galvanism and electro-magnetism are concerned; especially where the organic substance is converted into crystalline matter. The juxtaposition of mineral matters forms galvanic combinations, that produce the requisite currents.

3. *Means of determining the Nature of Organic Remains.*

The first requisite for determining the character of organic remains, is an accurate and extensive knowledge of zoology and botany. This will enable the observer to ascertain whether the species found in the rocks are identical with those now living on the globe.

The second important requisite is a knowledge of comparative anatomy; a science which compares the anatomy of different animals and the parts of the same animals.

This recent science reveals to us the astonishing fact, that so mathematically exact is the proportion between the different parts of an animal, "that from the character of a single limb, and even of a single tooth, or bone, the form and proportion of the other bones, and the condition of the entire animal, may be inferred."—"Hence, not only the framework of the fossil skeleton of an extinct animal, but also the character of the muscles by which each bone was moved, the external form and figure of the body, the food, and habits, and haunts, and mode of life of creatures that ceased to exist before the creation of the human race, can with a high degree of probability be ascertained."

CLASSIFICATION OF LIVING PLANTS AND ANIMALS.

It is essential that the learner should have some idea of the great Classes and Families of living Plants and Animals, in order to form an idea of those in a fossil state. For both groups are brought into the same great system of life. And since the living species are more numerous and perfect than any that have preceded them, the former are taken as the standard by which to arrange the latter.

A Flora consists of a species of plants that occupy any given district.

A Fauna consists of the species of animals in a district.

The following are the classes and leading families of living plants commencing with the most perfect, and terminating with the least perfect. We follow the arrangement adopted by Prof. Asa Gray.

VEGETABLE KINGDOM.

Series 1.—FLOWERING PLANTS, OR PHANEROGAMIA.

Plants which produce real flowers with stamens, pistils and seeds.

Class 1.—EXOGENS OR DICOTYLEDONS.

Plants increase by rings on the outside, the seed opening into two or more parts, or seed leaves, called cotyledons. Most common trees and herbs.

Class II.—ENDOGENS OR MONOCOTYLEDONS.

Eot increasing by external rings, but by threads or bundles of fibres from within. The leaves have parallel, not branching veins. Embraces most grasses, rushes, and bulbous plants; also palms, the most remarkable of plants. The seed has only one cotyledon.

Series II.—FLOWERLESS PLANTS, OR CRYPTOGAMIA.

Without flowers and propagating by spores instead of seeds.

Class 1.—ACROGENS.

1. *Filices or Ferns.* 3. *Equisetaceæ,* or Horsetail and Cattail family. 3. *Lycopodiaceæ* or Club Mosses. 4. *Marsileaceæ.*

Class 2.—ANOPHYTES.

Mosses and Liverworts.

Class 3.—THALLOPHYTES.

1. Algæ or Sea Weeds. 2. Lichenes or Lichens. 3. Fungi or Mushrooms; Puff Balls, Toad Stools, Mould, Rusts, Mildew, etc.

There is a great diversity among the most eminent zoologists in the classification of animals, often perhaps more in the name than in the grouping, We shall make no attempt to decide when such men disagree. Most of them, however, still follow Cuvier in dividing the whole Animal Kingdom into four great sub-kingdoms or provinces; though some have added others. We give below the systems of two of the most eminent living writers on this subject. In the course of the work some other systems, or parts of them, will come into view, because numbers which we wish to use are so connected with them that we can not separate them.

ANIMAL KINGDOM.

SUB-KINGDOM VERTEBRATA.

Class 1.—MAMMALIA or animals that nurse their young. The most usual orders of these are the following:

1. Bimana, or man. 2. Quadrumana, or monkeys. 3. Cheiroptera, or Bats. 4. Insectivora, or insect eaters, as the mole. 5. Carnivora, or flesh eaters. 6. Cetacea, the whale tribe. 7. Pachydermata, or thick skinned, as the horse, elephant, etc. 8. Ruminantia, the cud chewers, as the camel, deer, sheep, etc. 9. Edentata, as the sloth and armadillo. 10. Rodentia, the gnawers, as the mouse, squirrel, woodchuck, etc. 11. Marsupialia, as the opossum and kangaroo. 12. Monotremata, as the platypus, of New Holland.

Agassiz divides Mammalia into three orders. 1. Marsupialia. 2. Herbivora. 3. Carnivora. Owen makes fifteen orders. 1. Bimana. 2. Quadrumana. 3. Carnivora. 4. Artiodactyla. 5. Perissodactyla. 6. Proboscidea. 7. Toxodontia. 8. Sirenia.

9. Cetacea. 10. Bruta. 11. Cheiroptera. 12. Insectivora. 13. Rodentia. 14. Marsupialia. 15. Monotremata.

Class 2.—Aves, *or Birds.*

Agassiz gives four orders: 1. Natatores. 2. Grallæ. 3. Rasores. 4. Insessores.

Griffith and Henfrey give six. 1. Accipitres, the eagle, owl, etc. 2. Passerina, the swallow, etc. 3. Scansores, the cuckoo, parrot, etc. 4. Gallina, the fowl, pigeon, etc. 5. Grallæ, the ostrich and crane. 6. Palmipedes, the web-footed, as the duck and goose.

Class 3.—Reptilia, *or Reptiles.*

Agassiz divides them into two classes. 1. *Amphibians*, with three orders, 1, Cæciliæ; 2, Ichthyodi; 3, Anura. 2. *Reptiles*, with four orders, 1, Serpentes; 2, Saurii; 3, Rhizodontes; 4, Testudinata.

Owen divides the Reptiles into two classes. 1. *Amphibia*, with two orders, 1, Ganocephala; 2, Labyrinthodontia. 2. *Saurian Reptiles*, into eleven orders, 1, Thecodontia; 2, Cryptodontia; 3, Dicynodontia; 4, Enaliosauria; 5, Dinosauria; 6, Pterosauria; 7, Crocodilia; 8, Lacertilia; 9, Ophidia; 10, Chelonia; 11, Batrachia.

Class 4.—Pisces, *or Fishes.*

Agassiz divides them into three classes. 1. *Fishes proper*, with two orders, 1, Ctenoids; 2, Cycloids. 2. *Ganoids*, with three orders, 1, Coelacanths; 2, Acipenseroids; 3, Sauroids. 3. *Selachians*, with three orders, 1, Chimæræ; 2, Galeodes; 3, Batides.

Owen makes eleven orders.

Sub-Kingdom ARTICULATA.

Agassiz divides into three classes and ten orders. 1. *Worms*, with three orders. 2. *Crustacea*, with four orders. 3. *Insects*, with three orders, 1, Myriapods; 2, Arachnids; 3, Insects proper.

Owen divides the Articulates into six classes. 1. *Arachnida*, with four orders. 2. *Insecta*, with eleven orders. 3. *Crustacea*, with eleven orders. 4. *Epizoa*, with three orders. 5, *Anellata* with four orders. 6. *Cirripedia*, with three orders.

SUB-KINGDOM MOLLUSCA.

Agassiz makes three classes and nine orders. 1. *Acephala*, with four orders. 2. *Gasteropoda*, with three orders. 3. *Cephalopoda*, with two orders.

Owen divides into six classes and many orders. 1. *Cephalopoda*, with two orders. 2. *Gasteropoda*, with ten orders. 3. *Pteropoda*, with two orders. 4. *Lamellibranchiata*, with two orders. 5. *Brachiopoda*, subdivided into families only. 6. *Tunicata*, with two orders.

SUB-KINGDOM RADIATA.

Owen divides the Radiates into three sub-provinces (what are called sub-kingdoms above, he calls provinces), with numerous orders and families. 1. *Radiaria*, with five classes, 1, Echiondermata; 2, Bryozoa; 3, Anthozoa; 4, Acalephæ; 5, Hydrozoa. 2. *Entozoa*, with two classes, 1, Coelelmintha; 2, Sterelmintha. 3. *Infusoria*, with two classes, 1, Rotifera; 2, Polygastria.

Agassiz makes three classes of Radiates. Polypi, with two orders. 2. Acalephæ, with three orders. 3. Echinoderms, with four orders.

As to the infusoria, Agassiz says: "The infusoria as a class do not exist. It has been proved that a part of these are plants or their spores; others are the young of different animals, and the rest are perfect animals."

In his article on Palæontology in the eighth edition of the Encyclopedia Britannica, which has appeared since his classification above described, we find Professor Owen adopting a different view of some organisms which he had classed among the lower animals, as the following extract will show:

"The two divisions of organisms called plants and animals are specialized members of the great natural group of living things; and there are numerous organisms, mostly of minute size and retaining the form of nucleated cells, which manifest the common organic characters, but without the distinctive superadditions of true plants or animals. Such organisms are called *Protozoa*, and include the sponges, or *Amorphozoa*, the *Foraminiferæ*, or *Rhizopods*, the *Polycistinæ*, the *Diatomaceæ*, *Desmidiæ*, and most of the so-called *Polygastria* of Ehrenberg, or infusorial animalcules of older authors."—*Richard Owen, Ency. Brit., Art. Palæontology.*

In conformity with the above views, Professor Owen thus divides the *Protozoa :*

Class 1. Amorphozoa, Sponges.

Class 2. Foraminifera, Rhizopods.

Class 3. Infusoria, Animalcules.

GROUPING OF LIVING AND FOSSIL PLANTS AND ANIMALS INTO PROVINCES.

Existing animals and plants are arranged into distinct groups, each group occupying a certain district of land or water; and few of the species ever wander into other districts. These districts are called zoological and botanical provinces; and very few of the species of animals and plants which they contain can long survive a removal out of the province where they were originally placed; because their natures can not long endure the difference of climate and food, and other changes to which they must be subject.

Although naturalists are agreed in maintaining the existence of such provinces, they have not settled their exact number; because yet ignorant of the plants and animals in many parts of the earth. Besides, the provinces interfere with one another; and a single large province may embrace several minor ones. This is particularly the case with animals. So that zoologists divide them first into kingdoms, and these into provinces, as follows: 1. The first kingdom embraces Europe, which is subdivided into three provinces. 2. The second kingdom comprises Asia, divided into five provinces. 3. Australia, one kingdom and one province. 4. Africa, with the islands of Madagascar, Bourbon and Mauritius; one kingdom and one province. 5. America, one kingdom and four provinces. In all, five kingdoms and fourteen provinces.

Professor Schouw makes twenty-five regions of plants. The arrangement depends on the natural classification. Thus the region of Mosses and Saxifrages embraces the north polar regions as far south as the trees, and the upper part of the mountains of Europe. The region of Cactuses and Pepper embraces Mexico and South America to the river Amazon. The region of Palms and Melastomas embraces that part of South America east of the Andes between the equator and the tropic of Capricorn.

A few species seem capable of adapting themselves to all climates. This is eminently true of man, whose cosmopolite character is so marked, and his ability to adapt himself to different climates and circumstances so dependent upon his superior mental endowments, that the distribution of the different races of the human species can not be accurately judged of by that of any other species.

Sometimes mountains and sometimes oceans separate these districts on the land. In the ocean they are sometimes divided by currents or shoals. But both on land and in the water, difference of climate forms the most effectual barrier to the migration of species; since it is but a few species that have the power of enduring any great change in this respect.

In some instances, organic remains are broken and ground by attrition into small fragments, like those which are now accumulating upon some beaches by the action of the waves. But often the most delicate of the harder parts of the animal or plant are preserved; and they are found to be grouped together in the strata very much as living species now are on the earth.

From these facts it is inferred that, for the most part, the imbedded animals and plants lived and died on or near the spot where they are found; while it was only now and then that there was current enough to drift them any considerable distance, or break them into fragments. As they died, they sunk to the bottom of the waters and became enveloped in mud, and then the processes of consolidation and petrifaction went slowly on, until completed.

So very quietly did the deposition of the fossiliferous rocks proceed in some instances, that the skeletons and indusiæ of microscopic animals, as we have seen, which the very slightest disturbance must have crushed, are preserved uninjured; and frequently all the shells found in a layer of rock, lie in the same position which similar shells now assume upon the bottom of ponds, lakes and the ocean; that is, with a particular part of the shell uppermost.

In the existing waters we find that different animals select for their *habitat* different kinds of bottom; thus, oysters prefer a muddy bank; cockles a sandy shore; and lobsters prefer rocks. So it is among the fossil remains; an additional evidence of the manner in which they have been brought into a petrified state.

From the researches of Prof. E. Forbes in the Egean Sea, it appears, first, that increase of depth has the same kind of effect upon the marine animals, as increase of height has upon those on dry land, that is, the animals become more and more like those of a colder climate. Secondly, that most marine animals and vegetables inhabit particular localities, which at length become unfit for their abode, and they emigrate or die out. Thirdly, that species ranging widest in depth range furthest horizontally. Fourthly, below 300 fathoms, deposits of fine mud are going on without organic remains, because animals do not live there. These conclusions correspond to the manner in which organic remains occur in the rocks.

4. *Organic Remains arranged according to their Origin.*

Organic remains may be divided, according to their origin, into three classes: 1. Marine. 2. Fresh water. 3. Terrestrial.

The last class appear in most instances where they occur, to have been swept down by streams from their original situation into estuaries; where they were mixed with marine relics. Sometimes, perhaps, they were quietly submerged by the subsidence of the land.

The following table will show the origin of the remains in the different groups of fossiliferous rocks.

Cambrian and Silurian Systems.	Marine.
Old Red Sandstone.	Marine, Fresh Water and Terrestrial?
Carboniferous Limestone.	Do. Do. Do.
Coal Measures.	Terrestrial Estuary Deposits and submerged land. Rarely perhaps fresh water deposits.
New Red Sandstone Group.	Marine.
Oolitic Group.	Mostly Marine.
but in few instances,	Terrestrial.
Wealden Rocks.	Estuary Deposit.
Cretaceous Group.	Marine.
Tertiary Strata.	Marine and Fresh Water.
Alluvium.	Every variety of origin.

It appears from the preceding statements, that by far the greatest part of organic remains are of marine origin. Nearly all the terrestrial relics indeed, and many of fresh water origin, have been deposited beneath the waters of the ocean.

5. *Amount of Organic Remains in the Earth's Crust.*

The thickness in feet of the fossiliferous strata, as given in the tabular view of the stratified rocks, is as follows:

Alluvium,	500 feet,
Tertiary,	2,000 feet,
Chalk,	1,500 feet,
Wealden,	2,210 feet,
Oolite,	2,270 feet,
Lias,	1,160 feet,
Upper New Red,	3,100 feet,
Permian,	1,040 feet,
Carboniferous	3,000 feet, in Nova Scotia, 8,000 feet, in United States, 13,500 feet, in Europe,
Devonian,	8,950 feet, in United States, 10,000 feet, in Europe,
Upper Silurian,	5,100 feet, in United States, 8,400 feet, in Europe,
Lower Silurian,	20,000 feet.

The Cambrian or Huronian, which is from 10,000 to 26,000 feet thick, is not included in the above, because scarcely any fossils have yet been detected in it. The average thickness of the other fossiliferous strata is 57,035 feet, equal to about eleven miles.

Organic remains occur more or less in all the fossiliferous strata whose thickness has been given. As a matter of fact, they have been dug out several thousands of feet below the present surface. In the Alps, rocks abound in organic remains from 6,000 to 8,000 feet above the level of the sea; in the Pyrenees, nearly as high; and in the Andes and the Himalayas, at the height of 16,000 feet.

Prodigious accumulations of the relics of microscopic animals and plants are frequently found in the rocks.

EXAMPLE 1.—From less than 1.5 ounce of stone, in Tuscany, Soldani obtained 10,454 chambered shells;—400 or 500 of these weighed only a single grain; and of one species it took 1,000 to make that weight. These were marine shells. *Buckland's Bridgewater Treatise*, vol. 1, p. 117.

EXAMPLE 2.—In fresh water accumulations a microscopic crustaceous animal, called the cypris, often occurs in immense quantities; as in the Hastings sand and Purbeck limestone in England, where strata 1,000 feet thick are filled with them; and in Auvergne, where a deposit 700 feet thick, over an area twenty miles wide and eighty in length, is divided into layers as thin as paper by the exuviæ of the cypris. *Same Work*, p. 118.

EXAMPLE 3.—But perhaps the most remarkable example is that derived from microscopic animals and plants which have been regarded by Ehrenberg and others as exclusively animal, under the name of infusoria or animalculæ, but a part of them are doubtless plants. In one place in Germany is a bed fourteen feet thick, made up of skeletons so small, that it requires 41,000,-000,000 of them to form a cubic inch; and in another place, a similar bed is twenty-eight feet thick. In Massachusetts are numerous beds composed of these siliceous shields many feet in thickness; and similar beds occur all over New England and New York. Deposits of these *carapaces* or shields, have been discovered by Prof. William B. Rogers in the tertiary strata of Virginia, extending over large areas, and from twelve to twenty-five feet thick!

It is a moderate estimate to say, that two-thirds of the surface of our existing continents are composed of fossiliferous rocks; and these, as already stated, often several thousand feet thick.

This estimate might, without exaggeration, be confined to strata that contain marine exuviæ;—that is, such as were deposited beneath the ocean.

At the end of the next section we hope to be able to present a table of most of the fossil animals and plants known.

ICHNOLOGY.

This branch of Palæontology means literally, *the science of tracks*. It is tracks in stone, however, that fall within the province of geology, and hence we might call the science Ichnolithology.

But since this is less euphonic than Ichnology, and since fossil footmarks require for their elucidation the study of recent tracks, we prefer the latter term, proposed by Dr. Buckland.

This branch of Palæontology is of quite recent origin. The first scientific account of fossil footmarks, was that by Rev. Dr. Henry Duncan in the transactions of the Royal Society of Edinburgh, in 1828. They were probably the impressions of the feet of tortoises on what was then supposed to be the New Red Sandstone, but is now thought to be Permian Sandstone, of Corncockle Muir Quarry, in Scotland. In 1831, G. P. Scrope found numerous footmarks of small crustaceans on forest marble of the Oolite in England. In 1834, Professors Hohnbaum, Kaup, and Sickler published an account of tracks of the Cheirotherium, on New Red Sandstone, in Saxony. In 1836, the first description was given of the tracks in that most prolific of all localities, the valley of Connecticut river. Since that time numerous other descriptions of the same locality have appeared, by Dr. Deane, Sir Charles Lyell, Isaac Lea, Dr. J. Warren, and the authors of this work, and so many other localities have been discovered in Europe and America, that scarcely any fossiliferous formation is now without its footmarks. These will be described under the several rocks.

At first names were given to the different kinds of tracks, but now for the most part the animals that made them are named. Such animals are called *Lithichnozoa*, from the Greek words (λιθος, ιχνος and ζωον) signifying, ***stony track animals***, or ***track-discovered animals***. The following table gives a general view of their distribution up to the present time.

LITHICHNOZOA.

Formation.	*Class.*
Cambrian,	Annelids.
Lower Silurian,	
Potsdam Sandstone,	Crustaceans.
Hudson River Shales,	Crustaceans and Annelids.
Upper Silurian, Clinton Group	Fishes? Annelids. Gasteropods?
Devonian,	Batrachians, Saurians? and Chelonians.
Hamilton Group,	Crustaceans?
Carboniferous,	Batrachians, Saurians, Molluscs?
Permian,	Chelonians, Saurians.
Trias,	Batrachians, Saurians, Chelonians, Crustaceans.
Jurassic,	Marsupialoids, Birds, Lizards, Batrachians, Chelonians, Fishes, Crustaceans, Myriapods, Insects, Annelids.
Wealden,	Saurian. (Iguanodon?)
Alluvium,	Man, Carnivora, Ruminants, Birds, Batrachians, Annelids, Molluscs.

Under Alluvium we have mentioned only those animals whose tracks have been described by geologists, and of which specimens have been preserved in the cabinets.

FUNDAMENTAL PRINCIPLES OF ICHNOLOGY.

The question naturally arises, whether Ichnology has any principles at its foundation on which we can rely, or are its results conjectural; or, to give the inquiry a more scientific form, is there any such relation between the feet of animals and their general structure and character, that knowing the one, we can with strong probability infer the other, as we can determine the unknown quantity in an algebraic equation? We maintain the affirmative, for the following reasons:

1. Comparative anatomy and zoology teach us that an intimate relation exists between all the parts or organs of animals.

2. They teach us that the feet of animals are unusually characteristic, and their relations to other parts unusually clear, so as to furnish, in some instances, a basis of classification to the zoologist. Now a perfect track, especially one in relief, gives a complete model of the foot, and thus furnishes us with a better means of determining the nature of the animal than is sometimes used by the palæontologist, who frequently can obtain only a fragment of some other organ of the animal by which to judge of its nature.

3. We are able, very often, to determine the nature of living animals from their tracks. Who would confound the human track with that of any other animal? or the tracks of quadrupeds with those of birds? or of ruminants with those of the carnivora or marsupials? or among birds, those of the grallæ or waders with those of the web-footed or the pigeons? or those of the ostrich with those of the eagle or albatross?

4. We have the highest authorities for naming animals from their tracks. Such men as Professors Kaup and Richard Owen, Sir William Jardine and Isaac Lea, have done it. Cuvier, too, has said, that "any one who observes merely the print of a cloven hoof, may conclude that it has been left by a ruminant animal, and regard the conclusion as equally certain with any other in physics and morals. Consequently, the single footmark clearly indicates to the observer the forms of the teeth, of all the leg bones, thigh, shoulders, and of the trunk of the body of the animal which left the mark."

PERMANENT CHARACTERS IN THE FEET AND TRACKS OF ANIMALS BY WHICH DIFFERENT KINDS CAN BE DISTINGUISHED.

We have not space to draw out these characters in detail, nor even to enumerate but the most important. A full enumeration and description may be found in the Report on the Ichnology of New England made to the government of Massachusetts in 1858, page 24.

1. Tracks are of three kinds, 1, a simple trail such as serpents, annelids, molluscs, and perhaps some fishes might make; 2, trails accompanied by the impressions of feet, such as would be made by most kinds of reptiles, some fishes, crustaceans, and some in-

sects ; 3, impressions of feet only, such as might be left by most vertebral animals and all the invertebrate tribes that have feet.

2. Width of the trackway.

3. Angle made by the axis of the foot with the line of direction called the median line.

4. Distance of the tracks from the median line.

5. Number of feet.

6. Relative size and character of the feet before and behind.

7. Mode of progression ; by one row of tracks or two ; direct or oblique ; by steps or leaps, etc.

8. Length of step.

9. Size of foot.

10. Number of toes.

11. Whether thick-toed or narrow-toed.

12. Number and size of the phalangeal impressions.

13. Divarication or spread of the toes.

14. Character of the heel.

15. Claws and pellets.

16. Anomalous characters, such as indicate that the animal may have partaken of characters now found only in different classes or orders : like the icthyosaurus, pterodactyle and sauroid fishes.

A careful application of these and other less important characters will enable us, in most cases, but not in all, to decide from tracks whether the animal was vertebral or invertebral ; to which of the classes in these two great groups it belonged ; often to what family, genus and species. In this way have the examples of Lithichnozoa been determined, which will be given under the different formations in our next Section.

SECTION II.

PALÆONTOLOGICAL CHARACTERS OF THE ROCKS.

Organic remains are not thrown together confusedly in the rocks, but each of the great rock formations has its peculiar fossils, which are not found in the formations above or below. Usually the species are limited to a particular formation, although the genera have a wide range. It is desirable to give some idea, especially by drawings, of the leading and characteristic plants and animals that have successively peopled the globe since it became habitable. We begin with the lowest formation and pass upward.

1. Cambrian or Huronian System.

Lower Cambrian Sedgwick.

Fig. 148.

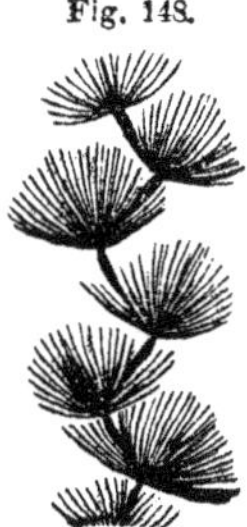

Notwithstanding the great thickness of this rock (12,000 feet in this country and 26,000 in Wales), not more than half a dozen fossils have been found in it. The most interesting is a zoophyte, found in Ireland, and called Oldhamia antiqua, Fig. 148, named after its discoverer, Professor Oldham. In this country and in Bohemia no fossils have been found in this rock.

"The reader," says Sir Roderick I. Murchison, "may look with reverence on this zoophyte; for, notwithstanding the most assiduous researches, it is the only animal relic yet known in this very low stage of unequivocal sedimentary matter."

Lithichnozoa, or animals made known by their tracks. The

Fig. 149.

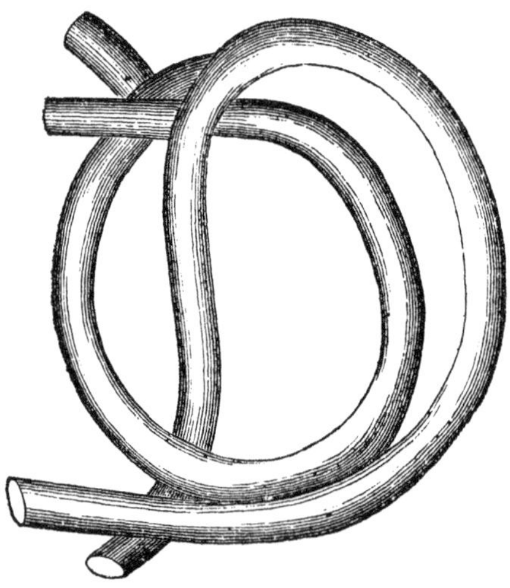

Palæochorda major.

trackway of a species of Annelid, called the Arenicolites didyma, occurs in what is called the Longmynd rocks, in Wales. The same formation contains also impressions of rain drops.

Fig. 149 is thought to be a plant; the Palæochorda major, from the Skiddaw slate, of England.

2. Lower Silurian System, *Murchison.*
Upper Cambrian, Sedgwick.

The plants of this vast system (20,000 feet thick,) are only a few, and mostly obscure sea weeds, which often go by the name of Fucoids, from their resemblance to the living genus, Fucus. Fig. 150 shows the Scolithus linearis which occurs in the Potsdam sandstone, as well as in quartz rock. But it is uncertain whether it is an animal or a plant.

Phytopsis tubulosum from the Black river limestone, of New York, is considered a plant and is shown on Fig. 151.

Fig. 150.

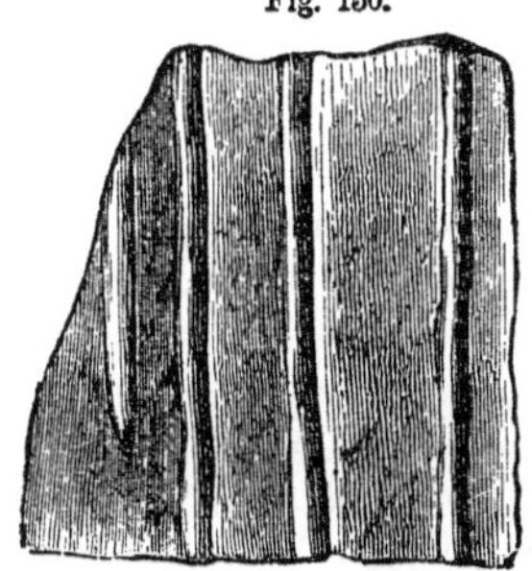

Fig. 151.

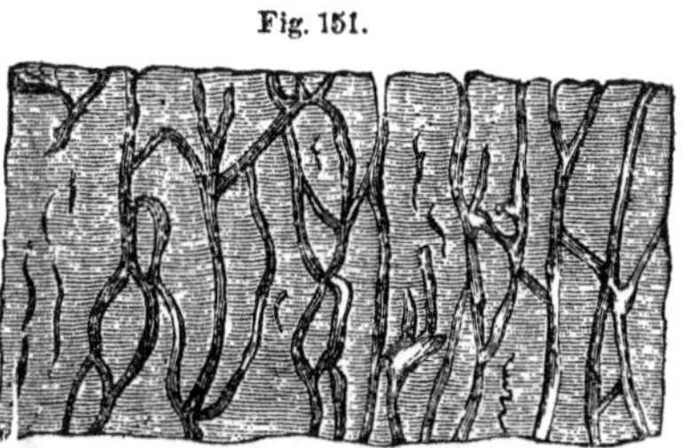

Phytopsis tubulosum.

Polypi.—The radiated animals are well represented in this system. Among these are the coral builders. These are minute radiated animals, called *Anthozoa*, that have the power of secreting carbonate of lime, and thus of building up large stony structures, called *Polyparias*, from the bottom to the surface of the ocean. They swarm in immense numbers in the seas of tropical climates, and form coral reefs which sometimes extend hundreds of miles. They seem to have existed in all ages, and to have formed similar deposits, which are now ranked among the limestones. Figs. 152, 153, 154 show several living species of these animals as they are attached to their stony habitations.

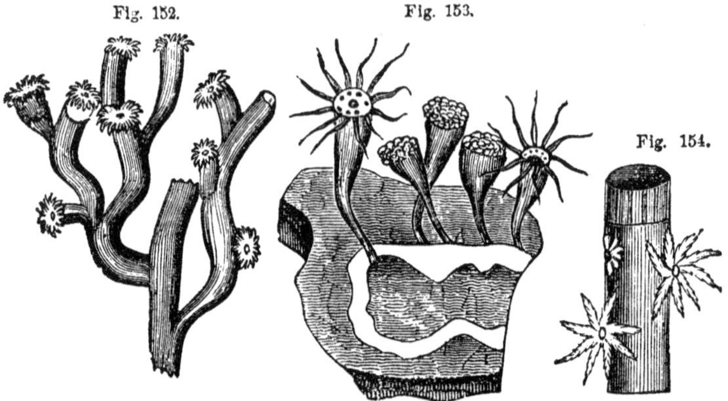

Polyparia.

The tentacles of these animals are provided with cilia or minute hairs on their margins, which are capable of being rapidly moved, so as to keep currents of water in motion, that food may be conveyed to their mouths. Immense numbers of the polypi unite in building up a single habitation, and they do this as if influenced by one instinct; so that the structure rises with the most symmetrical proportions. In the *Flustra carbasea* each polype has usually twenty-two tentacles; and on these, 2,200 cilia. An ordinary specimen of this species will contain 18,000 polypi; and of consequence, 396,000 tentacles, and 39,600,000 cilia. On the *Flustra foliacea*, Dr. Grant estimates 400,000,000.

These polypi mostly multiply by buds, called gemmules, which grow like the buds of plants from the parent, and after a time fall off and become distinct animals. A single polypi in this mode may produce a million of young in a month. They may also be multiplied by division, when each separate part becomes in a short time a whole animal. Different parts may also be made to grow together, and monsters of every form be produced. The Hydra is one of the genera of polypi; and by taking the heads of several individuals, and grafting them to one body, a Hydra with seven, or any other number of heads may be produced.

Fig. 155 shows the Columnaria alveolata from the Black river limestone. Fig. 156 represents the Favistella stellata, and Fig. 157, Chætetes lycoperdon; Fig. 158, shows the Cyathophyllum turbinatum which is found in the next three higher formations.

Graptolites.—These are another remarkable family of radiated animals that appeared in the Lower Silurian, and continued as high as the carboniferous system. Until the late researches of

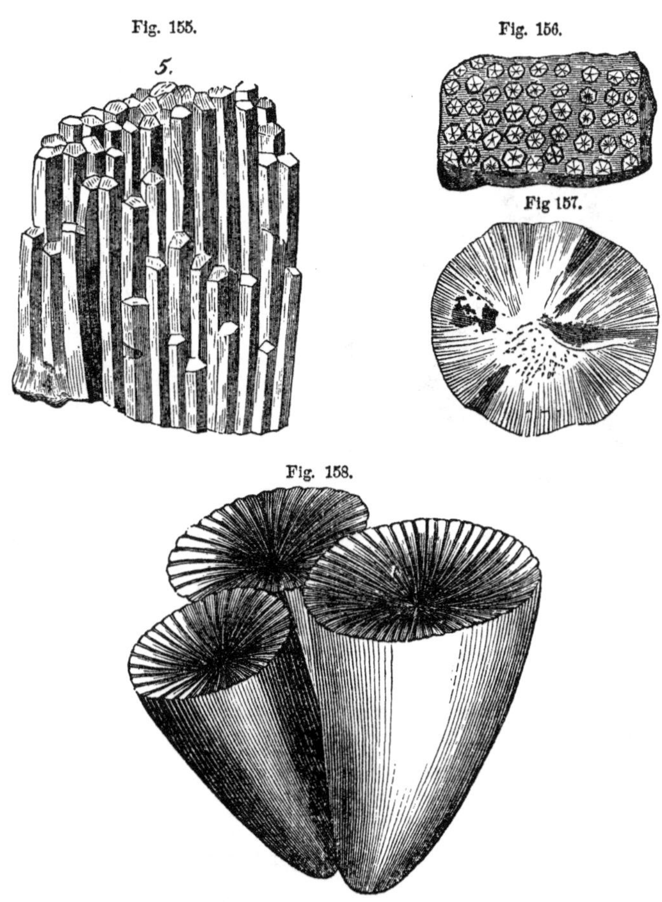

Cyathophyllum Turbinatum.

Professor Hall, little else had been described but the serrated arms of this peculiar animal. But he has traced it from its earliest development till it shows itself as a fixed (possibly free) animal, having a bilateral arrangement in its arms, as shown in Fig. 159, which is the Graptolithus Logani. The arms here are broken off; but in Fig. 160 they are extended far enough to bring the serratures into view. Generally these serrated arms are all that have been figured.

Fig. 159.

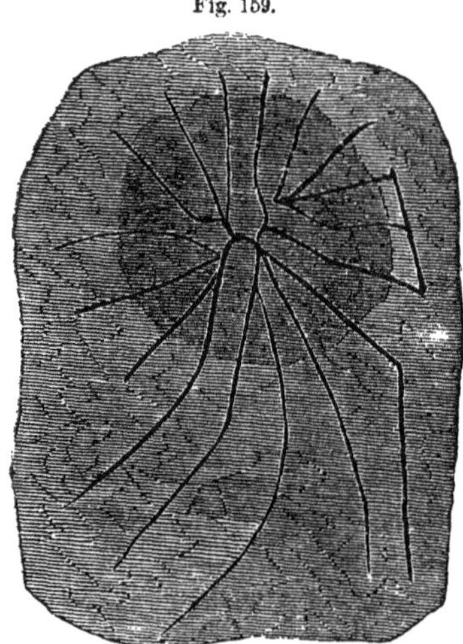

Fig. 160.

Brachiopods.—These are a class of bivalve shells with the valves unequal, and the arms of the animal long, that flourished abundantly in the lower Silurian Seas. Several of the genera also, (not the same species), have lived through all the changes of the earth's crust, and still inhabit the ocean.

The Lingula is one of these, on Fig. 161, from the Potsdam sandstone, and can hardly be distinguished from those now found alive. The Terebratula, Fig. 162 is one of the six genera of shells that have lived through all changes, and 30 species are still found in the ocean. Bronn gives a list of 410 species, 10 of which

Fig. 161.

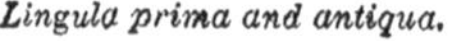

Lingula prima and antiqua.

Fig. 162.

Terebrabula.

occur in the Lower Silurian, 30 in the Upper Silurian, 72 in the Devonian, 38 in the Carboniferous, 11 in the Permian, 30 in the Trias, 83 in the Oolite, 127 in Chalk, and 34 in the Tertiary.

Another Brachiopod of the Silurian rocks is the Orthis (Fig. 163), of which Bronn gives 54 species in the Lower Silurian, 31 in the Upper Silurian, 43 in the Devonian, 12 in the Carboniferous Limestone, 3 in the Permian, and 2 in the Trias, where it died out.

Of the Spirifer, Fig. 163, characterized by a peculiar spire within its valves, shown in our figure, 16 species occur in the Lower Silurian, 18 in the Upper Silurian, 56 in the Devonian, 59 in Carboniferous Limestone, 7 in the Permian, 8 in the Trias, and 4 in the Oolite, where they terminate.

Fig. 163.

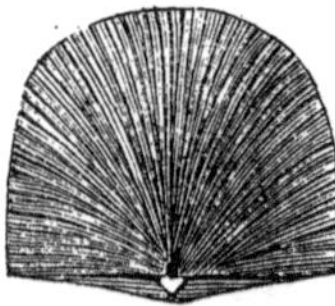

Orthis retrosistria.

Fig. 164.

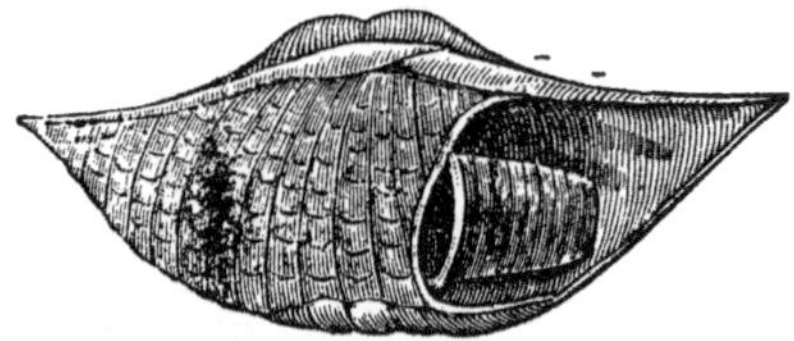

Spirifer.

The number and variety of the preceding Brachiopods that have been described are now so great, that the above genera are regarded as families, such as Terebratulides, Spririferides, etc., each embracing several genera. But details on this subject can not be here given. They will be found in the large works on Palæontology, such as those of Pictet, D'Orbigny, Hall, McCoy, etc.

Other interesting Brachiopods occur in the Lower Silurian; as for instance, Atrypa, Fig. 165.

Conchifera. Numerous representatives of this class of shells appeared early. For examples we give Ambonychia (Pterinæ), undata, Fig. 166, from the Trenton Limestone; Avicula demisa, Fig. 167; from the Hudson River Group, and Modiolopsis modiolaris, Fig. 168, from the same.

Fig. 165. Fig. 166. Fig. 167. Fig. 168.

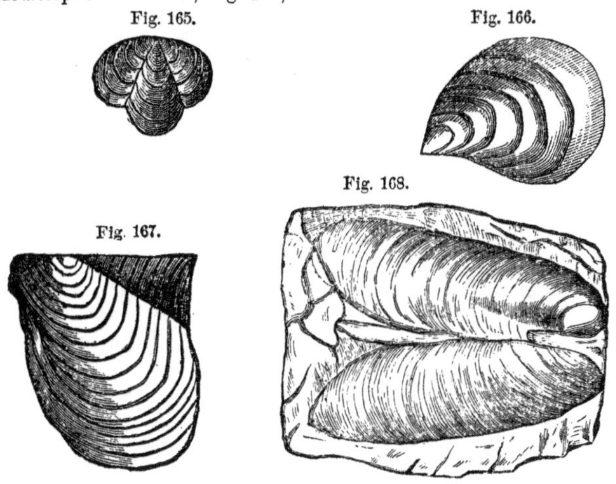

Gasteropoda. These shells are numerous in the Lower Silurian. Maclurea matutina occurs in the calciferous sandstone, Fig. 169. The M. magna is sometimes seven or eight inches in diameter. Murchisonia bellicincta is a delicate shell, Fig. 170. Fig. 171 shows Scalites angulatus of the Chazy limestone. Fig. 172 shows the Bellerophon bilobatus.

Fig. 169.

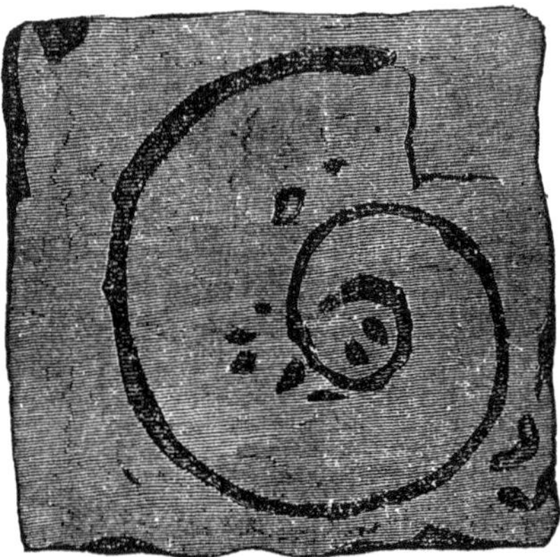

Maclurea matutina.

Fig. 171.

Fig. 170.

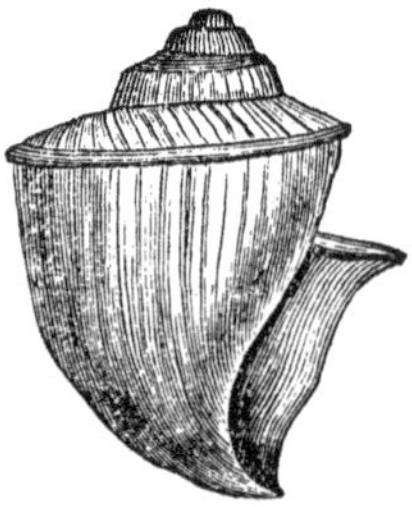

Fig. 172.

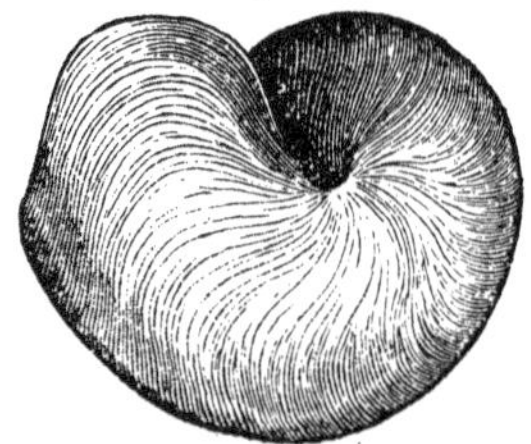

Fig. 173

Lituites cornu-arietis.

Cephalopoda. These were chambered shells, some of which were curved and some straight. We give an example of each. Fig. 173 shows the Lituites cornu-arietis.

Fig. 174 represents an Orthocera, which was straight, yet divided into chambers. This has been found six feet long and six inches in diameter; which must have had an animal around it larger than any living cephalopod. As many as seventy chambers have been counted in it. Bronn mentions 153 species; ten in the Lower Silurian, thirty-two in the Upper Silurian, forty-three in the Devonian, thirty-one in the Carboniferous limestone, eight in the coal measures and seven in the Trias, where it disappeared.

Fig. 174.

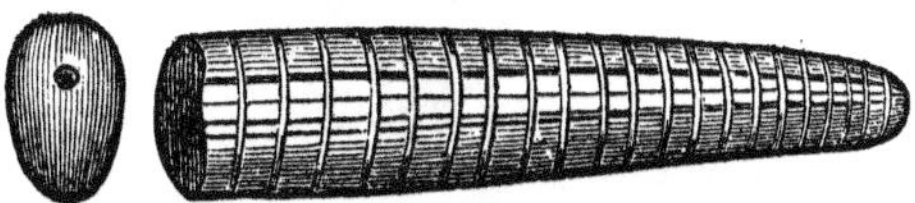

Orthocera.

Echinodermata.—The living animals of this class, are the star-fishes, which are found as low as the Lower Silurian, and extend through all the rocks. But the most remarkable animals of this class are the Crinoids, or Encrinites, so named from the resemblance of some of them to a lily (κριvov) of which the common stone lily (Fig. 263) is an example. The head was supported by a flexible column, that was made up of a vast number of bony rings, and at its lower end was fastened to the ocean's bottom, or to a piece of wood. The head was composed of five articulated arms, which were divided into fingers, and were used for obtaining food. The stem of this species was circular, but that of the Pentacrinite was five-sided, and its arms or tentacules vastly numerous. The number of little bones or joints composing the head of the lily Encrinite, was 26,000; but in the Briarean Pentacrinite, they amount to 100,000, and those of the side arms to 50,000 more. If each of these, as in the higher animals, required two muscles to move it, they would amount to 300,000; while the muscles in man amount to only 540.

Fig. 175 is a representation of the *Pear encrinite* or *Apicrinites* as it appeared when attached and in full life at the bottom of the ocean.

Prof. Pictet, in his great work on Palæontology, has grouped the Crinoids into nine families, as follows: 1. The Comatulideæ, or those free and without stalk; 2. The Pentremitideæ; 3. The Cystideæ; 4. The Cupressocrinideæ;

Fig. 175.

Pear Encrinite.

5. The Polycrinidæ; 6. The Haplocrinidæ; 7. The Anthocrinidæ; 8. The Cyathocrinidæ; 9. The Pycnocrinidæ. These families he divides into 105 genera. Of these, thirty-seven are peculiar to the Lower and Upper Silurian, sixteen to the Devonian, nineteen to the Carboniferous, one in the Permian; in the Triassic six, in the Jurassic fourteen, in the Cretaceous seven, in the Tertiary two, and two exist in our present seas, to which should be added a third, the Holopus, which is found only among existing animals. The genus Pentacrinus began its existence in the Triassic period, and has continued to the present time, though there have been several changes of the species.

Among the Crinoids the Cystideæ have attracted special attention, and those of Canada have been finely illustrated by Mr. E. Billings, palæontologist of the Canada Survey. According to him, there have been found in the lower half of the Lower Silurian (in Bohemia and nowhere else) four species, in the upper half sixty-three; in the Upper Silurian eighteen species, and perhaps some doubtful ones in the Devonian; above which none occur. Fig. 176 shows one of these little mailed animals from the Trenton limestone, the Pleurocystites filitextus.

Crustacea.—Crustaceans form the highest order of articulated animals. By far the most remarkable group of them in the earlier or palæozoic rocks are the Trilobites, so called because their shell or buckler is divided into three parts. So different are they from living crustaceans, that for a long time it was contended that they were molluscs or insects.

The shield or buckler of this animal covered its anterior part, while the abdomen had numerous segments that folded over each other like those on a lobster's tail. By this arrangement some of the species had the power to roll themselves up like the wood-

Fig. 176.

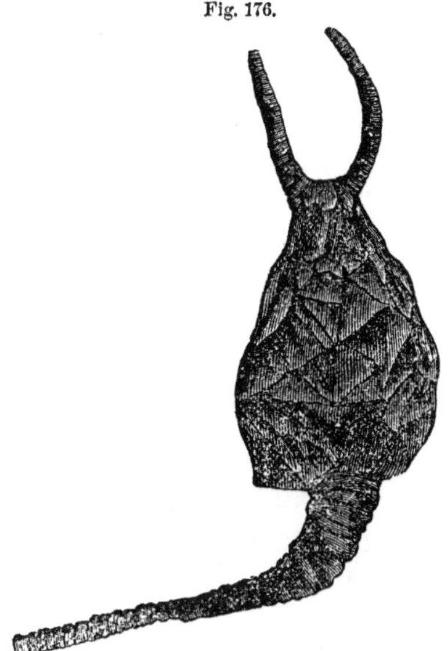

Pleurocystites filitextus.

louse and armadillo, and thus of defending themselves against enemies. They are from half an inch to six inches long, and longitudinal furrows divide them into three lobes.

It is well known that the eyes of many articulated animals are made up of a large number of facets, or lenses, placed at the end of tubes, which being arranged in a parallel position, form a compound eye, like a multiplying glass; which projecting from the head, enables the animal to see on all sides without turning the eye. The number of these little facets or lenses in the house-fly is 14,000; in the dragon fly, 25,000; in the butterfly, 35,000; in the Mordella, 50,000. In the Trilobite they vary from 400 to 6,000. Fig. 177 shows one of the eyes of this animal found in a fossil state.

Fig. 177.

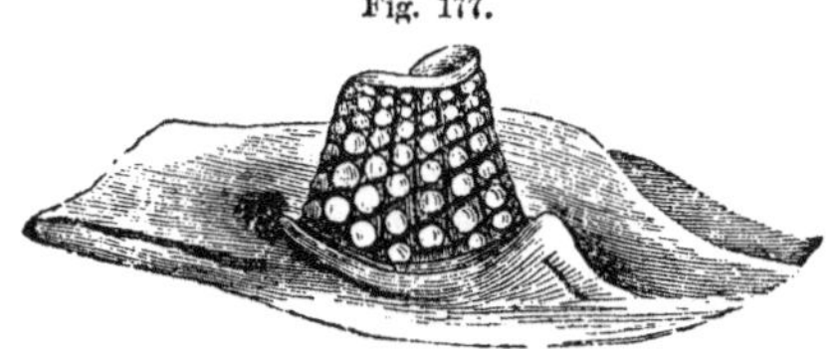

Pictet divides the Trilobites into twelve families, viz., the Harpidæ, the Paradoxidæ, the Calyminidæ, the Lichasidæ, the Trinucleidæ, the Asaphidæ, the Æglinidæ, the Illenidæ, the Odontopleuridæ, the Amphionidæ, the Brontidæ and the Agnostidæ. These he divides into forty-three genera, of which twenty-four are found in the Lower Silurian, half of which pass into the Upper Silurian, and eleven in this last formation that pass into the Devonian, while only one passes into the Carboniferous; above which none are found. But only in a very few cases is the same species found in any two of these formations. According to Prof. Owen, the whole number of species is 400 and of genera fifty; of which forty-six are Silurian, twenty-two Devonian, and four Carboniferous. Thirteen genera are peculiarly Lower Silurian, three Upper Silurian, one Devonian and three Carboniferous.

Fig. 178 exhibits one of the well-known forms of Trilobites from the Lower Silurian, the Paradoxides Tessini of Brongniart.

Fig. 178.

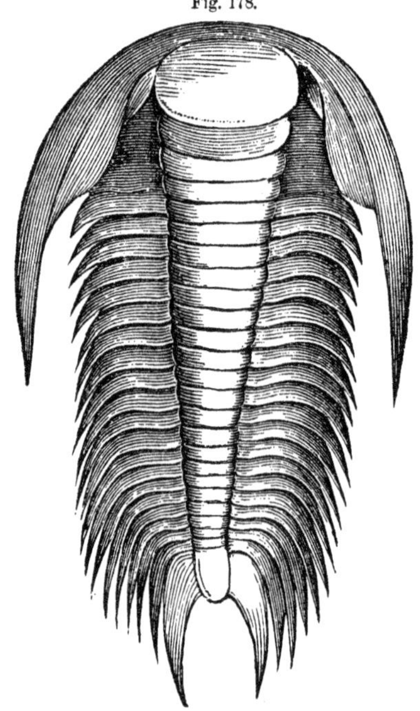

Fig. 179 is a top view and Fig. 180 a side view of the Sao hirsuta from the Lower Silurian of Bohemia.

Lithichnozoa.—In Potsdam sandstone, at Beauharnois, and other places in Canada, Sir William Logan has collected and described at least seven species of crustacean's tracks. Fig. 181 will give an idea of one species. A fine collection of them may be seen in the cabinet of the Canada Geological Survey at Montreal. Fig. 181, A, shows the tracks of a living crustacean, the Ocypode arenaria, sketched by Prof. Agassiz and kindly put into our hands.

Fig. 179. Fig. 180.

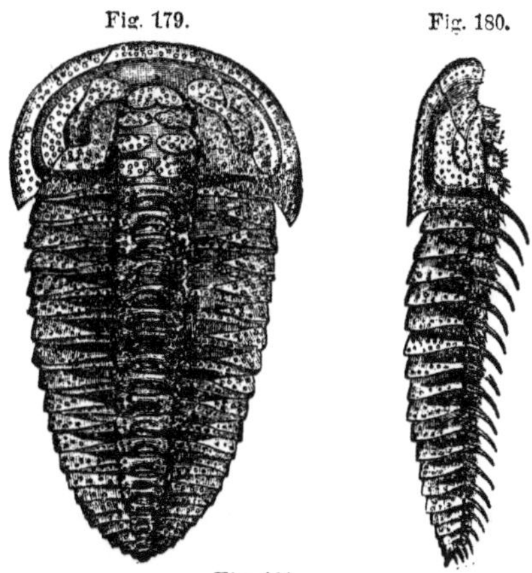

Fig. 181.

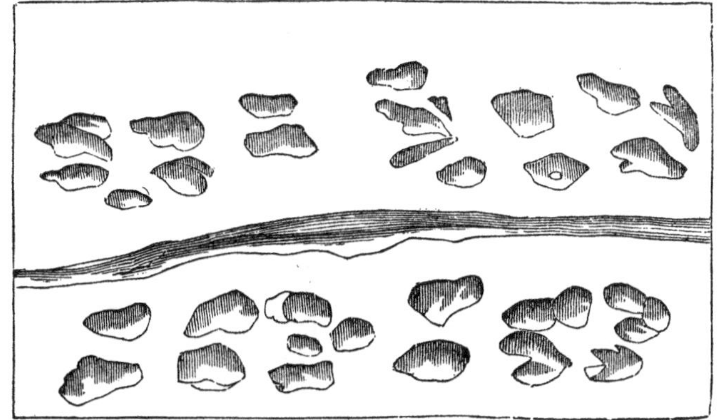

Prototichnites septem-notatus.

Fig. 181. A.

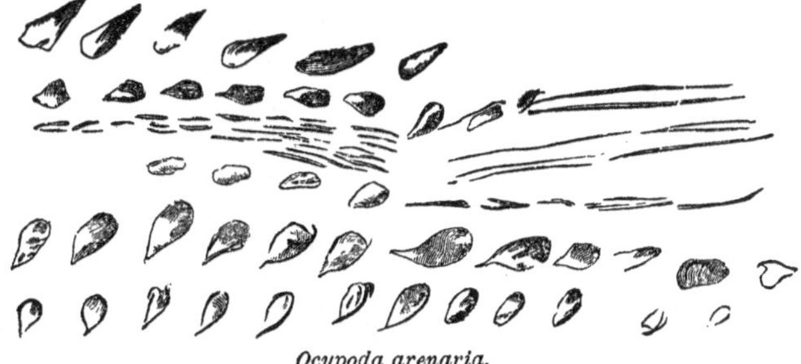

Ocypoda arenaria.

More recently analogous Prototichnites (the name applied to the Canada tracks by Prof. Owen) have been found in Scotland, called P. Scoticus.

Fig. 182.

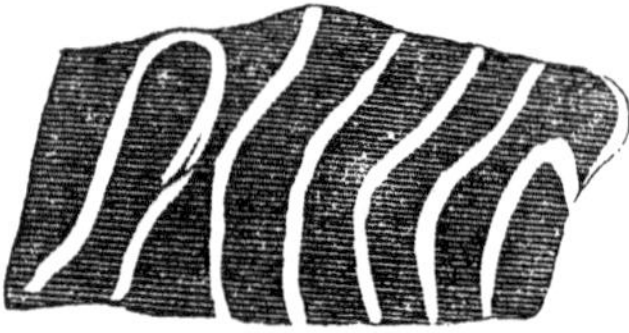

Crossopodia Scotica.

In the Llandeilo flags in Wales several species of Annelids occur which have sometimes left their trail, as represented on Fig. 182, which was made by Crossopodia Scotica.

On the Hudson River shales of Georgia in Vermont, we have found the trail probably of an annelid, as shown on Fig. 183.

Fig. 183.

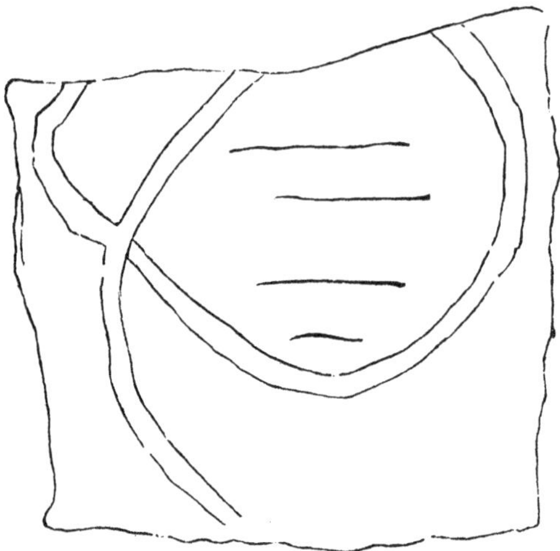

Trail of an Annelid.

3. Upper Silurian Period.

In the Lower Silurian rocks the plants hitherto discovered are all flowerless and the animals all invertebrate: crustaceans being the highest class, of which the crab and lobster are living examples. At the close of this period there seems to have been a considerable change in the state of things, and sometimes the higher strata are unconformable to the lower. For the most part the same classes of animals continue, though with new forms. Near the top a few vertebral animals appear, as the details below indicate.

Plants.—The plants of the Upper Silurian are very few and

mostly marine. In Europe the best writers mention only one or two species. But Professor Hall has described not less than ten species of Algæ or sea-weeds. Perhaps the most interesting of these is the Arthrophycus Harlani, shown on Fig. 184.

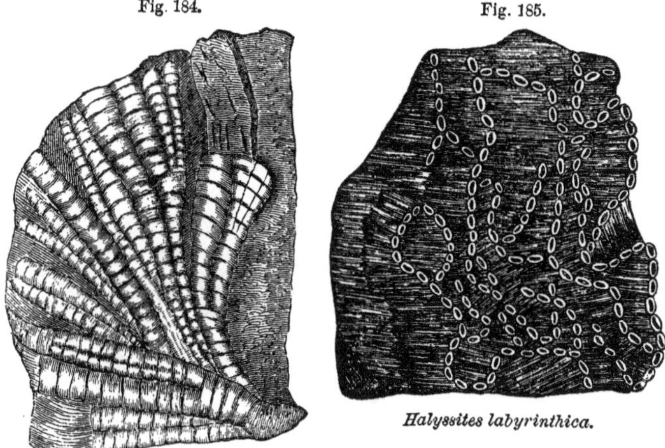

Fig. 184.

Fig. 185.

Halyssites labyrinthica.

It is not till we rise to the very top of this formation, and perhaps into the Devonian above, that we find any traces of land plants.

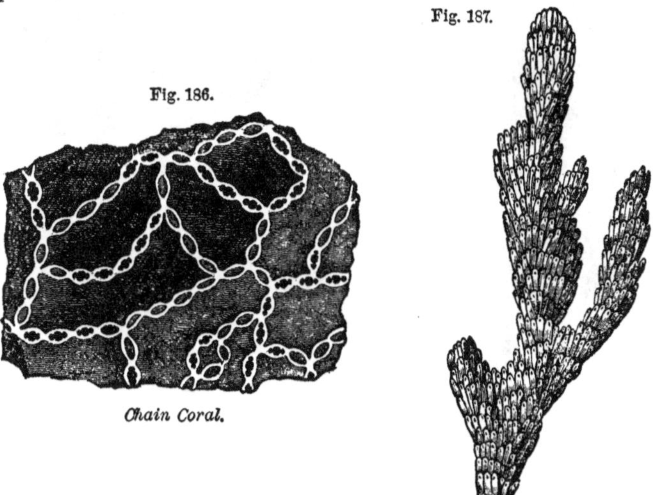

Fig. 186.

Chain Coral.

Fig. 187.

Animals. *Polypi.*—Of the zoophytes one of the most striking species is the chain coral. We give two specimens in Figs. 185 and 186.

Favosites polymorpha, another genus, is shown in Fig. 187.

Fig. 188.

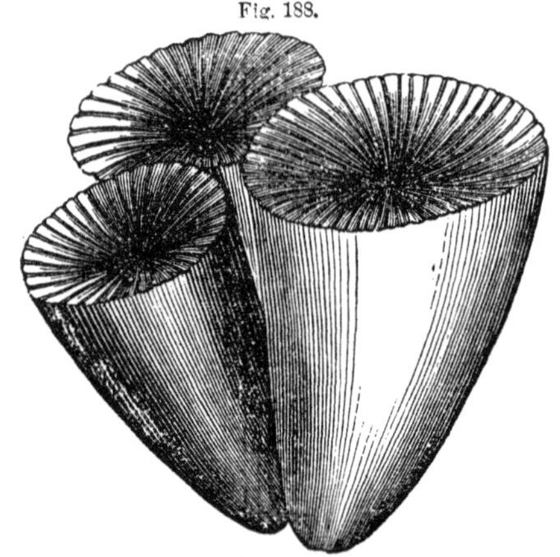

Cyathophyllum turbinatum.

Fig. 189.

Fig. 190.

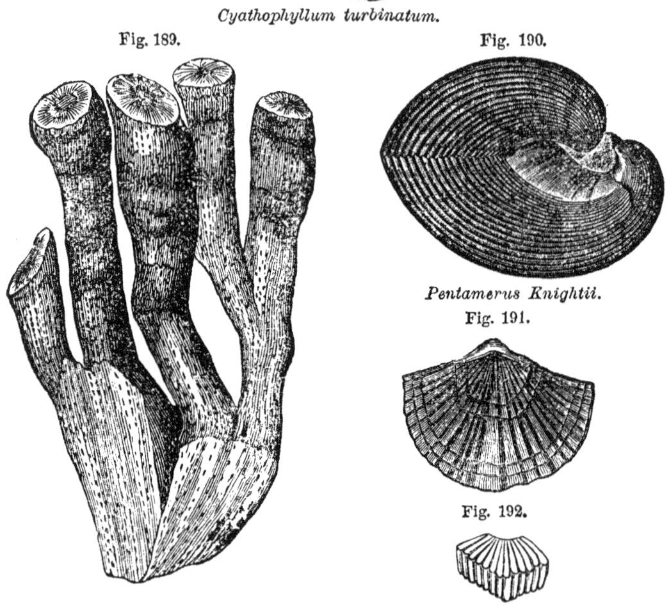

Pentamerus Knightii.

Fig. 191.

Fig. 192.

Another genus of these old corals is the Cyathophyllum, often mistaken in our country for the horns of deer, etc. Fig. 188 shows one species of this genus, the C. turbinatum. Fig. 189 represents the Cyathophyllum cæspitosum.

Brachiopoda.—New species of these shells abound in the Upper Silurian, belonging to the same genera, as in the Lower Silurian most frequently, but

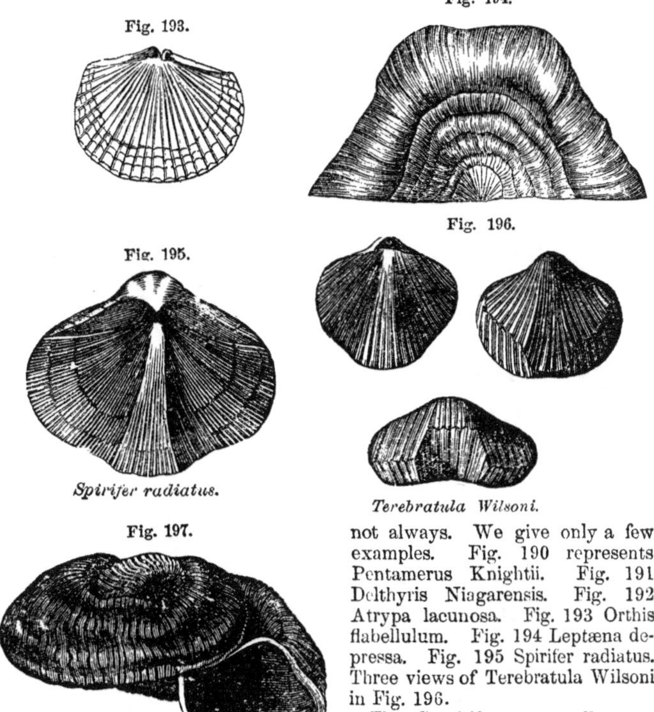

Fig. 193.

Fig. 194.

Fig. 195.

Spirifer radiatus.

Fig. 196.

Terebratula Wilsoni.

Fig. 197.

not always. We give only a few examples. Fig. 190 represents Pentamerus Knightii. Fig. 191 Delthyris Niagarensis. Fig. 192 Atrypa lacunosa. Fig. 193 Orthis flabellulum. Fig. 194 Leptæna depressa. Fig. 195 Spirifer radiatus. Three views of Terebratula Wilsoni in Fig. 196.

The Conchifera are well represented. Fig. 197 shows a Gasteropod, the Euomphalus rugosus. Fig. 198 shows a Cephalopod, the Conularia Niagarensis, from the Niagara group of New York.

The Crinoids are abundant; but we have room to present only three. Fig. 199 represents the Caryocrinus ornatus from the Niagara group. Fig. 200 shows the Ichthycrinus lævis, from the same formation. The sculpture on many of these crinoids is often extremely beautiful. Fig. 201 shows the Hypanthocrinus decorus, allied to the lily.

Of the other Echinoderms, we show in Fig. 202, the star fish Ophiura constellata, and in Fig. 203, the Palæaster Niagarensis, from the Niagara limestone.

We regret not having room for more of the beautiful Trilobites found in this formation. Fig. 204 exhibits the Calymene Blumenbachii. Fig. 205 the same rolled up.

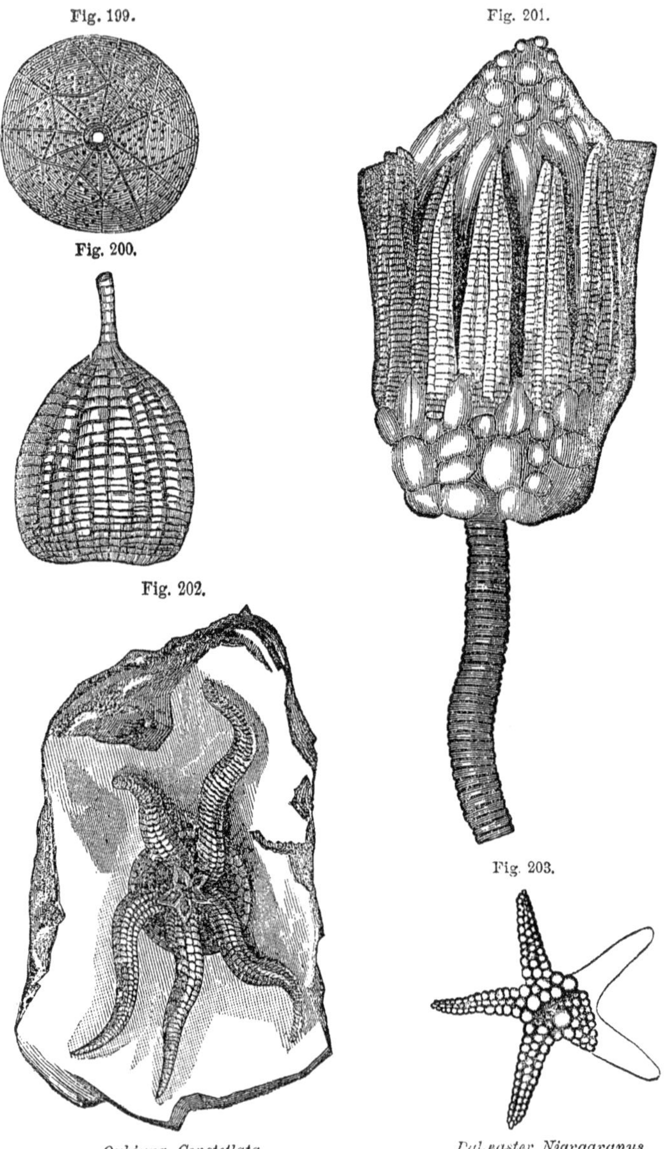

Fig. 199.

Fig. 200.

Fig. 201.

Fig. 202.

Fig. 203.

Oyhiura Constellata.

Palæaster Niargaranus.

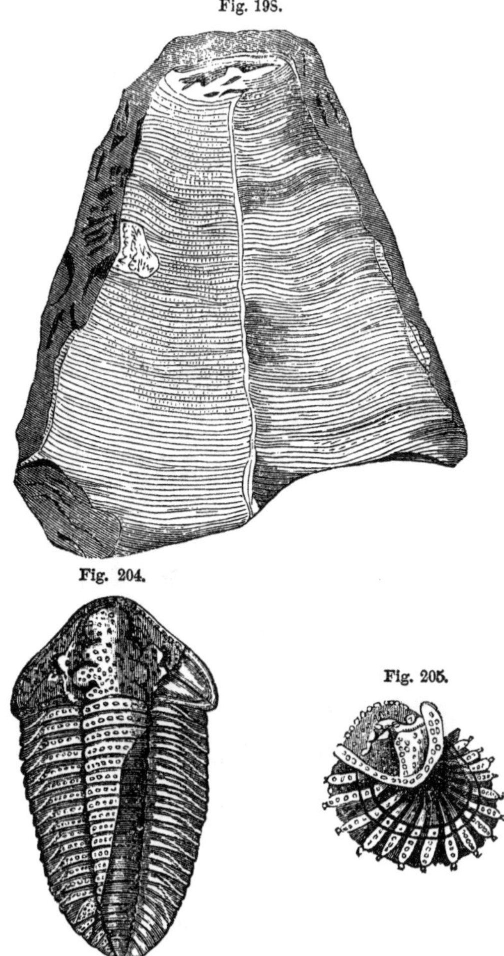

Fig. 198.

Fig. 204.

Fig. 205.

Fishes.—Near the top of this formation a few fishes have been found; of some of them, however, we know but little, as only broken fragments occur. Pictet mentions only three genera, and four species as certain; but Agassiz adds four other genera from the broken fragments: these were from the Upper Ludlow rocks; but in 1859 a very perfect fish was found in the Lower Ludlow rocks, which carries down this class of animals 1,000 or 2,000 feet. These few fishes seem to be only an anticipation of the great development of this class of animals in the Devonian group, and as none of those in the Silurian are of any special interest, we defer a description of this important class of animals to the next group.

Lithichnozoa.—Professor James Hall in the second volume of the Palæontology of New York, has described as many at least as six species of tracks on the Clinton group of the Upper Silurian, in that State. He suggests that these were made by Molluces (Gasteropods), Annelids and Fishes, as shown on figures 206, 207, and 208.

Fig. 206.

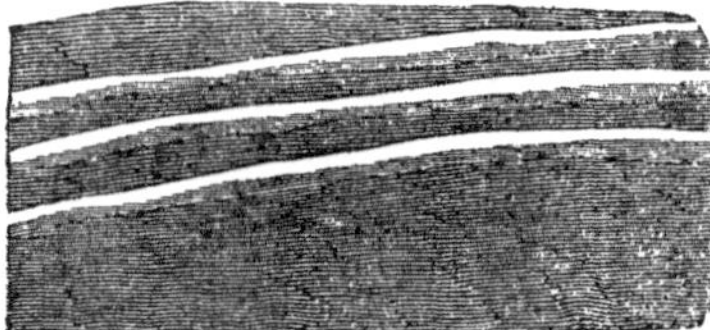

Annelid Track.

Fig. 207.

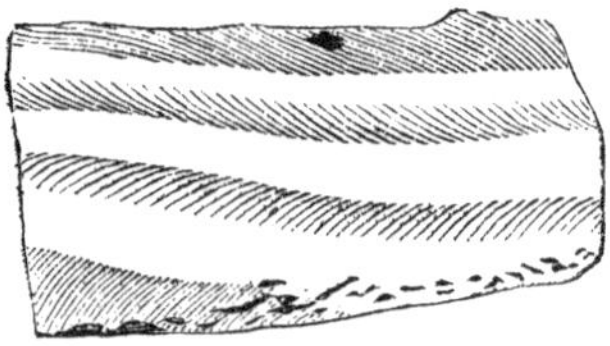

Annelid Track.

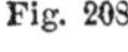

Fig. 208.

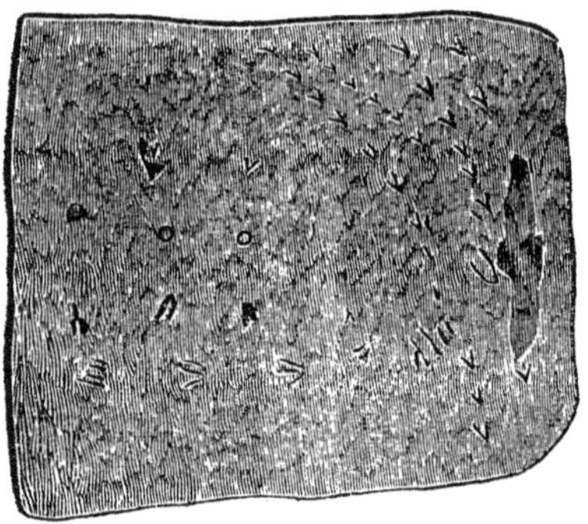

Fish Tracks.

4. DEVONIAN PERIOD.

Plants.—These are few and badly preserved, so that great uncertainty still exists as to their character. Some of them, however, were sea weeds, and some land plants of as high organization as the coniferous or pine tribe. A few years ago, quite a number of plants were referred to the Devonian Period. In 1849, Bronn enumerates nearly fifty species of monocottyledons, as well as half a dozen less perfect flowerless plants. But these probably occur in rocks which are now placed higher in the series. We give a sketch of only one species from this formation, which seems well determined by Professor Hall, from the Chemung group, and which resembles a good deal, plants in the coal measures above the Devonian. It is the Sphenopteris laxus, Fig. 209.

Fig. 209.

Sphenopteris laxus.

ANIMALS.—*Polypi.*—Of the corals we present three examples: Fig. 210,

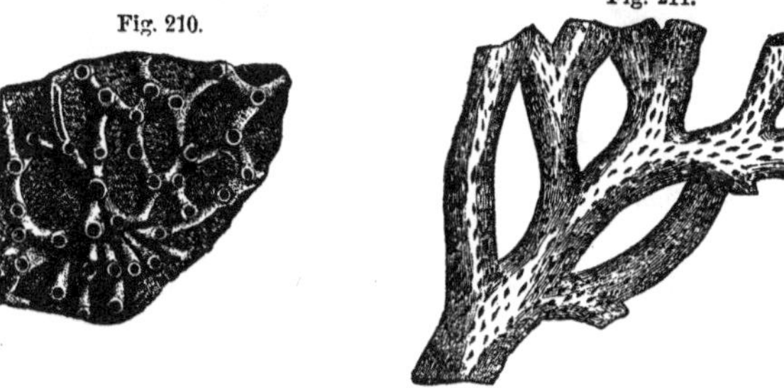

Fig. 210.

Fig. 211.

represents a fragment of Aulopora serpens.: Fig. 211, Favosites fibrosa, and Fig. 212, Astrea rugosa.

Fig. 212.

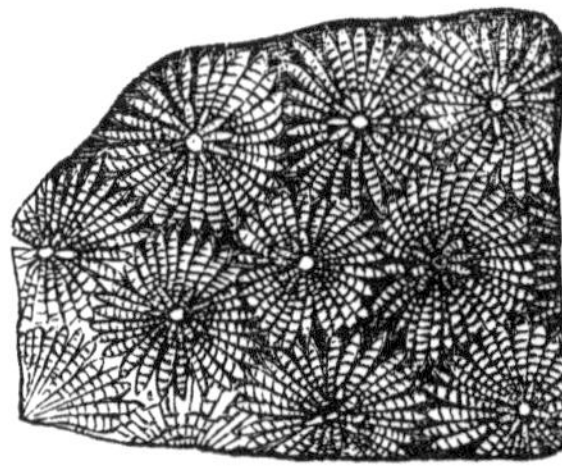

Fig. 214.

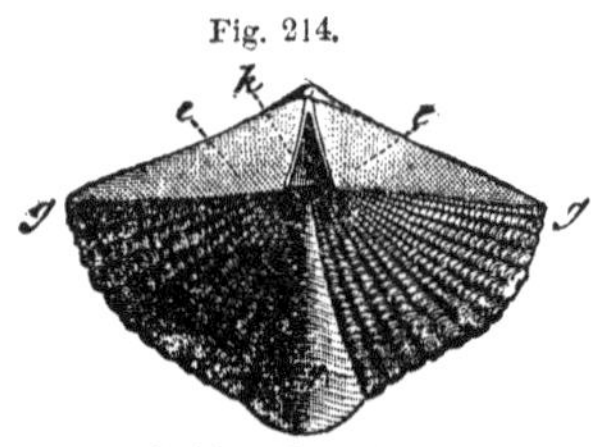

Spirifer hystericus.

Of the Brachiopoda we show in Figs. 213, 214, 215, two interesting species of Spirifer; one of which, Fig. 213, goes with late authors under the

Fig. 213.

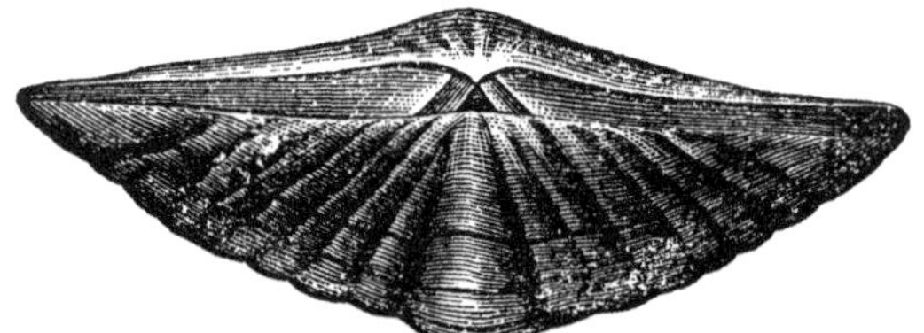

Trigonotreta speciosa.

name of Trigonotreta, and Fig. 214 is Spirifer hystericus. Fig. 215 gives a good idea of the two spirals found in this shell.

Fig. 215.

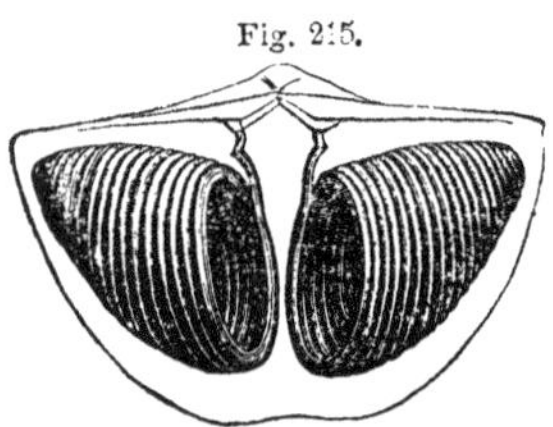

Spirals in Spirifer.

Fig. 216.

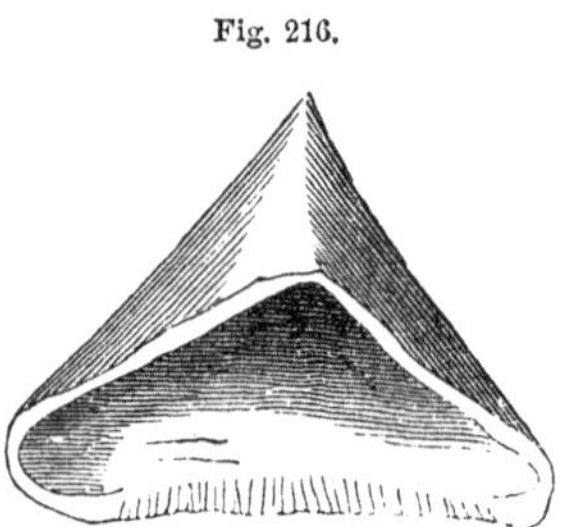

Calceola sandalina.

A quite peculiar genus of Brachiopods is the Calceola sandalina, named from its resemblance to a shoe, and shown on Fig. 216.

Of the Conchifera or Acephala, we give only one species in Fig. 217, the Posidonomya Beecheri. We give two Cephalopoda. Fig. 218 shows the Goniatites costulatus, and Fig. 219, the Cyrtoceras undulatus of the Corniferous Limestone.

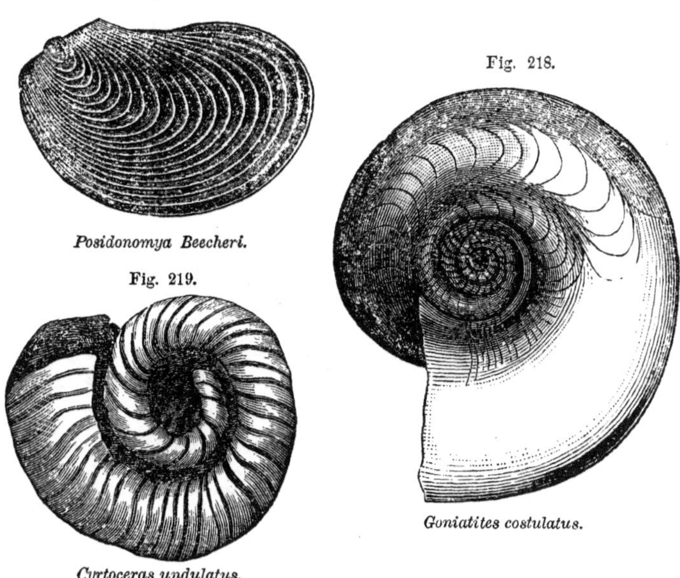

Fig. 217. *Posidonomya Beecheri.*

Fig. 218. *Goniatites costulatus.*

Fig. 219. *Cyrtoceras undulatus.*

Fig. 220 shows a Devonian trilobite, the Acidaspis elliptica.

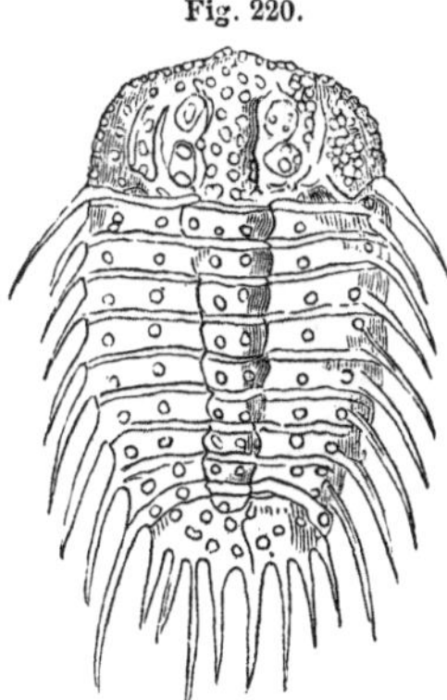

Fig. 220.

Fishes.—The most characteristic feature of the Devonian rocks is the great development of fishes which they present. With the exception of a few species near the close of the Upper Silurian Period, this is their earliest appearance; and they present many strange forms, and are largely developed in all the subsequent formations. This is not true of any other class of vertebral animals, and hence the history of the fossil fishes is peculiarly instructive. It is said that not less than 1700 species are found fossil, and 10,000 are now living. Professor Agassiz estimates the number of species that have lived in all past periods, at not less than 30,000 species.

This eminent writer in his great work entitled, *Recherches sur les Poissons par L'Agassiz*, divided all fishes into four classes, distinguished by the character of their scales; the important discovery having been made by him, that there is such a relation between the form of the scale and the organization of the fish, that if those having similar scales be brought together, they will be found to correspond closely in their nature. The following are the forms of the scales in the four classes:

Fig. 221, No. 1, shows one of the enameled plates that belongs to the *Placoids;* No. 2, the plates covered with enamel, identical in structure with the teeth, covering the *Ganoids;* No. 3, the toothed or comb like scales of the *Ctenoids;* and No. 4, the circular plates without enamel of the *Cycloids*.

Fig. 221.

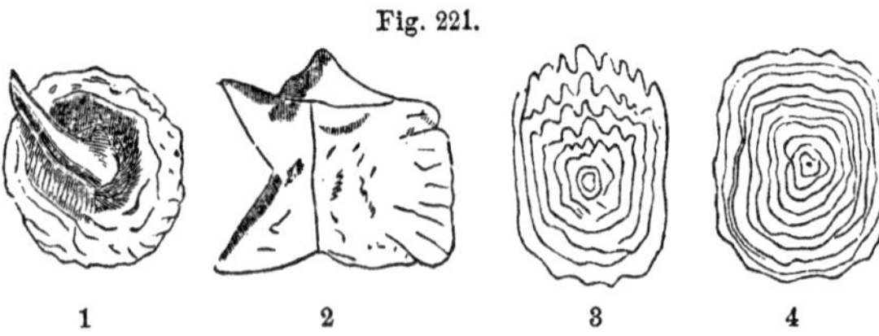

From the classification of animals which we have given on a previous page, it would appear that Agassiz has given up the above arrangement. He will doubtless explain fully his new system in his great work on the Natural History of the United States. J. Muller, according to Pictet, had proposed desirable changes in the system founded upon the scales.

He proposes six classes: 1. The *Sirenoid* fishes, that have both lungs and gills; 2. the *Teleosteans*, or fish with proper bones; 3. the *Ganoids;* 4. the *Elasmobrancheans*, or the Placoids of Agassiz; 5. the *Cyclostomeans*, or cartilaginous fishes; 6. the *Leptocardians*, or fishes without a heart, being only one genus. Of these the first, fifth, and sixth classes are not found fossil.

Naturalists have described 504 genera of fossil fishes, distributed as follows:

In the Upper Silurian	7
In the Devonian	56
In the Carboniferous	70
In the Permian	16
In the Trias	23
In the Jurassic	66
In the Cretaceous	78
In the Tertiary	188

Of all fossil animals the fishes cast the most light upon the laws of palæontology. In their different formations they are found to be separated from one another by the most distinct characters. Those of a particular formation seem to have been created for that formation only, and rarely do the genera extend beyond it as they do in most of the lower animals. Not a single genus below the chalk has survived to the present day, and above that point the number of extinct genera is quite large, amounting to two thirds of those in the chalk, and one third of those in the lower tertiary.

Those of the carboniferous strata disappear with the deposition of the new red sandstone, and those in the Oolite suddenly vanish with the appearance of the chalk. Not one species has yet been found that is common to any of the two great geological formations, or that is now living in the ocean.

We cannot but remark here, how entirely opposed are these facts to a prevalent hypothesis that the different sorts of animals in the rocks, as we ascend, have been slowly changed from one into the other by a natural process. Here, on the contrary, we find such great and entire changes in the successive groups as can be explained only by new creations.

"We find in the history of fishes," says Pictet, "many arguments against the hypothesis of the transition of species from one into the other. The Teleosteans could not have had their origin in the fishes which existed before the cretaceous epoch, and it is impossible to derive the Placoids and Ganoids from the Teleosteans. The connection of faunas, as Agassiz has said, is not material, but resides in the thought of the Creator." It is well to take heed to the opinions of such masters in science, when so many, with Darwin at their head, are inclined to adopt the doctrine of gradual transmutation in species.

The Devonian fishes had great peculiarities. Indeed, to the close of the Jurassic period no fish had the horny scales such as now cover four-fifths of them. The Devonian fishes were many of them covered with bony plates forming a buckler, and were also of peculiar form. Fig. 222 shows the under side of the Cephalaspis Lyellii.

Fig. 222.

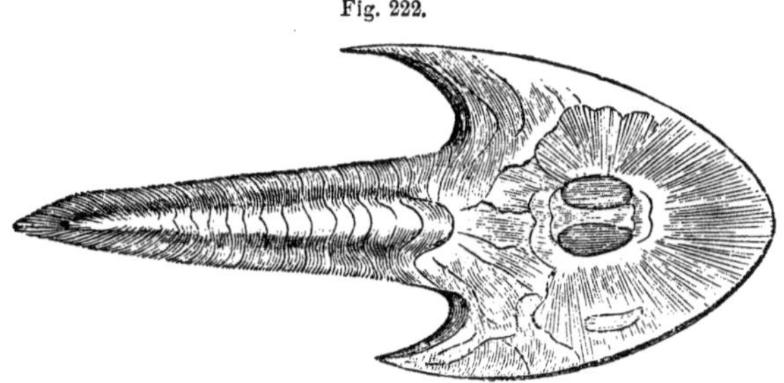

Cephalaspis Lyellii.

Fig. 223 is a side view of the same fish.
In Fig. 224 is represented another of these fishes, the Pterichthys cornutus.

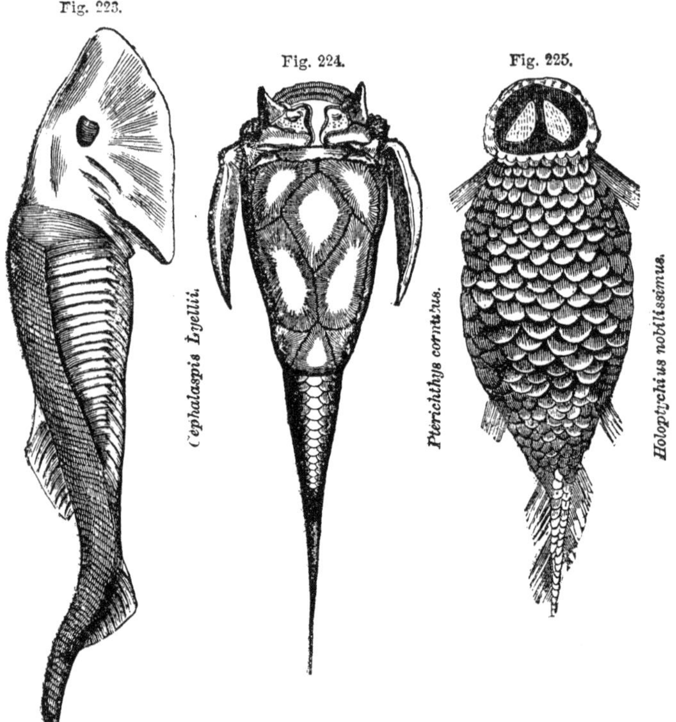

Fig. 223. Fig. 224. Fig. 225.

Another fish of peculiar aspect, which is common in the Devonian rocks of this country, is the Holoptychius nobilissimus, of which Fig. 225 will give some idea.

Reptiles.—From at least two sources of evidence we find that reptiles began to appear as early as the Devonian period, though some geologists suspect that the Upper Old Red Sandstone of Great Britain belongs to the Carboniferous group; and if so, it would raise the reptiles into that formation. But one skeleton of a reptile, Fig. 226, called the Telerpeton Elginense, and Leptopleuron lacertinum, by Professor Owen, has been found in the Old Red Sandstone of Morayshire, which Dr. Mantell regards either as a fresh water Batrachian, or a small terrestrial lizard. Another, a true Saurian reptile, has been found in the same formation. In another part of the same rock small bodies have been discovered, which have been referred with some probability to the eggs of the frog.

Fig. 226.

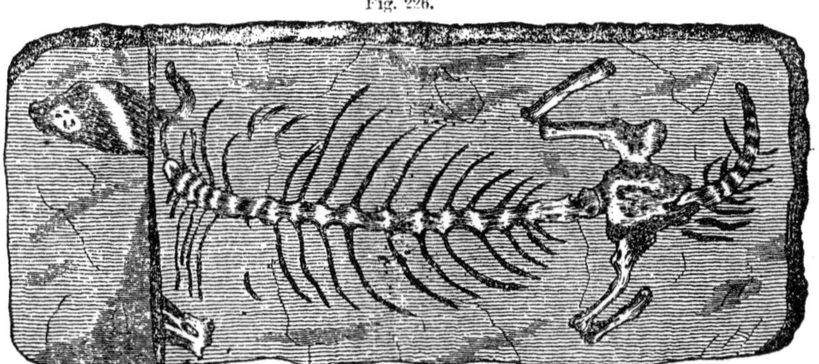

Telerpeton Elginense. From the Old Red Sandstone.

Fig. 227.

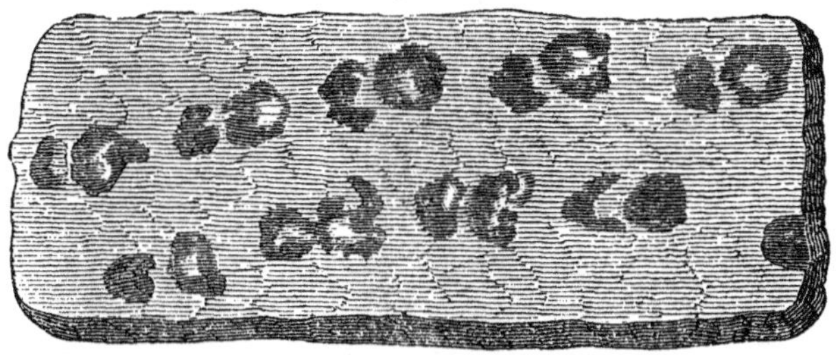

Fig. 228.

Lithichnozoa.—In the Old Red Sandstone of Scotland, Captain Brickenden has described tracks which agree best with the Chelonians or Tortoises, as shown on Fig. 227.

In the red sandstone near Pottsville, in Pennsylvania, Isaac Lea has described the tracks of an animal which he calls Sauropus primævus. They are shown on Fig. 228.

Professor H. D. Rogers, in opposition to Mr. Lea's opinion, places these tracks in the lower part of the carboniferous formation. So he does, also, those other analogous species which he found 1500 feet lower in the series. He also found what he thinks may be the trail of a mollusc in his Umbral Series.

5. Carboniferous Period.

The two very distinct parts into which this formation is divided, differ widely in their palæontological characters. The lower part, called the carboniferous or mountain limestone, is rich in marine relics. But it is the remains of terrestrial plants, which so abound in the upper part, called the coal measures, that gives the name and the highest interest to the formation. Marine remains abound in this part also; but the land plants predominate. In the brief space which we can devote to the fossils of this formation we shall dwell chiefly upon the plants and upon the higher tribes of animals.

Plants.—Compared with the formations below there was an immense development of plants in the carboniferous rocks. Previously it would seem that not much dry land existed, certainly not in a condition for producing vegetation; for the plants in these lower rocks are almost entirely marine, and of course flowerless. But in the carboniferous era land plants were introduced abundantly, more than 683 species having been described in that formation, according to Prof. Unger. But they were mostly flowerless plants, chiefly such as form the class of Acrogens, such as the Ferns, the Equisetaceæ and Lycopodiaceæ. Yet many of them were large trees. Take the ferns, for example. In tropical regions, at the present day, these sometimes grow as high as forty or fifty feet, and the trunks are covered with the scars of the leaves that have fallen off. Fig. 229 exhibits some of these gigantic ferns.

Most of the fossil ferns probably belonged to species no larger than the ferns now growing in temperate latitudes; but some of them were tree ferns. Fig. 230 shows Neuropteris ovata from the coal, and it can hardly be distinguished from species now common in this country.

Fig. 229

Tree Ferns.

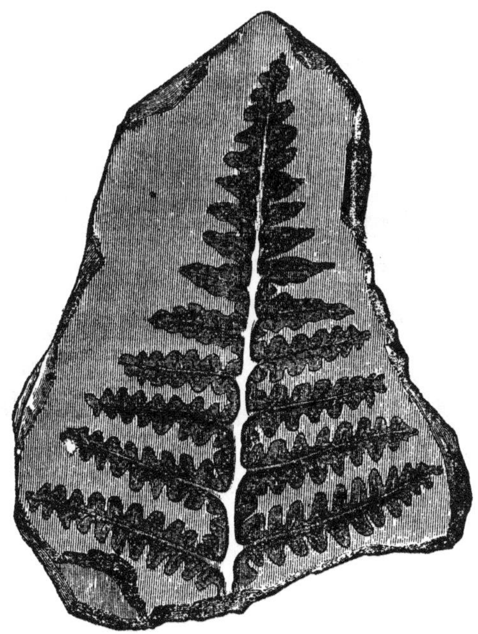

Neuropteris Ovata.

More than 250 species of ferns have been already dug from the coal strata in Europe, and it is an interesting fact that at present not more than fifty species of this tribe of plants are natives of Europe. They are far more abundant in tropical regions; and hence it seems a fair conclusion that the climate in Europe and the United States, during the coal period, was tropical.

Professor Lindley made some experiments to determine what sorts of plants would longest resist the action of water. The leaves and bark of most dicotyledonous plants, that is, our present forest trees and flowering plants generally, were destroyed in two years. The monocotyledons, such as the palms, were more enduring, but grasses perished. Funguses, mosses and most of the lowest forms of vegetables, soon disappeared; but ferns were the most enduring of all. In short, those plants most abundant in a fossil state endured the best. Hence it is inferred that the frailer sorts may have been much more abundant in early times than their number found fossil would indicate.

Stigmaria.—Immediately beneath every bed of coal (and sometimes twenty or thirty beds lie above one another in the same basin) is a layer of arenaceous shale, from six inches to ten feet thick, called *under clay* or *fire clay*. In this, and here only, is found the peculiar fossil called Stigmaria, Fig. 231. It is ascer-

Fig. 231.

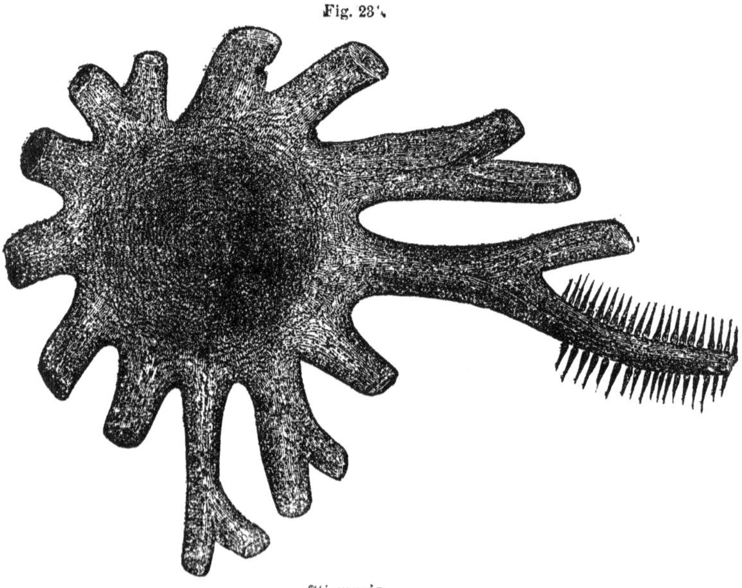

Stigmaria.

tained to be the root of another kind of fossil which is abundant in the coal, and in the rock immediately above it, called Sigillaria.

Sigillaria.—These are large flattened trunks, from thirty to sixty, and even seventy feet long. They are covered by scars or

Fig. 232.

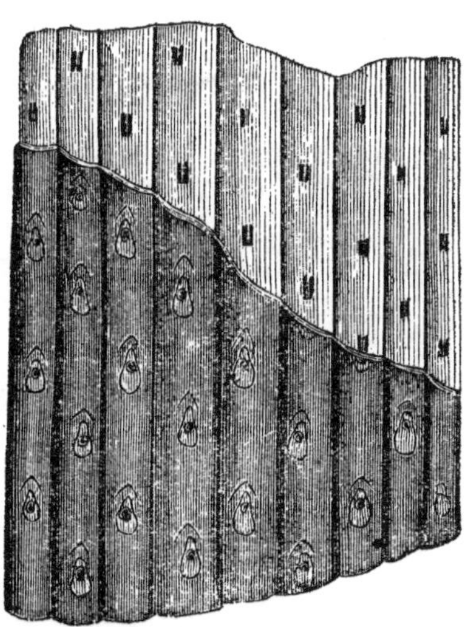

Sigillaria.

Fig. 233

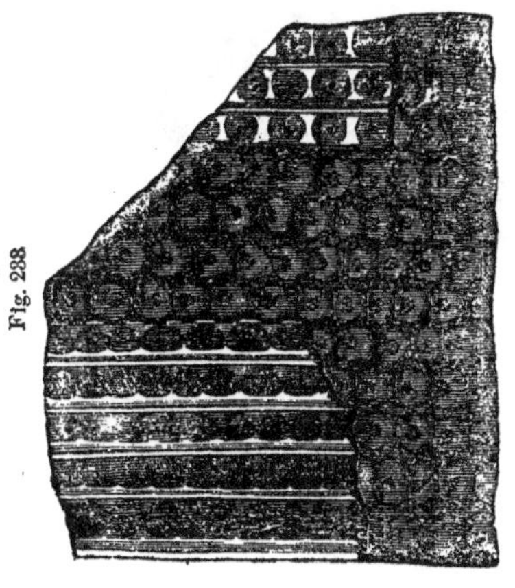

Sigillaria elegans

Fig. 234.

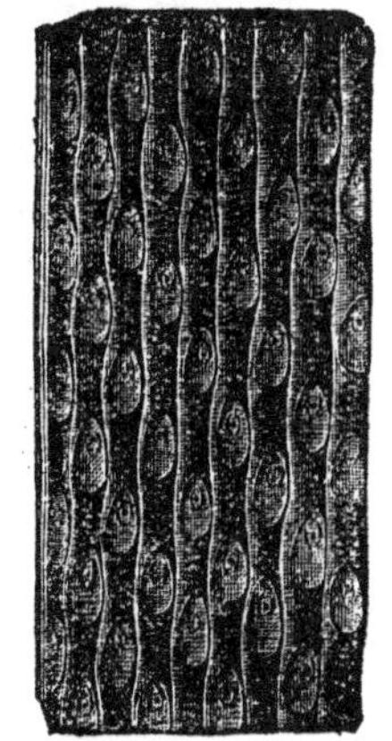

Sigillaria Groesseri.

cicatrices, showing the points to which leaves were attached while growing. They were probably hollow trunks, or became so before falling, and hence are so much flattened. Doubtless they formed the source of most of the beds of coal.

More than thirty-five species are known, of which Figs. 232, 233, 234 will give an idea. Fig. 235 represents a trunk standing.

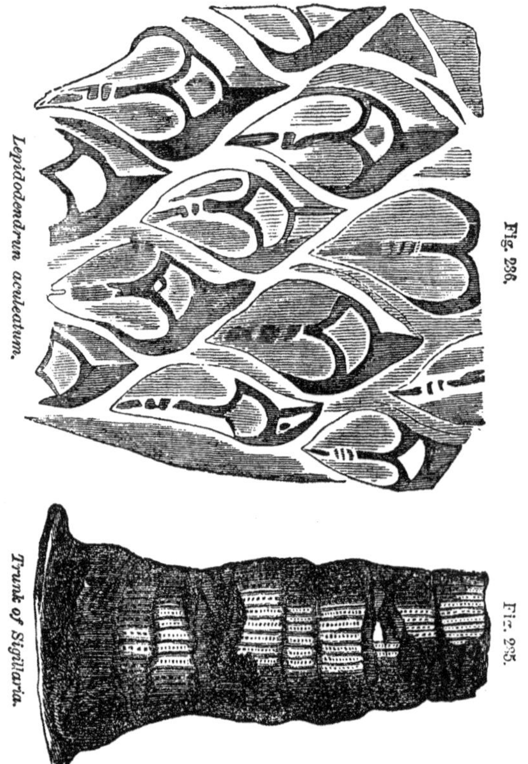

Fig. 236. *Lepidodendrum aculeatum.*

Fig. 235. *Trunk of Sigillaria.*

Lycopodiums, or *Club Mosses.*—The living plants of this family, about 200 species, are small, rarely in temperate climates exceeding a few inches, and in tropical climates never more than three feet in height. But the Lepidodendron, which is an allied fossil plant, grew forty to fifty feet high. The trunk is beautifully tessulated or scarred like the Sigillaria, as is shown in the L. aculeatum, on Fig. 236.

Fig. 237 shows one of these large trunks. Another genus of this family is the Ulodendron, shown in Fig. 238.

Fig. 237.

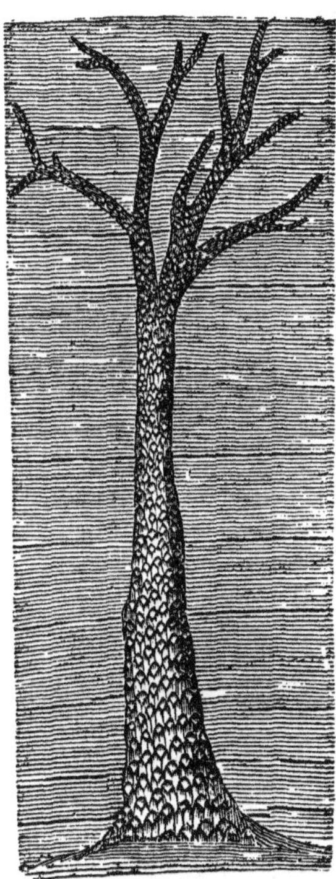

Lepidodendron.

Fig. 238.

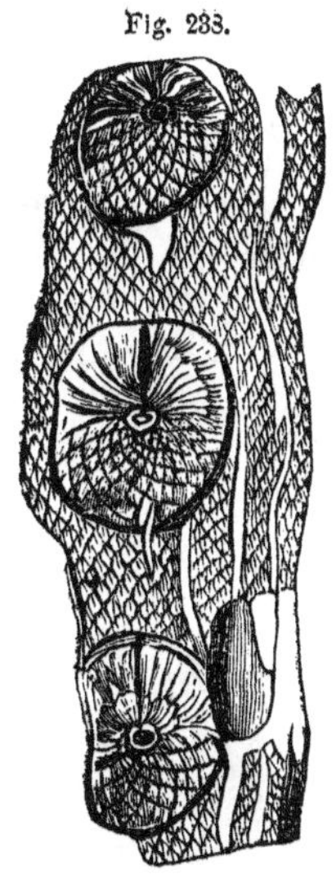

Ulodendron.

Equisetaceæ.—The living plants of this family go by the name of horsetails, cattails, rushes, etc., and are of diminutive size. But not so the fossil species. The most remarkable and common is the Calamites. These were large jointed reeds, attaining sometimes the size of trees. A sample is given in Fig. 239.

The large trunks that have been described are sometimes found standing upright in the mine and penetrating the sandstone layers. Fig. 240 shows a succession of vertical stems of this sort in the coal measures at the head of the Bay of Fundy in Nova Scotia. The mass of sandstone containing the stems is 2,500 feet thick, and the length of the upright trunks six or eight feet. Only ninety-two feet of the beds is represented.

Fig. 239.

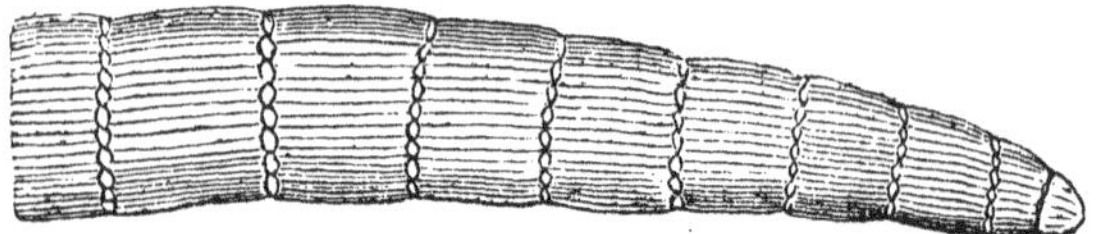

Calamites.

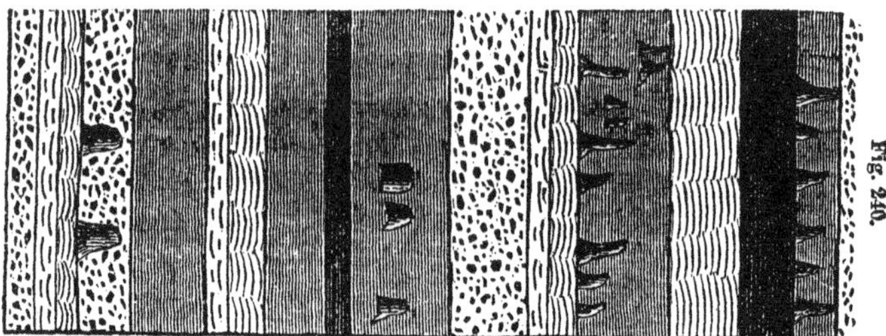

Fig. 240.

Fig. 241.

Annularia.

Asterophyllitece.—These plants belong to the Acrogens, but have the aspect of aster flowers, and hence the name of the family. They were not numerous, but quite peculiar. Fig. 241, exhibits a species of Annularia. Fig. 242 shows the Spenophyllum emarginatum, another genus of these plants.

Fig. 242.

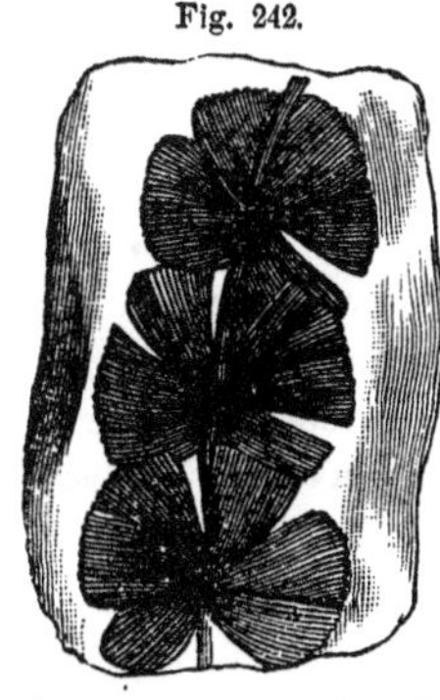

Flowering plants began to appear in the coal formation. Several species of palms and of grasses are reckoned among the monocotyledons. The dicotyledons are mostly reckoned as belonging to the Coniferæ or pine tribe, especially to the Araucaria, a species of which now grows on Norfolk Island, on the west coast of America, and is cultivated frequently in conservatories. Trunks of the fossil trees have been found in several quarries in Scotland, penetrating the strata obliquely, and being sixty to seventy feet long, and from four to six feet in diameter at the base. Coniferous plants have a peculiar microscopic structure, which is retained in true petrifactions, and can be made manifest by polishing thin plates.

Animals.—This is the first formation in an ascending order in which any of the Protozoa have been found of much size, although they occur as deep as the Lower Silurian. Here we have in great abundance in the Carboniferous Limestone, the Fusulinæ, which belong to the class Foraminifera. These are organisms mostly microscopic, of a simple structure, protected by a shell, and bearing a considerable resemblance to chambered shells. The number found fossil amounts to 73 genera and 657 species, increasing in number and variety as we ascend, and attaining their maximum in the present seas. We give only one example here, the Fusulina cylindrica, considerably magnified in Fig. 243.

A tertiary limestone, the "Calcaire grossier," is used at Paris as a building stone, and so abounds in Foraminifera, that we may almost regard the capital of France as constructed of these minute shells.

Of the Bryozoa, we give one example from the Carboniferous group, which is Coral, the Archimedipora Archimedea (Fig. 244), from Kentucky. Its name is derived from its resemblance to the Archimedean screw. The Bryozoa are now put among the Molluscs.

Fig. 243.

Fig. 244.

The fossil Bryozoa already described amount to 1676 species, 213 of which are found in chalk.

Fossil shells, both bivalve and univalve, are abundant in the carboniferous limestone, but we pass by them all for want of room except the Chambered shells, which belong to the Cephalopods.

Fig. 245.

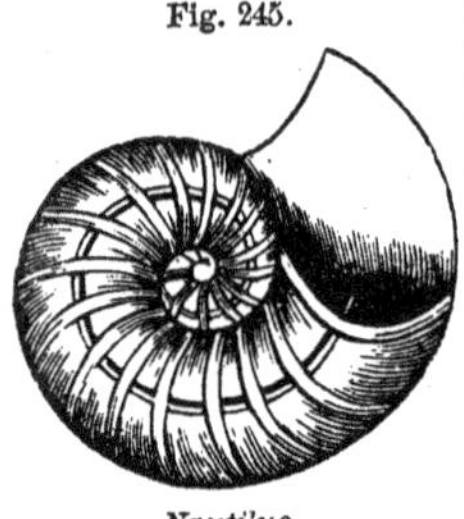

Nautilus.

The two principal families of them, the Nautilidæ and Ammonitidæ, are divided into numerous chambers, connected by a tube called a siphuncle, both which facts are shown in Fig. 245. The strait and partially unrolled cephalopods we have already described under Orthocera in the Silurian rocks, and the Ammonitidæ, are not developed till we reach the secondary strata. But the Nautilidæ are multiplied in the carboniferous strata which contain not less than forty species.

The extinct species of cephalopod molluscs amount to 1500, divided into 50 genera. 1400 of these, divided into 30 genera, belong to shells similar to the pearly Nautili, of which only five or six species exist in the present seas.

The Cephalopods possessed horny mandibles, or beaks, which are frequently found fossil, and have been called Rhyncholites. Figs. 246 and 247 show two of them.

Fig. 246.

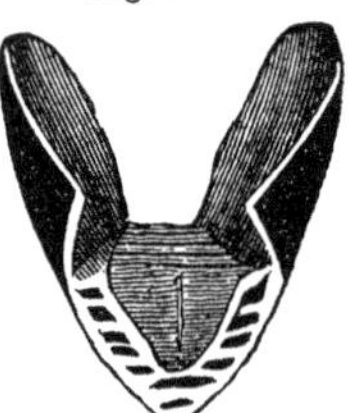

Fig. 247.

Among the Crinoids found in this formation one of the most beautiful is the Pentremite, of which Figs. 248 and 249 show the Pentremites conoideus from Indiana, as described by Professor Hall.

Fig. 248.

Fig. 249.

Fossil insects has been found as low as the coal measures. Of the Arachnida (spiders, scorpions, etc.), 131 species are described in the rocks. Of these, the most interesting is the scorpion, found in Bohemia, and shown on Fig. 250, the Cyclophthalmus Bucklandi.

Fig. 250.

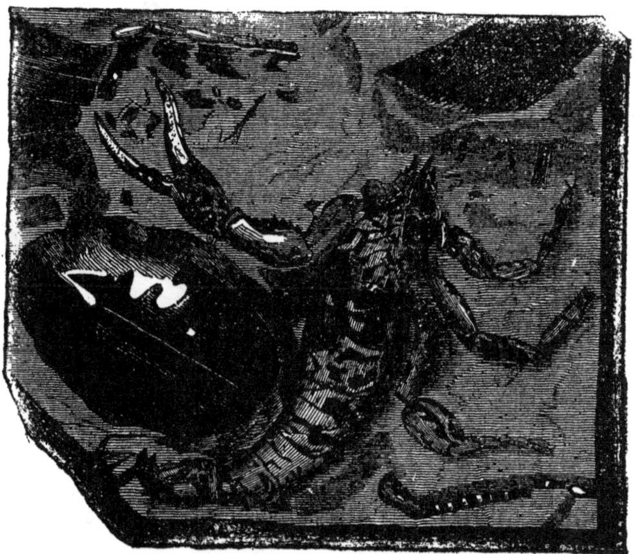

Cyclopthalmus Bucklandi.

According to Bronn, in 1848, eleven species of insects had been found in the carboniferous strata, thirty-one in the lias; forty-six in the oolite; fifty-seven in the wealden; two in the cretaceous, 1545 in the tertiary, and one in the alluvium, making 1699 in all. This embraces the Myriapods, the Arachnida, and Hexapods.

Not less than 70 genera and over 150 species of fishes have been described in the carboniferous formation. They begin to have a much nearer resemblance to living fishes than those of Devonian age, as the sketch of Palæoniscus Duvernoi (Fig. 251), and of Amblypterus macropterus (Fig. 252), will show.

All the fishes below the Trias, however, have one remarkable peculiarity. The vertebral column, or backbone, is prolonged far

Fig. 251.

Fig. 252.

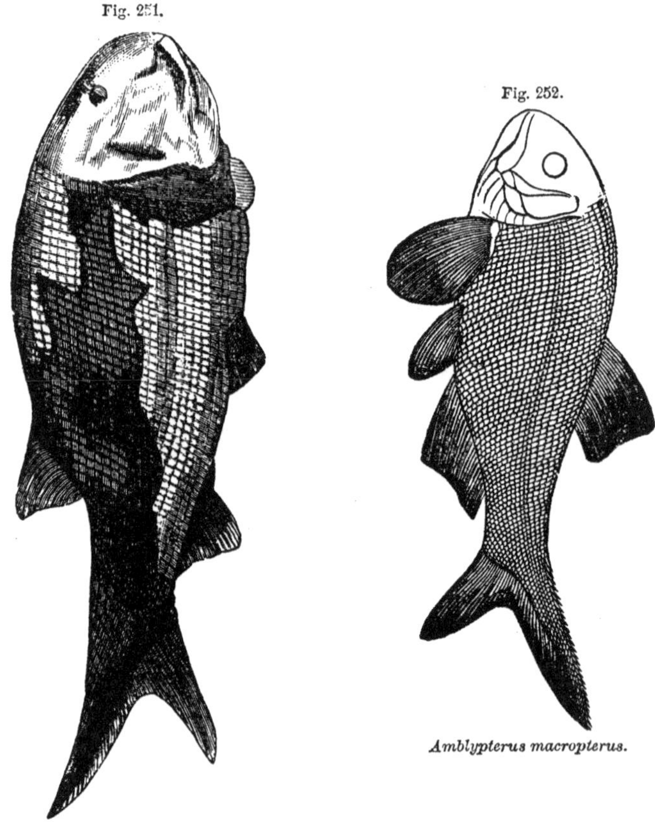

Amblypterus macropterus.

Palæoniscus Duvernoi.

into the upper lobe of the tail, as may be seen in the above figures. This makes the tail unsymmetrical, or as it is usually slyled, *hetero-cercal.* Above the Permian this peculiarity is rarely seen, though among living fishes it is possessed by the sharks, the dog-fishes, and sturgeons. But most living species, as well as the fossil, from the Permian upward, have symmetrical or *homocercal* tails; that is, the vertebral column terminates at the middle of the base of the tail, as an examination of some of the figures of fishes we shall present in the higher formations (see Figs. 289, 290) will show.

This mark often enables the geologist to determine from what part of the rock series a fossil specimen was obtained, and thus helps to fix the age of the rock in which it occurs.

Fig. 253.

A few living fishes, such as the Port Jackson shark, have strong dorsal spines, covered with small teeth, as weapons of defense. These have been found abundantly with the fossil species, and have been described by the name of *Ichthyodorulites.* The finest examples of these which we have met in European works is shown on Fig. 253, which we introduce here, although it belongs to the Wealden formation. It belongs to the genus thefodus.

We give in Fig. 254 a most remarkable and beautiful example of what is most probably an ichthyodorulite from the coal formation of our own country. It was found by Dr. S. B. Bushnell, of Montezuma, in Indiana, and presented by him, through Rev. John Hawks, to the senior author of this work, by whom it was deposited in the cabinet of Amherst College. It has the aspect of a shark's jaw, but was most probably dorsal. It has been referred to Prof. Agassiz for description. If it be an ichthyodorulite, it is the most beautiful that has ever been discovered. It was found a foot above a bed of coal.

Agassiz denominates some of the old fossil fishes, *Sauroid*, or like Saurian reptiles, because their anatomical structure, especially their large striated conical teeth, resemble those of saurians. They are an example of what we shall find common, viz., a union of characters in some fossil animals now found only in different families.

Reptiles.—We have presented decided evidence that reptiles began to exist as early as the Devonian period. As we should expect, we find them, though not very abundantly, in the Carboniferous strata. Professor Owen has paid great attention to this class of animals, and some

Fig. 254.

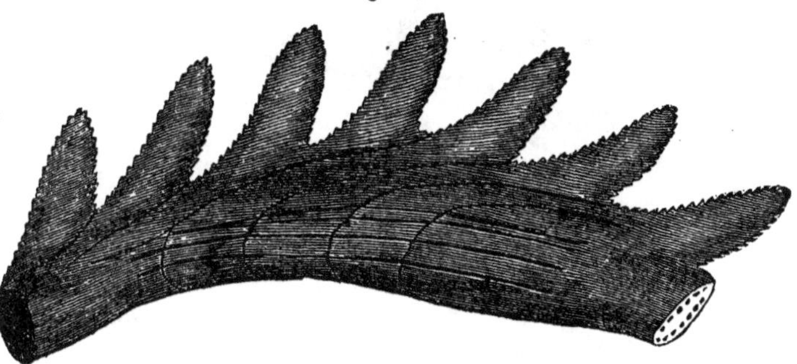

details respecting his classification, already briefly stated, may be desirable.

He divides the Amphibious Reptiles into two orders: 1. *Ganocephala*, animals allied to the living Proteus and Lepidosiren, being intermediate between fish-like batrachia and lizards and crocodiles. 2. *Labyrinthodontia*, animals between batrachians and lizards and fishes.

The Saurian Reptiles, Owen divides into eleven orders: 1. *Thecodontia*, embracing the Protosaurus and other genera, among which is the *Bathygnathus*, described by Dr. Leidy, from Prince Edward's Island. 2. *Cryptodontia*, between lizards, tortoises and birds. 3. *Dicynodontia*, combining characters found in crocodiles, tortoises, lizards and mammalia. 4. *Enaliosauria*, embracing most remarkable fossil Saurians which will be noticed further on. 5. *Dinosauria*, great land Saurians. 6. *Pterosauria*, or flying Saurians. 7. *Crocodilia*, crocodilians. 8. *Lacertilia*, lizards. 9. *Ophidia*, serpents. 10. *Chelonia*, tortoises. 11. *Batrachia*, frogs and salamanders. Quite recently he has made some change in this plan.

Prof. Jeffries Wyman has described, under the name of Raniceps, a fossil batrachian, reckoned by Owen with his Ganocephala, in the carboniferous rocks of Ohio, where are, also, two other allied species. Wyman also suggested the reptilian character of the Dendrerpeton Acadianum from Nova Scotia. But the most interesting of the carboniferous reptiles is the Archegosaurus, the head of which is shown in Fig. 255. This, according to Prof.

Fig. 255.

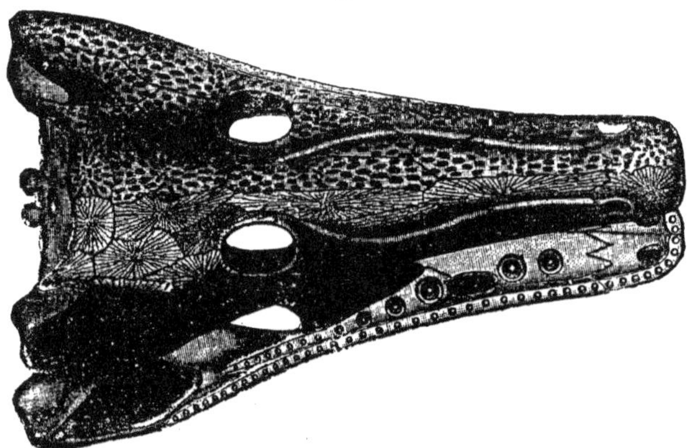

Archegosaurus.

Owen, belongs to the Ganocephala, differing from Batrachians in some important respects, and allied to the living Proteus and Lepidosiven. Owen has also described a Labyrinthodont reptile from the coal of Pictou, in Nova Scotia, which he calls Baphetes planiceps.

Lithichnozoa.—In the western part of Pennsylvania Dr. A. T. King has described tracks in the carboniferous formation which are doubtless those of a Batrachian. We give in Fig. 256 a slight sketch of some of them. It is called Batrachopus primævus.

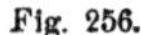

Fig. 256.

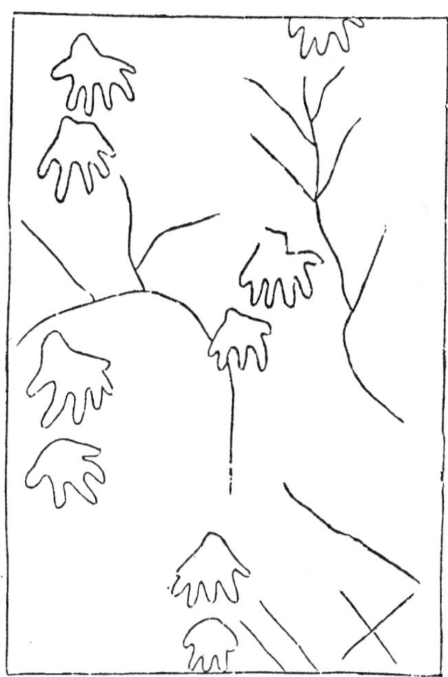

Footmarks in Pennsylvania.

Dr. Buckland has given an account of Ichthypodulites, or fish tracks, from the coal formation.

Hugh Miller has described abundant tracks on the coal measures of Scotland which are reptilian; but he does not decide whether they were those of a Batrachian or Lizard.

6. Permian Period.

Both the plants and animals of this period are so much like those of the preceding, though differing specifically, that we shall give but a few examples.

Fig. 257 represents a peculiar plant, the Nöeggerathia expansa, found in Russia.

Fig. 257.

Fig. 258.

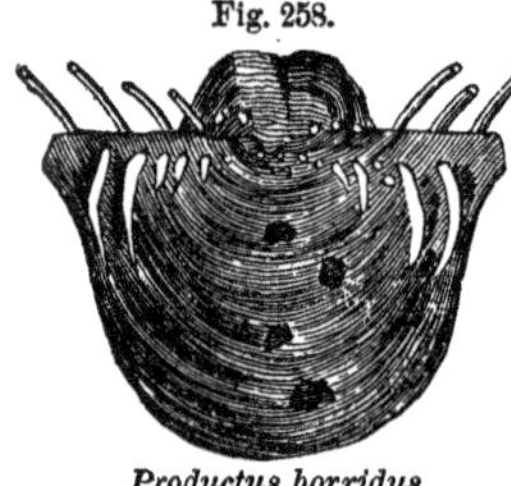

Productus horridus.

Fig. 258 shows a somewhat peculiar brachiopod shell, the Productus horridus.

Several new genera and many new species of fishes, and especially of reptiles, appear in this formation, and clearly distinguish it from the carboniferous. In England, Russia and Thuringia, several thecodont saurians have been found, such as the Protosaurus, Thecodontosaurus and Palæosaurus. In this country Professor Emmons has found in the sandstone of North Carolina the Palæosaurus, the Clepsiosaurus, and one which he names Rutiodon Carolinensis, and he considers these fossils as proving that sandstone to be Permian. Under Lithichnozoa we shall see that other reptiles are found in this rock, as shown by their tracks.

Lithichnozoa.—The first tracks discovered and described by Dr. Duncan in Scotland, are now thought to be in Permian rather than Triassic sandstone. The sketch, Fig. 259, evidently of a tortoise, was given by Dr. Buckland.

Lately Sir William Jardine has described these tracks in a splendid folio. He has given nine species: five of them he refers to Chelonians, two to Saurians, one to Batrachians, and one is left doubtful.

Having now reached the top of the Palæozoic deposits, it may be well to state certain leading facts as to the organic remains common to the whole.

1. These deposits are characterized by the entire absence, so

Fig. 259.

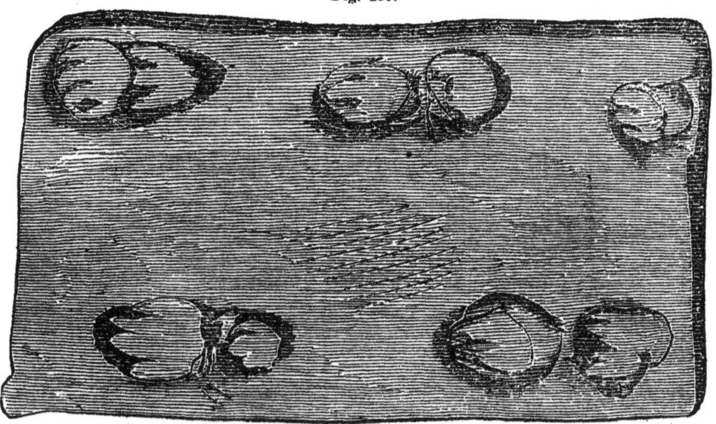

far as has yet been discovered, of birds and mammiferous animals, and by the rarity of all other vertebral animals except fishes.

2. Also among the molluscs, by the existence of numerous cephalopods with simple divisions between the chambers—such as are found no more among the newer rocks; also by brachiopods, such as are found rarely afterwards.

Fig. 260.

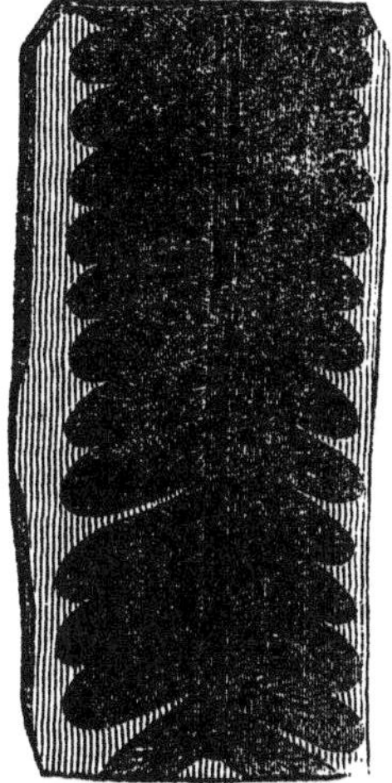

3. By the existence of numerous trilobites, of which there is no trace at a later period.

7. TRIASSIC PERIOD.

The fossils of the trias are less numerous than those of several formations below and above; but some of them are peculiar, and possess unusual interest.

Plants.—The general aspect of the scanty Triassic Flora differs much from that of the Permian and Carboniferous group, but some plants are very similar, as the Neuropteris elegans, shown in Fig. 260.

A more characteristic plant is the Voltzia heterophylla figured below Fig. 261.

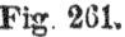

Fig. 261.

Voltzia heterophylla.

Animals.—Of the Brachiopods we give one fine form, the terebratula vulgaris, in Fig. 262.

Fig. 262.

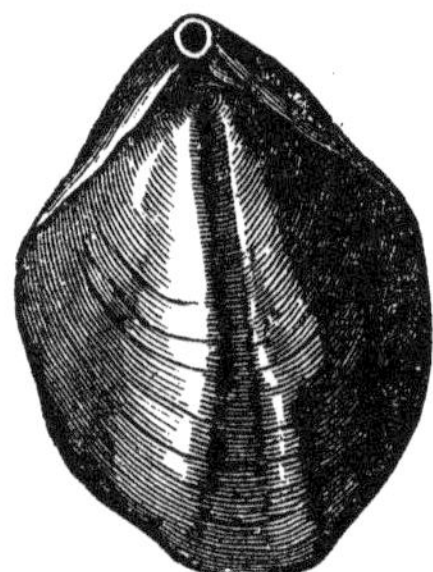

Among the Crinoids we have in the Muschelkalk of the trias, the well known and beautiful Lily Encrinite (Encrinus liliiformis) which is shown in Fig. 263, with a cross section of the stem beneath.

The fishes of the trias are all *homocerques.* The genera known are twenty-three. We give only the sketch of the upper jaw of one, the Placodus Andryani, Fig. 264. It will be seen that the almost entire roof of the mouth is covered with large flat teeth. Sometimes it is entirely covered.

This animal is regarded, and probably with good reason, as a reptile, by Prof. Owen. Prof. Pictet places it among the fishes.

In the palæozoic rocks the reptiles have been few. But in the trias they begin to show themselves in large numbers and of peculiar characters; a fit commencement of their enormous development in the next higher formation.

Fig. 263.

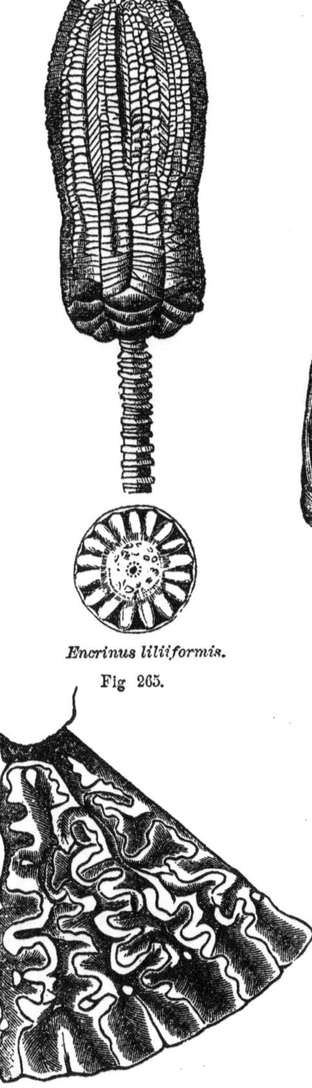

Encrinus liliiformis.

Fig. 265.

Tooth of the Labyrinthodon.

Fig. 264.

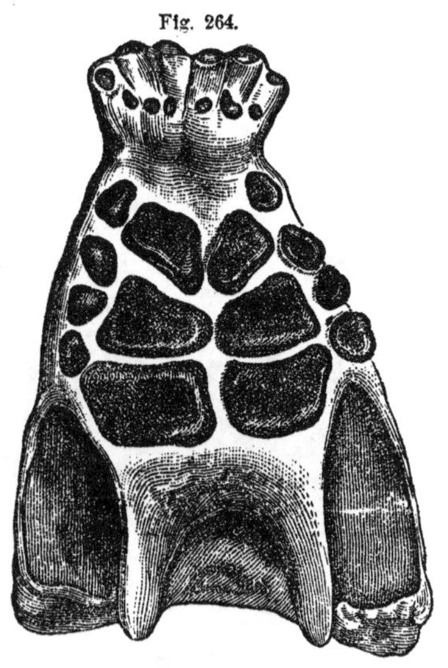

Placodus Andryani

The Labyrinthodonts are perhaps the most interesting. They are so named from the labyrinthine character of their teeth, when viewed upon a cross section, as in Fig. 265, which shows a portion of the tooth only when cut across and polished. Professor Owen describes them as reptiles having the essential bony characters of the *Batrachia*, but combining these with other bony characters of crocodiles, lizards and ganoid fishes, and exhibiting all under a bulk which rivaled that of the largest crocodiles of the present day. The form of the largest Labyrinthodonts, if we may judge by the great breadth and flatness of the skull, must have more resembled that of the toad or the land salamander.

Prof. Owen describes five British species of the Labyrinthodon, one of which is identical with the Mastodonsauus found in Germany. Fig. 266 shows the skull of this species, some of which have been found from thirty to forty-eight inches long.

Fig. 266.

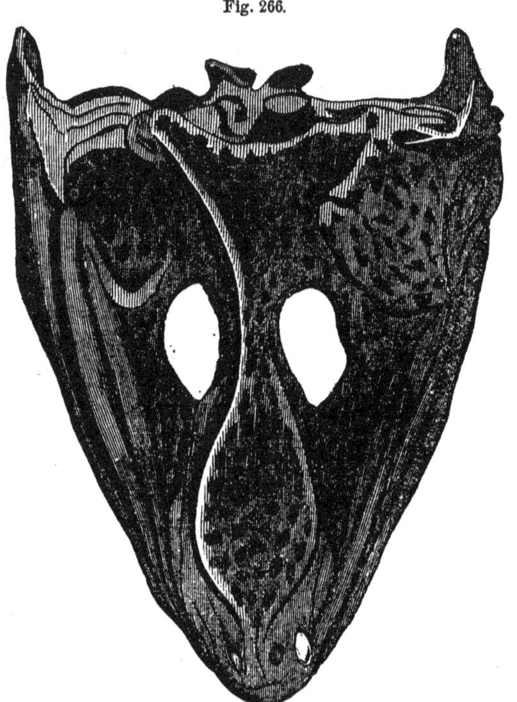

Head of the Labyrinthodon.

Prof. E. Emmons has lately found in the sandstone of North Carolina, a Labyrinthodont named by Leidy, *Dictyocephalus elegans*. He has also found in the same rock three genera, and four or five species of Thecodont reptiles. Dr. Leidy has described likewise a Thecodont Saurian, the Bathyguathus borealis, from the sandstone of Prince Edward's Island.

Another remarkable family of reptiles, represented by the Rhynchosaurus articeps, has been described by Professor Owen, from the trias of England. The tracks found in connection with the bones seem to indicate that the animal had feet resembling those of birds: three toes pointing forward, and sometimes one pointing backwards. Owen says that the "formation of the skull has brought to light modifications of the lacertine structure leading towards Chelonia and Birds which before were unknown."

Two well-marked examples of mamiferous animals have at length been found as low as the upper part of the trias, or certainly not higher than the lower part of the lias. One is the Microlestes, a small insectivorous quadruped, found both in Germany by Professor Plieninger, and in England by Charles Moore, though determined by Professor Owen. Among living mammals the small Myrmecobius, an insectivorous marsupial, comes nearest to the Microlestes. The other genus is the Dromatherium, sylvestve discovered, and named by Professor E. Emmons in the North

Fig. 267.

Tracks of Cheirotherium Barthii.

Carolina sandstones, with the reptilian remains already described. This, also, comes nearest to the Myrmecobius among living animals. Prof. Emmons is inclined to place it in the lower part of the trias, or even in the Permian, and it is probably the oldest known mammal.

Lithichnozoa.—Early in the history of footmarks some were found in the new red sandstone of Hildburghausen in Saxony, having such a resemblance to the human hand that Professor Kaup gave to the animal that made them the name of *Cheirotherium* or *hand animal.* The fore and hind feet were quite unequal, as shown below in Fig. 267, which is Cheirotherium Barthii.

Similar tracks were subsequently found in Cheshire, England; also those of the three-toed reptile, Rhynchosaurus. Crustacean tracks were likewise found in Cheshire; also some resembling a horse-shoe by Dr. Cotta in Saxony, which may have been made by Chalonians. Prof. Owen suggests that the Cheirotherian tracks may have been made by the Labyrinthodon above described. But such an animal would leave two rows of tracks, whereas those of the Cheirotherium form only a single row, as in the above figure, and, it would seem, must have been made by an animal with narrow body and long legs like some marsupials, and not by such an animal as that in Fig. 268, which is the Labyrinthodon as restored by Prof. Owen.

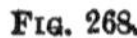

FIG. 268.

Labyrinthodon pachygnathus.

8. JURASSIC OR OOLITIC PERIOD, EMBRACING THE WEALDEN AND THE LIAS.

This formation is very prolific of fossils. Among so many that are interesting we find it difficult to make a selection.

Plants.—The vegetation of this period was not remarkable as to quantity; but it was characterized by the predominance of Coniferæ, or the Pine tribe, and of Cycadaceæ, both of which are Gymnosperms, or with naked seeds. While only two genera and twenty species of the latter are found among living plants, thirty-four species occur in the Oolite and four in the chalk of Great Britain, where no living species is found. Fig. 269 will give an idea of a living species, the Cycas revoluta.

Fig. 269

Cycas revoluta.

Fig. 270.

Cycadoidea megaphylla.

In Fig. 270 is given a representation of a fossil Cycad, the Cycadoidea megaphylla.

Large petrified trunks of Coniferæ occur in several formations. In the isle of Portland, on the coast of England, is a remarkable subterranean forest of these trees, or rather their stumps, standing perpendicular to the strata and rooted in a black vegetable mould, the whole now converted into stone. It is represented in Fig. 271.

Among the ferns of this period is a remarkable one, of which

Fig. 271.

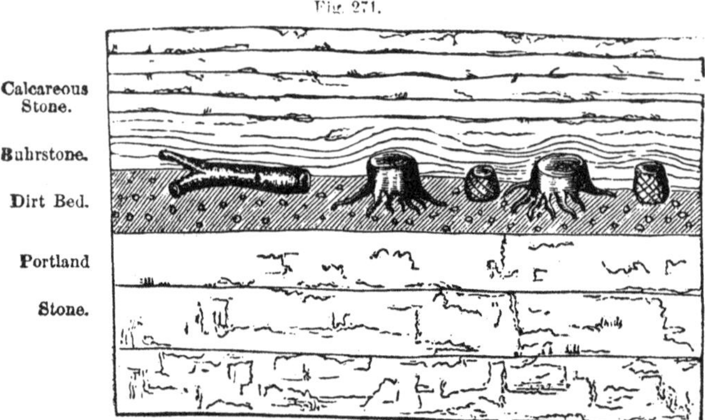

Subterranean Forest, Isle of Portland.

Fig. 272.

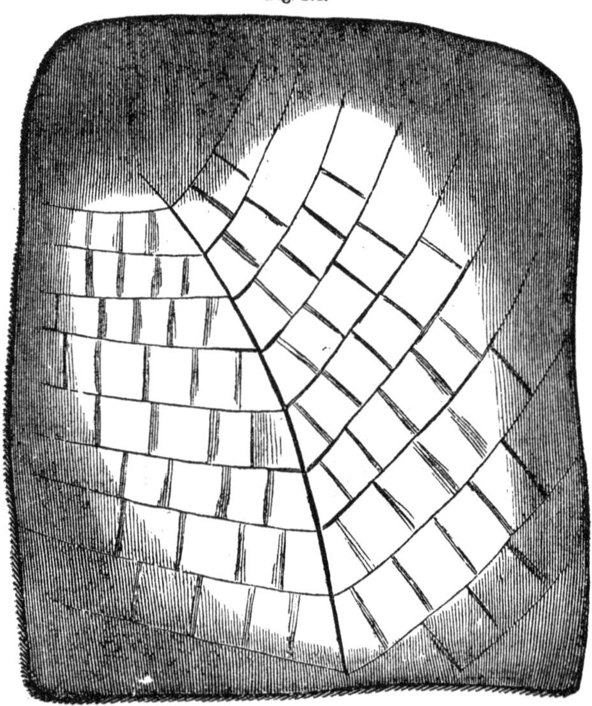

Clathropteris rectiusculus, East Hampton.

there seems to be only one species, and that found both in the lower part of the lias and upper part of the trias. It had quite large reticulated fronds radiating from a center, like some tropical ferns of the present day; an example of which may be seen in Fig. 229. In Fig. 272 we present a small portion of a frond of this Clathropteris found in East Hampton, Massachusetts, by Edward Hitchcock, Jr.

Fig. 273 shows a fern, the Coniopteris Murrayana, with a part of the frond magnified, showing fruit—a very unusual occurrence. This is an oolitic plant.

Fig. 273.

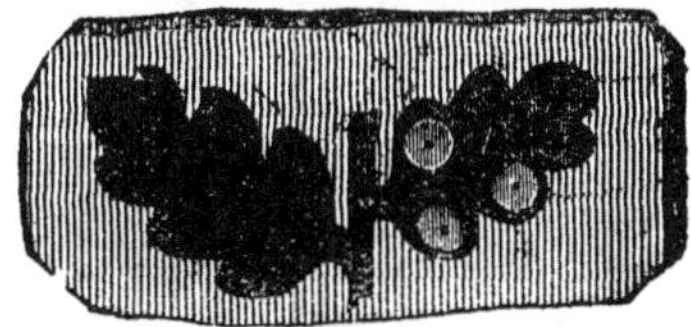

Coniopteris Murrayana.

Animals.—Of corals we present only one. Fig. 274 shows the Prionaster oblonga.

The bivalves and univalve shells are very abundant in this formation. We pass by all except the Cephalopods which have an immense development in the Oolite. These have already been

Fig. 274.

Pr.onastr Oblonga.

Fig. 275.

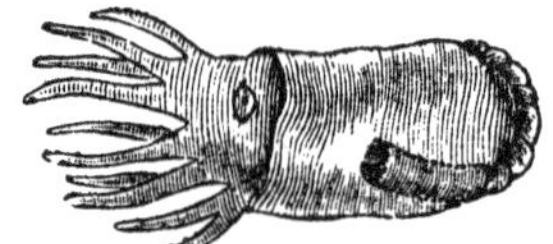

Spirula Peronii.

described in part, but some of the families need farther elucidation. Some of the Cephalopods have the shell within the body, as is shown in Fig. 275, which represents a living species, the Spirula Peronii.

The Orthoceratite, Lituite, Baculite, Hamites, Scaphite, Turrilite and Belemnite seem to have belonged to this description of shells. Fig. 276 shows the Hamites attenuatus from the Gault, which lies a little above the oolite in the cretaceous system, but is introduced here for the sake of illustration.

Fig. 276.

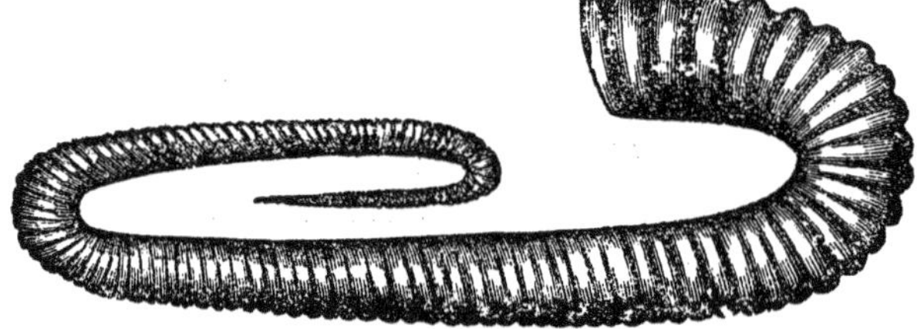

Hamites attenuatus.

The ordinary appearance of a Belemnite is that of a conical arrow head, as shown in Fig. 277. At the blunt end it is usually hollow, and if one half be split off, the section, as seen in the figure, will show a conical cavity.

Fig. 277.

This was the main part of the internal shell; but its structure was more complicated. Besides this cone-shaped shell, there was a conical, thin, horny sheath, extending outwards and enlarging. This part contained an ink-bag like the cuttle fish of the present day, which produces the Sepia, or India Ink. There was also a thin, conical internal chambered shell, placed within the hollow cone above described, having a construction analogous to that of the Nautilus and Orthocera. Fig. 278 is an imaginary restoration of the Belemnosepia, a family of belemnites proposed by Buckland and Agassiz.

Fig. 273.

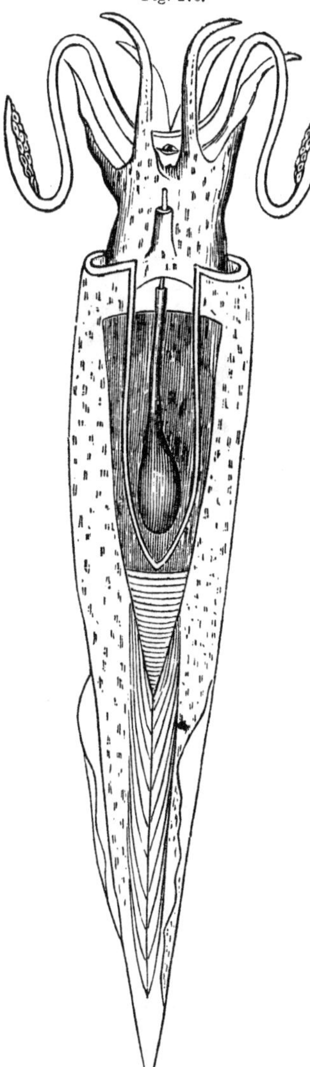

Belemnosepia.

The geological cabinets of three ladies in England (Misses Anning, Phillpots and Baker) are mentioned by Dr. Buckland as furnishing the specimens of belemnites containing petrified ink bags; the ink of which was pronounced by the best artists to be of superior quality. Thus were these ancient cephalopods identified with the modern cuttle-fishes.

Belemnites are mostly confined to the oolite and the chalk, where at least 100 species have been obtained. Species of Sepia occur in the tertiary, as well as in the present seas.

Of the Ammonites more than 500 species have been described. Bronn enumerates 2 in the Upper Silurian, 22 in the Trias, 317 in the Oolite, and 211 in the Chalk. Many of these, as well as the other chambered shells, are beautifully figured and embossed. Some of them are three feet in diameter, looking like a carriage wheel.

Fig. 279.

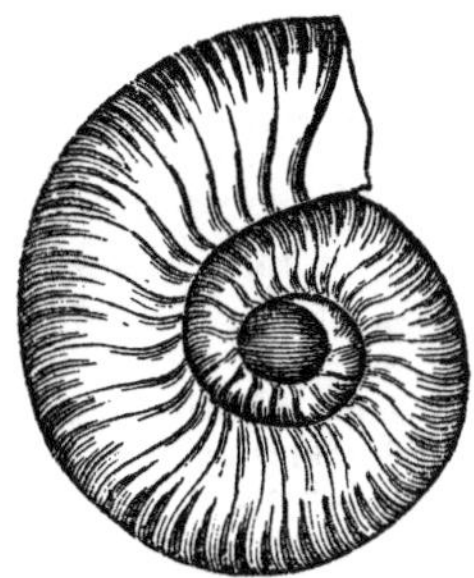

Ammonites planulatus.

Fig. 279 shows the Ammonites planulatus. Fig. 280, the A. Humphresianus. Fig. 281, A. nodotianus. Fig. 282, A. catena.

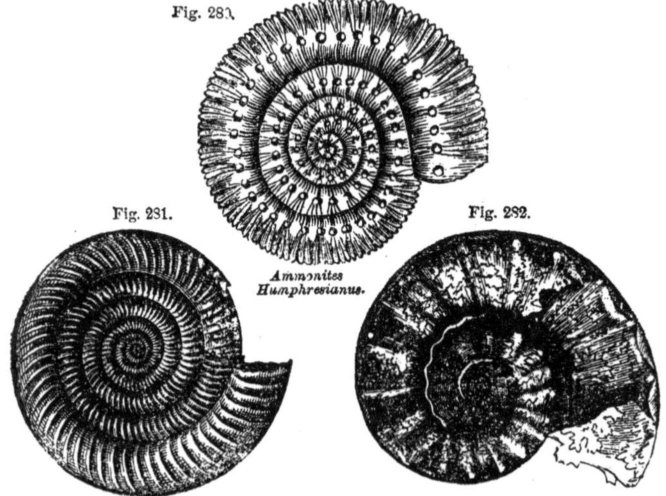

Fig. 280.

Ammonites Humphresianus.

Fig. 281.

Ammonites nodotianus.

Fig. 282.

Ammonites catena.

Fig. 283.

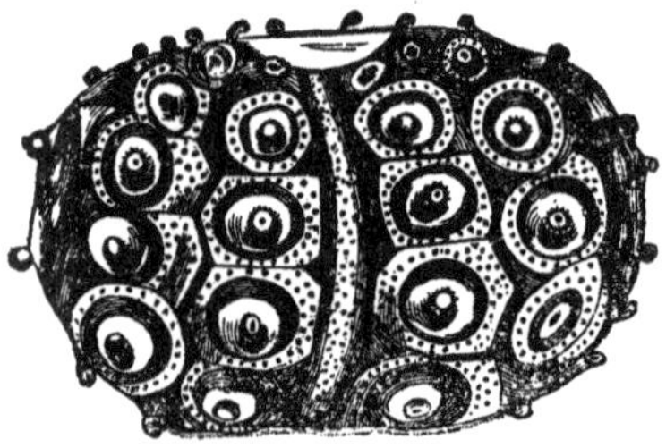

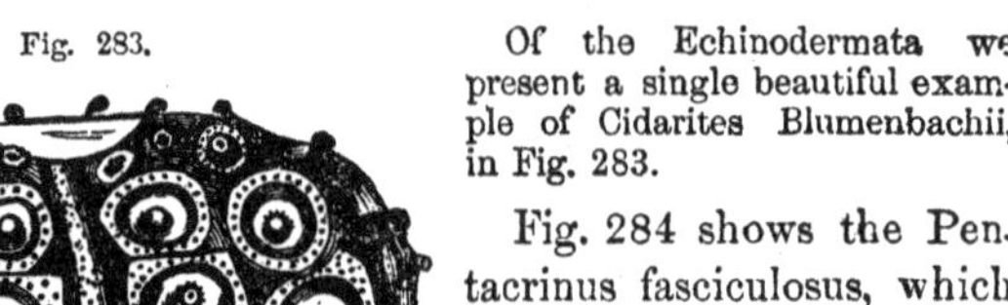

Cidarites Blumenbachii.

Of the Echinodermata we present a single beautiful example of Cidarites Blumenbachii, in Fig. 283.

Fig. 284 shows the Pentacrinus fasciculosus, which is called the Briarean Pentacrinite on account of the great number of its arms—the fabled Briareus being supposed to have a hundred hands. The bones and the fingers in this pentacrinite were 100,000, and those of its side arms 50,000 more. If there were two muscles, as in the higher animals, to each bone, it would require 300,000 in this echinoderm.

Fig. 285 shows another elegant species, the Apiocrinus Rossyanus.

In Fig. 286, we present a single example of an Oolite Annelid, the Serpula flagellum, which is a calcareous tube once occupied by a worm.

Insects and Myriapods.—In 1849, Bronn enumerated seventeen species of fossil Myriapods, such as the centipede, two of

Fig. 284.

Fig. 285.

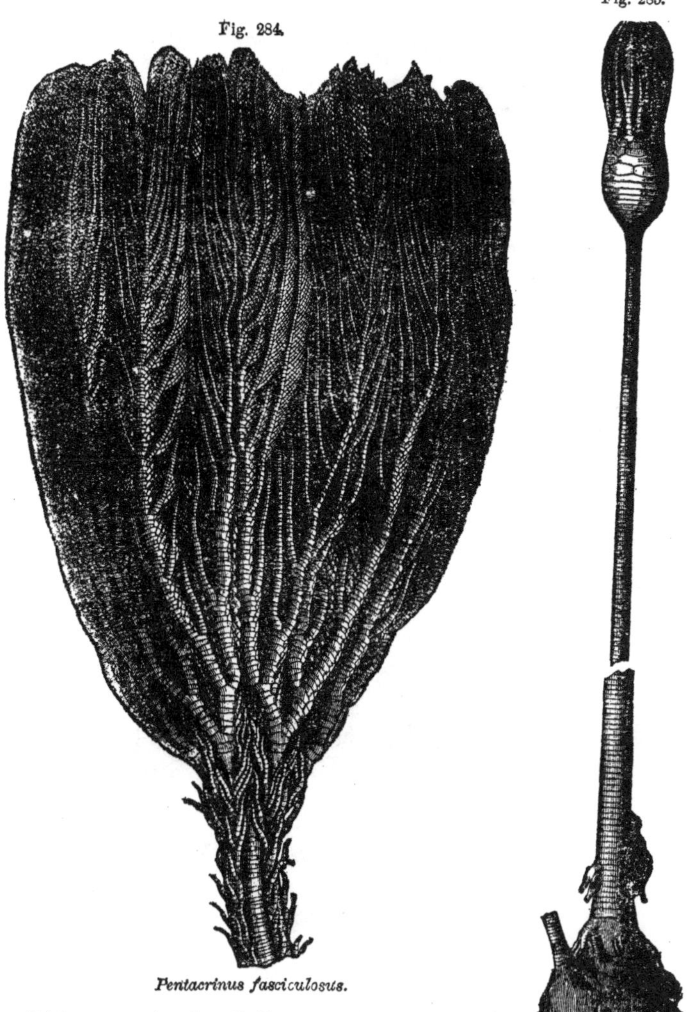

Pentacrinus fasciculosus.

Apiocrinus Rossyanus

which occur in the Oolite, and the rest mostly in the tertiary; also 131 species of Arachnoidea, or scorpions and spiders, of which two species are found as low as the coal measures, one in the Oolite, and the rest mostly in the tertiary; also 1551 species of hexapod insects,

Fig. 286.

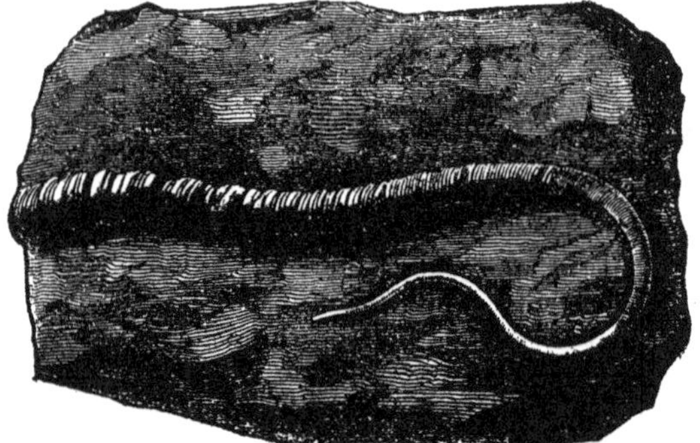

Serpula flagellum.

of which 847 are coleoptera. Nine species of these insects are found in the carboniferous formation, 120 in the Oolite, two in the chalk, and the rest mostly in the tertiary. Fig. 287, shows a species of Libellula from the Oolite.

Fig. 287.

Libellula.

In the sandstone of the Connecticut river, in Massachusetts, a fossil occurs, which at first was thought to be a Myriapod, but it seems rather to be the larva of an insect. A view of it greatly enlarged is given in Fig. 288.

Fig 288

Fishes.—The genera of fishes already described in the Oolitic series, is 66; larger than in any formation below, except the carboniferous. They are homocercal. We give a sketch in Fig. 289. of the Dapedius punctatus from the lias, and in Fig. 290, a restored Tetragonolepis from the same formation.

Fig. 289.

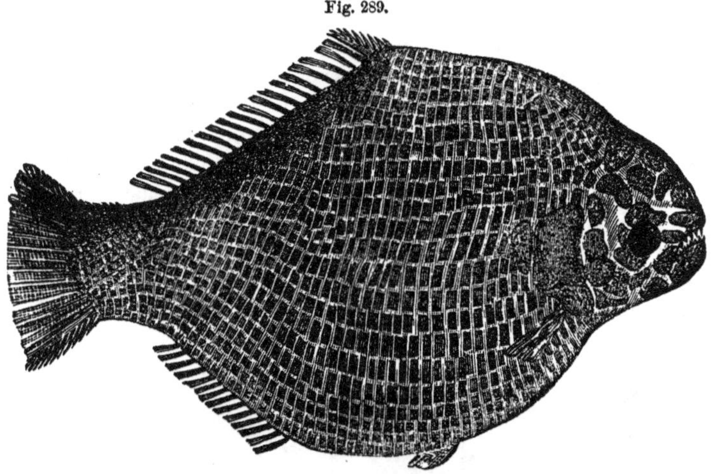

Dapedius punctatus.

Fig. 290.

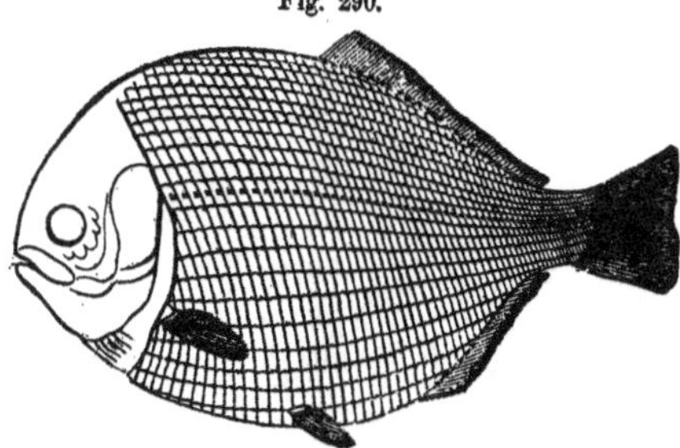

Tetragonolepis.

Reptiles.—This was the age of reptiles, remarkable both for their peculiar forms and formidable dimensions. We shall try to give some idea of a few of the most important.

Ichthyosaurus.—This animal, sometimes more than thirty feet long, and of which thirty species are known, had the snout of a porpoise, the teeth of a crocodile (sometimes amounting to 180), the head of a lizard, the vertebræ of a fish, the sternum of an ornithorhynchus, and the paddles of a whale: uniting in itself a combination of mechanical contrivances which are now found among three distinct classes of the animal kingdom. One of its paddles was sometimes composed of more than 100 bones; which gave it great elasticity and power, and enabled the animal to urge its way through the water with a rapid motion. Its vertebræ were more than 100. Its eye was enormously large; in one species, the orbital cavity being fourteen inches in its longest direction. This eye also, had a peculiar construction to make it operate both like a telescope and a microscope: thus enabling the animal to descry its prey in the night as well as day, and at great depths in the water. The length of the jaws was sometimes more than six feet. Its skin was naked, some of it having been found fossil; its habits were carnivorous, its food, fishes and the young of its own species; some of which it must have swallowed several feet in length. This fish-like lizard was an inhabitant of the ocean. Fig. 291 exhibits a restored ichthyosaurus.

Fig. 291.

Ichthyosaurus communis.

The head of the ichthyosaurus, with its enormous eye, is shown on Fig. 292.

Fig. 292.

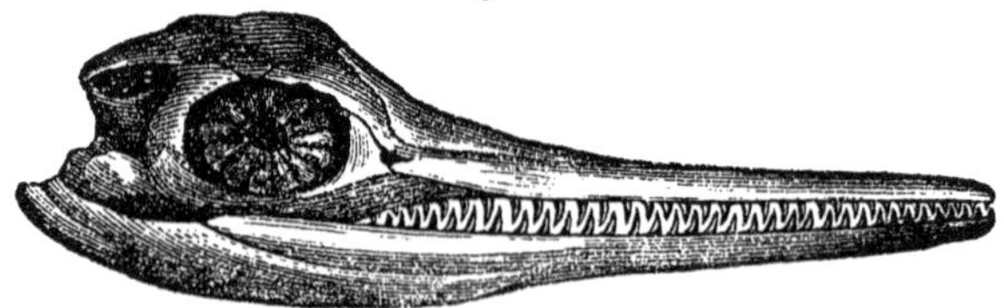

Head of Ichthyosaurus.

Plesiosaurus.—This animal, of which twenty species have been found, has the general structure of the ichthyosaurus. Its most remarkable difference is the great length of the neck, which has from twenty to forty vertebræ; a larger number than in any known animal; those of living reptiles varying from three to six, and those of birds from nine to twenty-three.

The largest perfect specimen yet found is eleven feet long, with about ninety vertebræ. Its paddles were proportionally larger than in the ichthyosauri. It was carnivorous; an inhabitant of the ocean, or rather of bays and estuaries, where it probably used its long neck for seizing fish beneath, and perhaps flying reptiles above the waters. Fig. 293 exhibits a restoration of one of the most remarkable species, the *P. dolichodeirus.*

Fig. 293.

Plesiosaurus dolichodeirus.

In Fig. 294 we give a sketch of one of the most perfect skeletons of Plesiosaurus macrocephalus as it lay in the rock.

The preceding were carnivorous reptiles that lived in the sea. But during the same period the land was tenanted by others, called Dinosaurians, of still more gigantic size, whose teeth indicate that they were mostly vegetable feeders, or possibly sometimes living on a mixed diet. We give a few examples.

Megalosaurus.—This name (meaning a *great saurian*) has been given by Dr. Buckland to a gigantic terrestrial reptile, thirty feet long, allied to the crocodile and monitor in structure, and found in the oolite. The animal was carnivorous; and in the structure of its teeth are combined the knife, the saw, and the sabre. Its principal food was probably crocodiles and tortoises. It had a Dinosaurian companion, called the Hylæosaurus, about twenty-five feet long.

Fig 294.

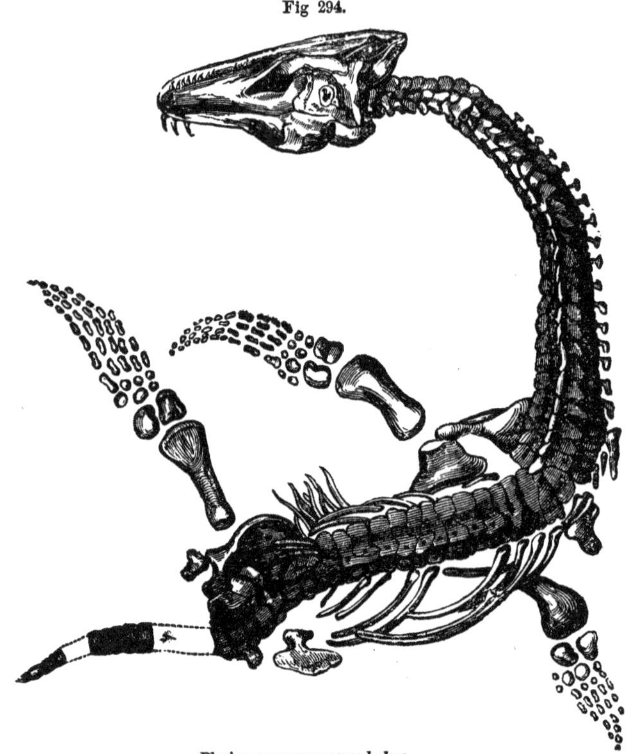

Plesiosaurus macrocephalus.

Iguanodon.—This animal approaches nearest in its structure, especially that of the teeth, to the living iguana; a reptile of the warmer parts of this continent; and hence its name; signifying an animal with teeth like the iguana. Its average length was about thirty feet; circumference of the body, 14.5 feet; length of the hind foot, 6.5 feet; circumference of the thigh, more than seven feet! The form of the teeth shows it to have been herbivorous, like the living iguana. It had a horn four inches long upon the snout, like some species of iguana. Fig. 295 will give some idea of the iguanodon.

The Pterosaurians, or flying reptiles come next. They are divided into several genera but a description of Pterodactylus crassirostris will give a good idea of the whole, which are probably the most heteroclitic of all fossil reptiles. Fig. 296 shows a perfect

Fig. 295.

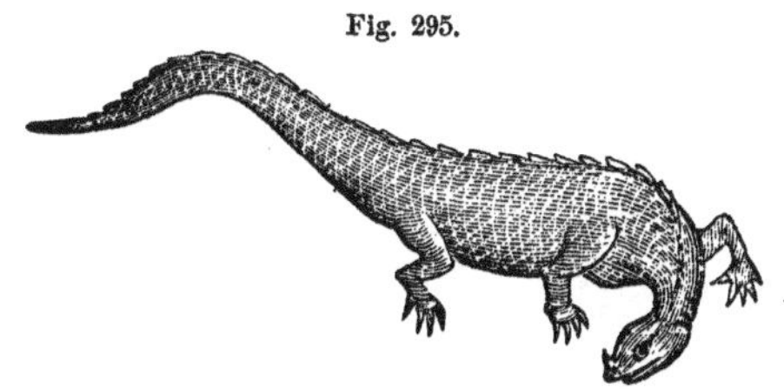

Iguanodon.

skeleton, with an enormous extension of the fifth or innermost digit or finger of the pectoral limbs. This could be only for the attachment of a membrane for flying, as in the bat.

Fig. 296.

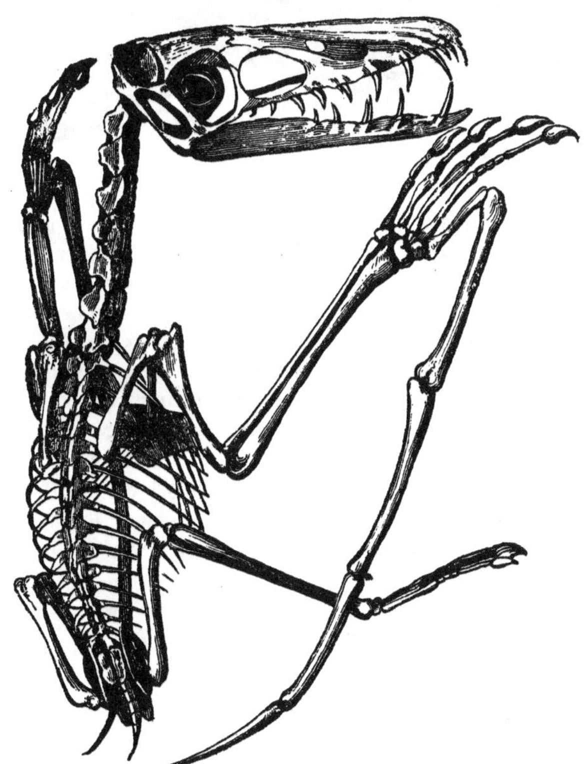

Pterodactylus crassirostris.

In Fig. 297, the membrane as it is snpposed to have existed, is attached to the elongated finger. Some of them were so large that the distance from tip to tip of their wings, when spread, was eighteen or twenty feet.

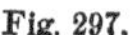

Fig. 297.

Pterodactylus crassirostris.

This animal had the head and neck of a bird, the mouth of a reptile, the wings of a bat, and the body and tail of a mammifer. Its teeth, as well as other parts of its structure, show that it could not have been a bird; and its osteological characters separate it from the tribe of bats. But in many respects it had the characters of a reptile. These animals were doubtless able to fly like the bat, while the fingers with claws projecting from their wings enabled them to creep or climb. When their wings were folded, they could, perhaps, walk on two feet; and it is most likely, also, they could swim. Their eyes were enormously large; so that they could seek their prey in the night. They probably fed on insects chiefly; though perhaps, also, they had the power of diving for fish.

"Thus," says Dr. Buckland, "like Milton's fiend, all qualified for all services, and all elements, the pterodactyle was a fit companion for the kindred reptiles that swarmed in the seas, or crawled on the shores of a turbulent planet."

"The Fiend,
O'er bog, or steep, through straight, rough, dense, or rare,
With head, hands, wings, or feet, pursues his way,
And swims, or sinks, or wades, or creeps, or flies."
Paradise Lost, Book 2, line 947.

"With flocks of such-like creatures flying in the air, and shoals of no less monstrous ichthyosauri and plesiosauri swarming in the ocean, and gigantic crocodiles and tortoises crawling on the shores of the primeval lakes and rivers; air, sea, and land must have been strangely tenanted in these early periods of our infant world." *Bridgewater Treatise*, vol. i. p. 224.

Fossil crocodiles are quite numerous from the lias to the chalk inclusive. The earlier species had great peculiarities. Many of them had long and slender jaws like the gavial of the Ganges. Of the twelve living species, three are alligators, eight true crocodiles, and one gavial. Fig. 298, shows the head of the Mystriosaurus Tiedmanni, from the lias; a good example of long and slender jaws.

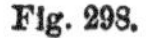

Fig. 298.

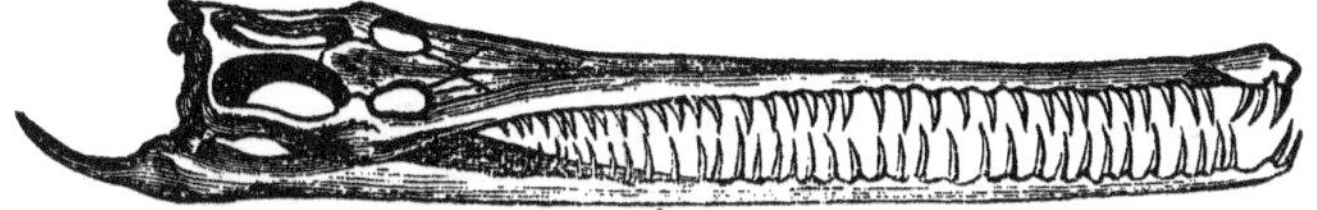

Mystriosaurus Tiedmanni.

Excepting their tracks, as already detailed, we have no evidence of the existence of Chelonians or tortoises, below the lias. In general these are not larger than those now living; but in some of the higher deposits they are found from eight to twenty feet in diameter.

Birds.—The only evidence of the existence of birds as early as the lias, or even perhaps the later periods of the trias, depends upon their tracks in New England. For no trace of their skeletons has been found, nor of any thing connected with them, save a few coprolites. But so clearly do some of the tracks correspond to the feet of birds in their form, the number of their toes, and especially in the number of phalanges, that Professor R. Owen, the most eminent of European comparative anatomists, seems fully satisfied that they were formed by this class of animals. See his arguments on the subject in his admirable work on Palæontology published in 1859. The case is argued also in the Ichnology of New England, where full details of the facts are given. In what follows in this work upon the oolitic Lithichnozoa the facts are also briefly stated.

Mammalia.—These are warm-blooded, air-breathing, viviparous, vertebrate animals. The lowest group on the scale of organization and character are the marsupials, like the kangaroo and opossum; and these, as we might expect and as we have seen, began to appear towards the close of the triassic period. Not less than seven genera have been found in the oolitic series, viz., the Amphitherium, Amphilestes, Phascolotherium, Stereognathus, Spalacotherium, Triconodon and Plagiaulax. They were mostly small animals, and some of them were insect-eaters. They have been found chiefly in England.

Lithichnozoa.—The most remarkable locality on the globe for fossil footmarks, so far as yet known, is the Connecticut valley in Massachusetts and Connecticut. Till of late the rock has been regarded as new red sandstone, and that formation is perhaps present in the series; but the belt that contains the footmarks seems more probably to be the equivalent of the lower part of the oolite, say liassic, or possibly it is the upper part of the trias. In Hitchcock's Report on the Ichnology of New England, published by the government of Massachusetts, the tracks of 119 species of Lithichnozoa are figured and the species described, all of whose tracks are preserved in the Ichnological Cabinet of Amherst College. These tracks vary in size from those twenty inches to those one twentieth of an inch long, and it would require nearly half a million of the latter to cover as much space as one track of the former. The animals are divided in that Report into the following groups, with more or less probability. These we propose, as the subject is one of novelty and interest, to illustrate by several drawings.

Group 1. *Marsupialoids.*—One of these, the Anomœpus major, is shown on Fig. 299. Fig. 300 shows the Anisopus gracilis. Fig. 301 shows a row of the tracks of Anisopus Deweyanus. There are five species of this group.

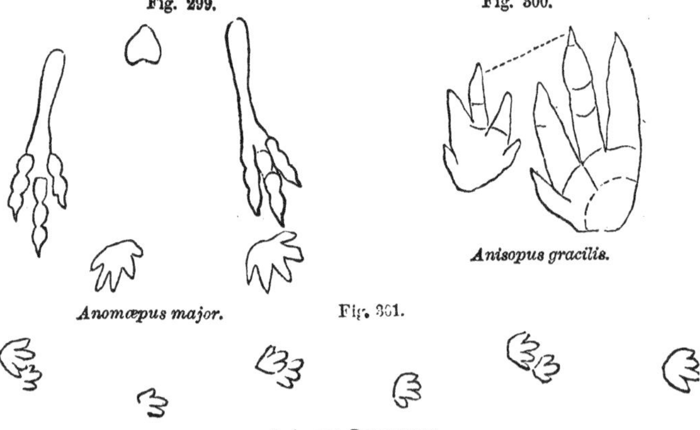

Fig. 299.

Anomœpus major.

Fig. 300.

Anisopus gracilis.

Fig. 301.

Anisopus Deweyanus.

Group 2. Pachydactylous, or Thick-toed Birds.—This is the most distinct and important group. The toes often show the impressions of the phalanges or joints most distinctly, and their number corresponds exactly with those of birds. This is shown on Fig. 302, and also on Fig. 303, which shows a part of a very perfect specimen in the cabinet, some five feet long. The track of the largest species, Fig. 302, is eighteen inches long—fourteen species in the group.

Fig. 302.

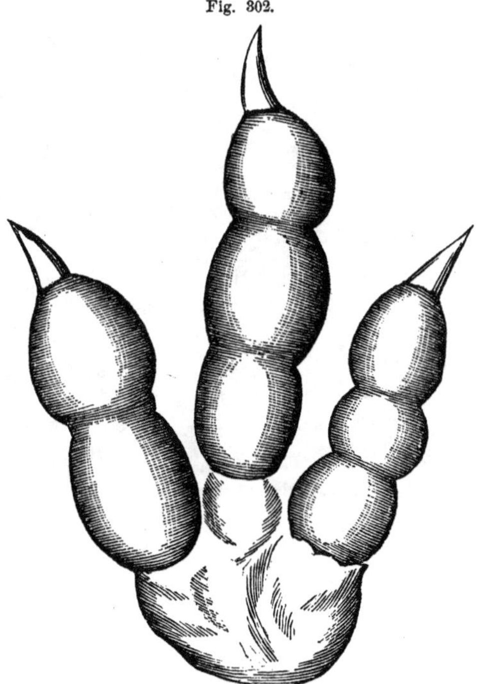

Brontozoum giganteum.

Group 3.—*Leptodactylous or Narrow-toed Birds.*—Of the 17 species of these, Fig. 304 will give an example; and Fig. 305 shows some rows of what seems to have been a biped, yet it is placed among the Ornithoid Batrachians for reasons that can not be here given. It is Apatichnus circumagens.

Group 4.—*Ornithoid Lizards or Batrachians.*—That is, animals which, though upon the whole, we must regard as lizards and batrachians, still have some characters that ally them to birds.

Fig. 303.

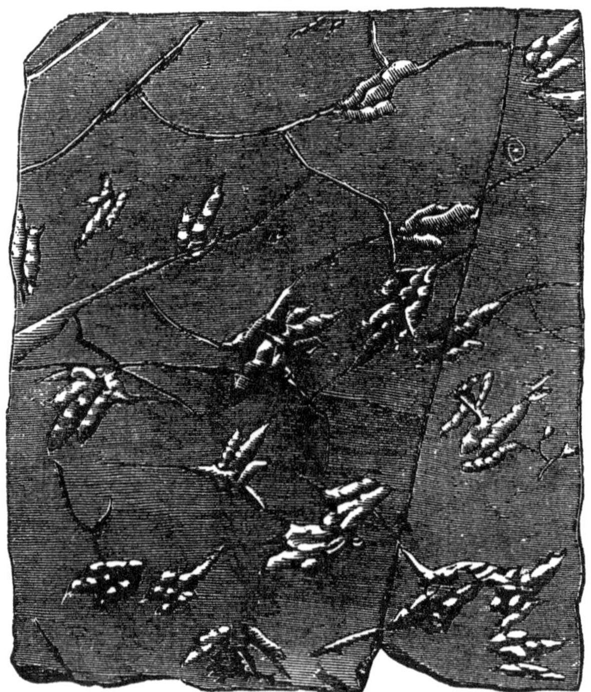

Brontozoum Sillimanium.

Fig. 304.

Ornithopus gracilior.

Fig. 305.

Apatichnus circumagens.

The most remarkable of the twelve species is shown on Fig. 306, the Gigantitherium, it being so far as yet discovered, an enormous two-legged animal, yet dragging a tail. Its foot is seventeen inches long.

Group 5.—*Lizards.* Of the 17 species of this group, Fig. 307 shows the hind foot, fifteen inches loug, of the largest species, the Polemarchus gigas.

Fig. 307.

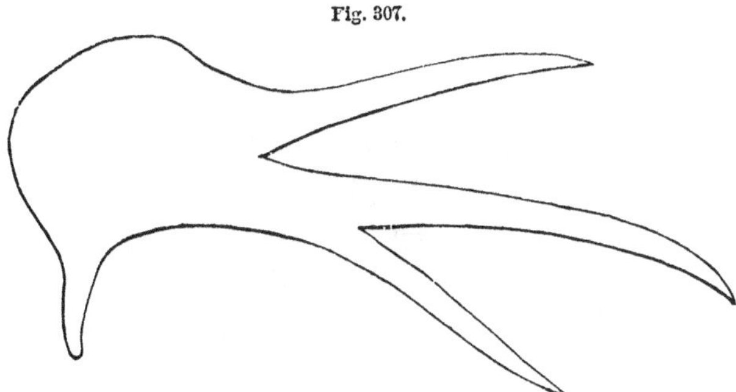

Polemarchus gigas.

Fig. 306.

Gigantitherium caudatum.

14

Fig. 308 is the track of a small lizard, the Orthodactylus floriferus.

Fig. 308.

Orthodactylus floriferus.

Group 6.—*Batrachians.*—Of the sixteen species of this group, the Otozoum Moodii is the most remarkable—the track is twenty

Fig. 309.

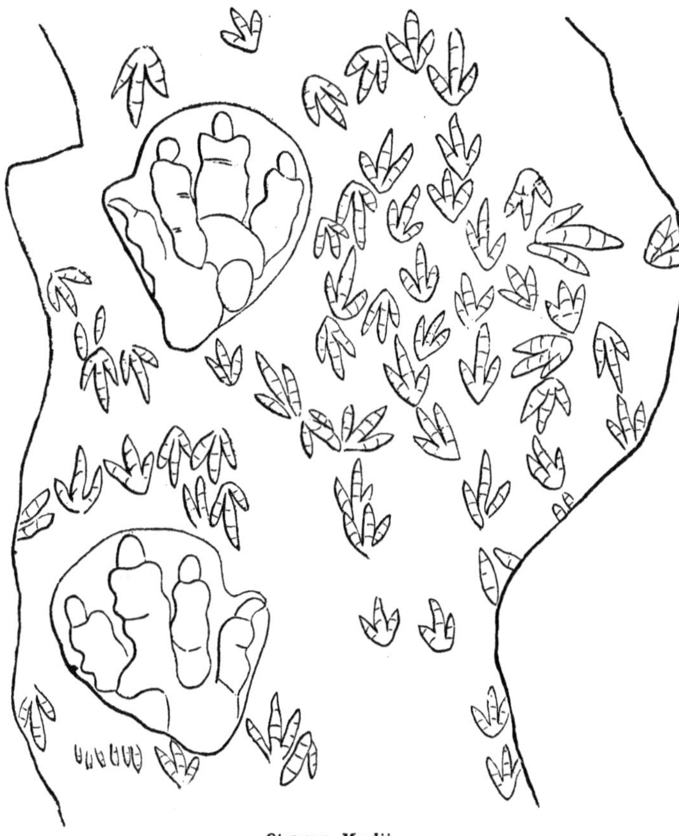

Otozoum Moodii.

inches long, and covers more than a square foot of surface. The drawing Fig. 309, exhibits two hind tracks of this four-footed web-footed animal, with numerous smaller tri-digitate tracks on the same slab.

Fig. 310.

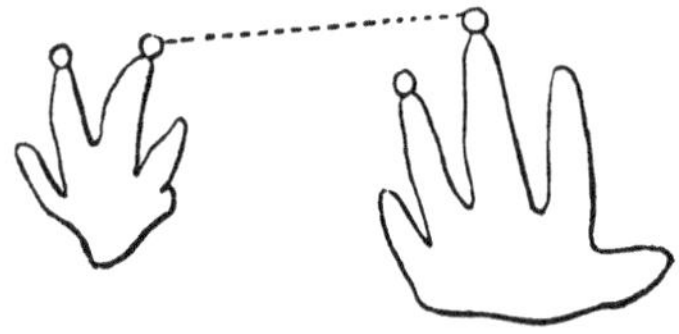

Fig. 310 shows tracks of Cheirotheroides pilulatus, with pellets on only a part of the toes, like some living frogs.

Group 7.—*Chelonians or Tortoises.*—The fore and hind feet of one small species out of the eight in this group, are given on Fig. 311. It is the Ancyropus heteroclitus.

Fig. 311.

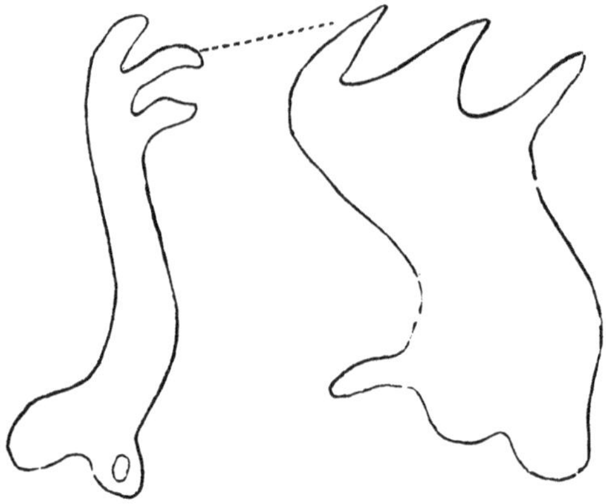

Ancyropus heteroclitus.

Group 8.—*Fishes.*—The tracks of this class are so peculiar that we omit a figure. And yet it is an undoubted fact that fishes do sometimes come out of the water, and walk, or rather hobble, over the land. Four species are given in the Report.

Group 9.—*Crustaceans, Myriapods, and Insects.*—Perhaps it is not possible to distinguish between these classes in many cases by their tracks, and, therefore, they are grouped together. The following sketches are copied from slabs in the cabinet, and are of the natural size. Those with six legs were most likely insects; the the others perhaps crustaceans, or myriapods. Fig. 312 shows Hamipes didactylus. Fig. 313 two trackways of Bifurculapes laqueatus, and one of Hexapodichnus horrens. Fig. 314 was made by Copeza triremis (*the three oared oar-foot.*) Fig. 315 by Acanthichnus cursorius, and Fig. 316 by Lithographus cruscularis

Fig. 312.

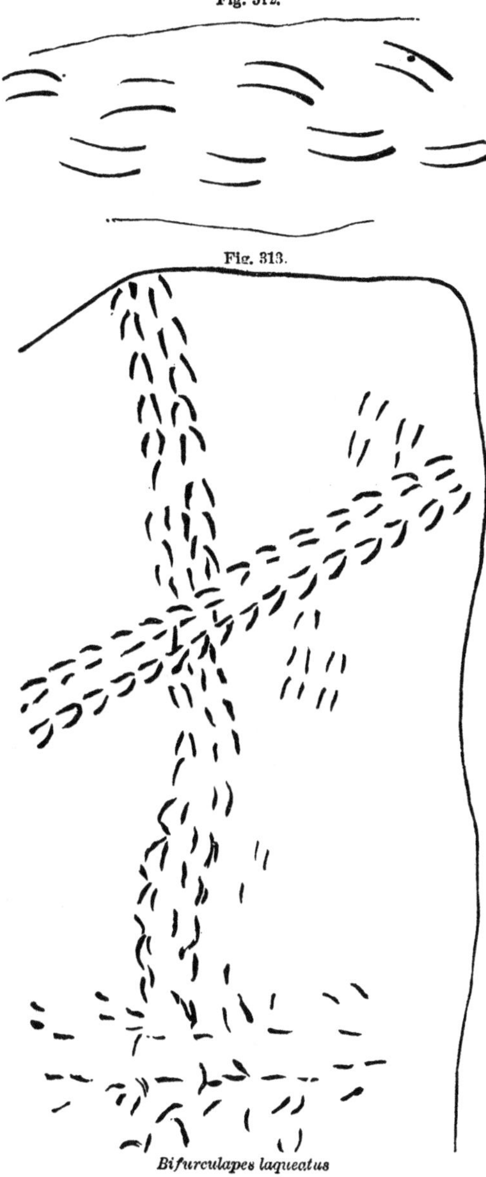

Fig. 313.

Bifurculapes laqueatus

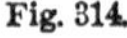

Fig. 314.

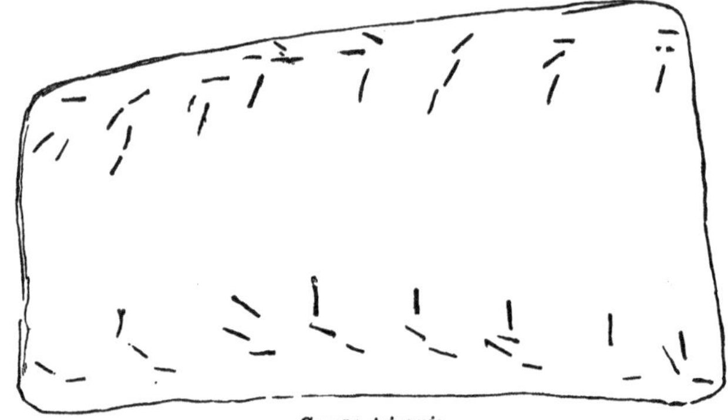

Copeza triremis.

and Bifurculapes elachistotatus; the latter the smallest of all tracks yet discovered, requiring half a million of them to fill a space equal to the foot of Otozoum Moodii.

Fig. 315.

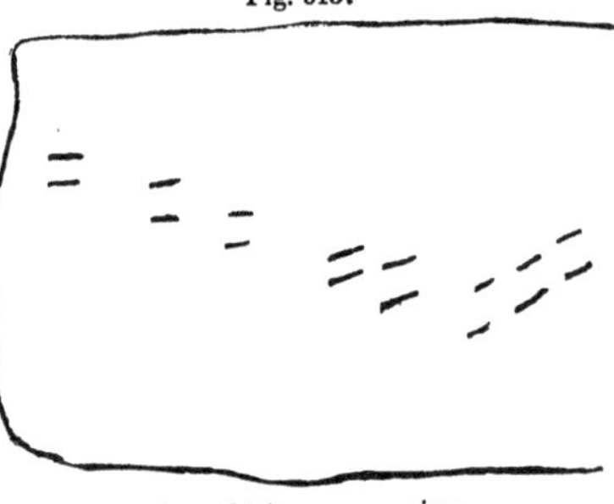

Acanthichnus cursorius.

We give two species of Annelid tracks. Fig. 317 shows a single furrow like that made by the earth-worm in summer after a shower, and is called

Fig. 316.

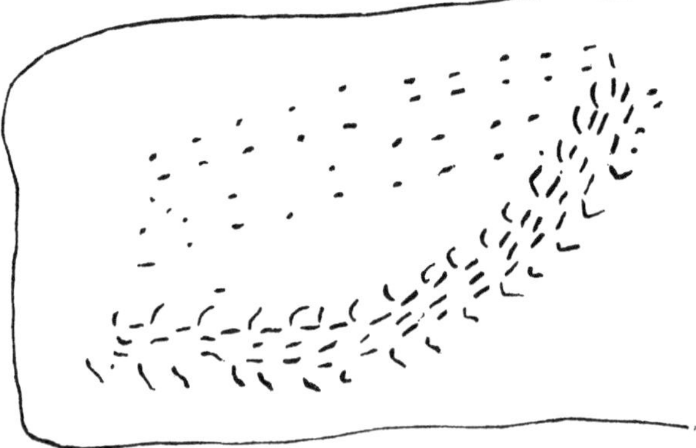

Bifurculapes elachistotatus.

Unisulcus intermedius. Fig. 318. represents a worm which moved by fixing its head upon the mud and drawing up its posteriors, then advancing its head to find another fulcrum. This is called Halysichnus laqueatus.

Fig. 317.

Unisucus intermedius.

This, although a very meagre account, discloses a very remarkable fauna in the Connecticut valley, in sandstone days. Yet with the exception of one or two skeletons of rather small reptiles, and a few coprolites, the tracks are all the evidence we have of the former existence of so many huge and strange beings. Well may we say with Hugh Miller, "they are fraught with strange meanings, those footprints of the Connecticut."

Fig. 318.

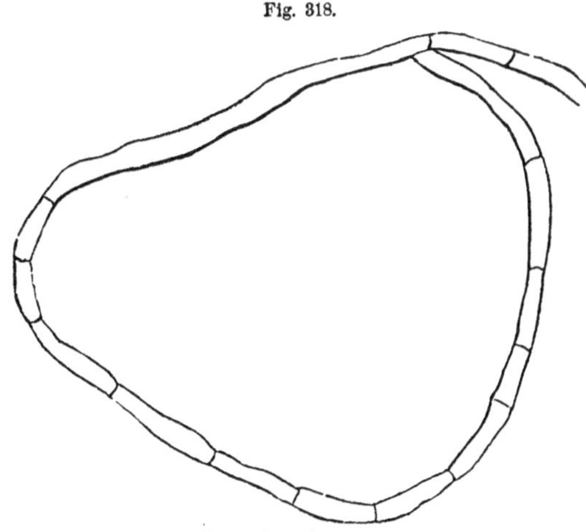

Halysichnus laqueatus.

There is another curious fact generally connected with these footmarks. The same surfaces are frequently covered with small pits, which can not be distinguished from those made on mud and clay by rain drops; and such they are now regarded. Even the direction of the wind at the time they were made can sometimes be determined by the parallel elongation of the rain impressions. Fig. 319 will give an idea of the usual appearance of the fossil rain drops.

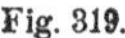

Fig. 319.

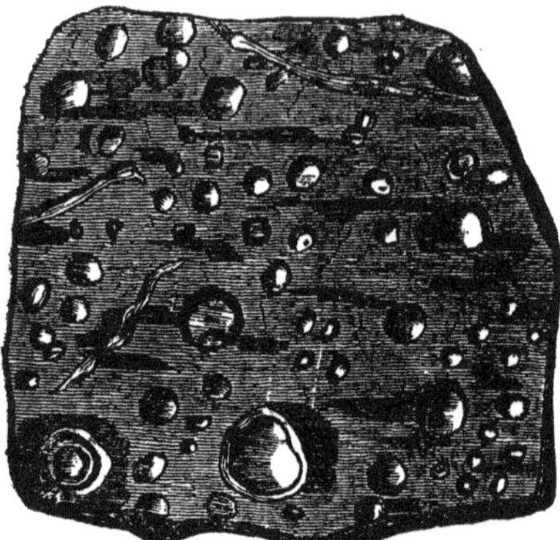

Fossil Rain Drops.

These fossil rain drops are found on all the aqueous deposits as far back as the Cambrian.

Mode of Formation of the Tracks.—A child hardly needs any help in forming a theory as to the origin of fossil footmarks. He will say they must have been made by animals walking over the surface while the rock was soft, which was subsequently hardened. The rain drops, which are frequently on the same surface, show that it was out of water when trodden upon, though some cases prove it to have been sometimes under the water. The whole surface must have been subsequently covered in order to bring mud over the tracks to form the rock above them and preserve them. Thus would the tracks be slowly filled up, and not entirely disappear till several layers of mud had accumulated above them; and besides, the weight of the animal would bend downward several layers of mud beneath the tracks, so that when the rock was afterwards split open, we should find the same tracks, more or less perfectly exhibited, on several layers. Putting these layers together by hinges, we get a *fossil book* of great interest. Fig. 320 represents the most remarkable volume of this sort ever put together. Two tracks are here shown passing through five layers of rock. The specimen (in the Amherst Ichnological Cabinet) is nineteen inches long by eight, and five inches thick when shut.

Fig. 320.

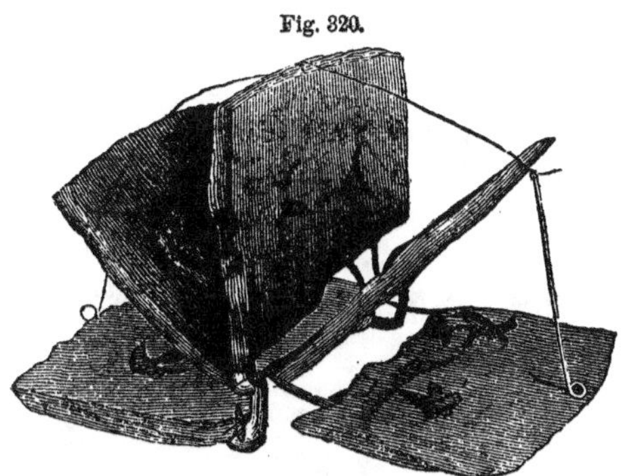

The Massachusetts Ichnological Report groups the 119 species of Lithichnozoa hitherto discovered in the Connecticut valley as follows:

Marsupialoid animals	5
Thick-toed birds	14
Narrow-toed birds	17
Ornithoid Lizards or Batrachians	10
Lizards	17
Batrachians, the Frog and Salamander family	11
Chelonians or tortoises	8
Fishes	4
Crustaceans, Myriapods and Insects	18
Annelids, or naked worms	8
Of uncertain place	6

Lithichnozoa in the Wealden.—Mr. Beckles has obtained from the Hastings sand, a middle member of the Wealden formation in England, impressions of enormous size, which are three toed, and the animal apparently a biped, Fig. 321. Yet Prof. Owen is inclined to regard them as made by the Iguanodon, and supposes the tracks of the fore feet were always covered by the hind feet. The largest of these tracks are twenty-eight inches long and twenty-five broad, and the stride sometimes reaches forty-six inches.

9. Cretaceous Period.

The plants of the cretaceous system, including the green sand, are not very numerous or important; and we shall pass them by. The animals, however, are very abundant. The Protozoa are largely developed, especially the Amorphozoa, or organisms allied to sponges, and the Foraminifera. Of the first we present two examples. Fig. 322 shows the Ventriculites radiatus and Fig. 323 the Cœloptychium lobatum, both from the European chalk.

Fig. 322.

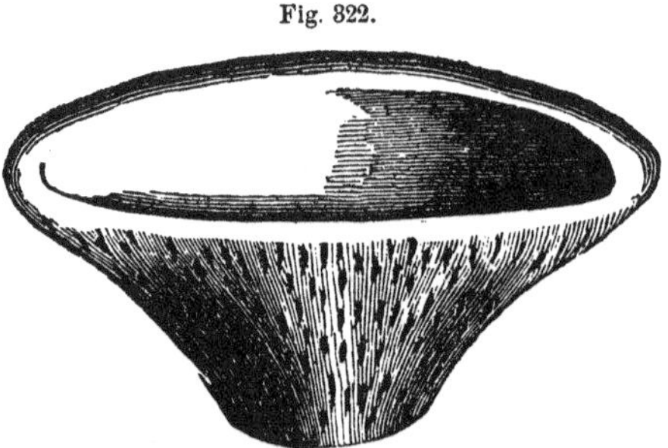

Ventriculites radiatus.

Fig. 323.

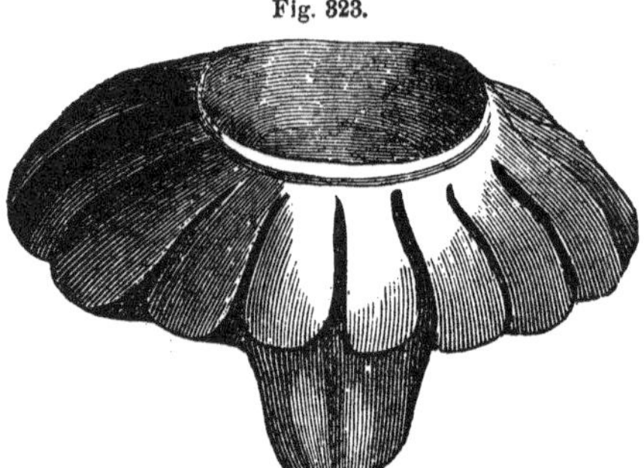

Cœloptychium lobatum.

We have elsewhere spoken of the great difficulty naturalists have experienced in disposing of the sponges, both living and fossil. They are certainly organic; but whether animals or plants, or to be regarded as an intermediate group, as Owen supposes, remains to be decided. Pictet divides them into three families; 1, the Spongides; 2, the Clionides; 3, the Petrospongides: the last of which is exclusively fossil. They commence with the silurian, where are three genera. One is added in the Devonian, four in the Permian, five in the Trias, nine in the Oolite, and nineteen in the chalk.

The Foraminifera are mostly microscopic, and they form often a large part of rocks. Nearly half of the chalk of northern Europe is composed of them. We give below, in Figs. 324, 325, 326, 327, 328, 329, six species, greatly magnified, from the chalk formation.

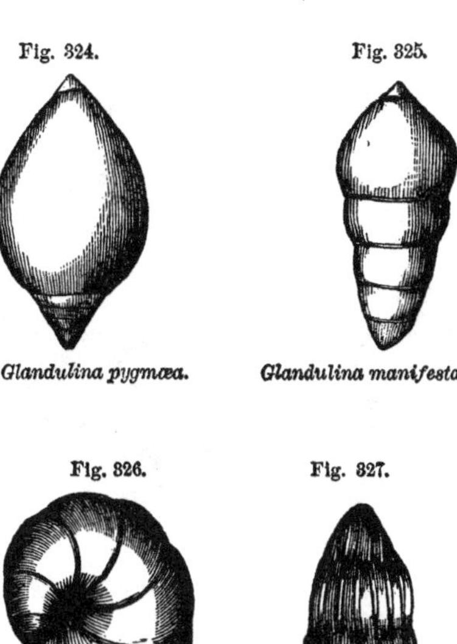

Fig. 324. *Glandulina pygmæa.*

Fig. 325. *Glandulina manifesta.*

Fig. 326. *Nodosaria proboscidea*

Fig. 327. *Cristellaria Spacholtzi.*

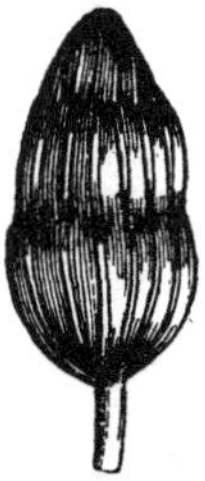

Fig. 328.

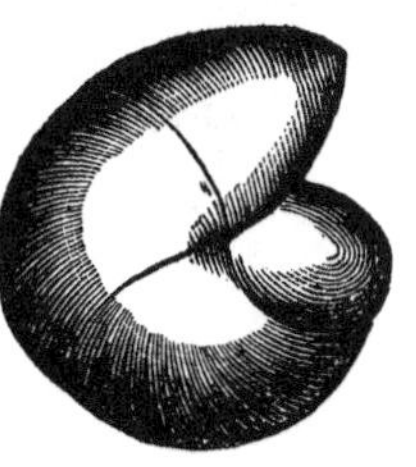

Nonionina quaternia

Fig. 329.

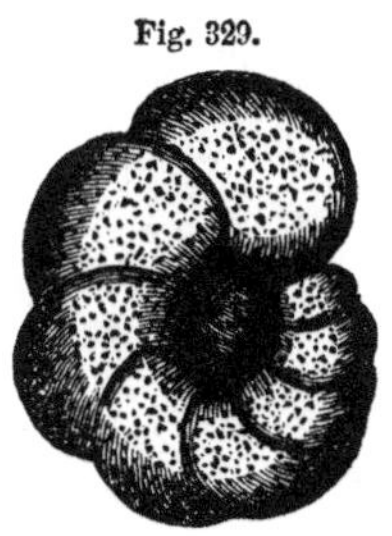

Rotalina involuta.

In the Reticulipora obliqua, Fig. 330, we give a single example of a fossil Polyzoa, or Bryozoa; animals regarded by some as a branch of the molluscs, coming under the division Molluscoidea.

Fig. 330.

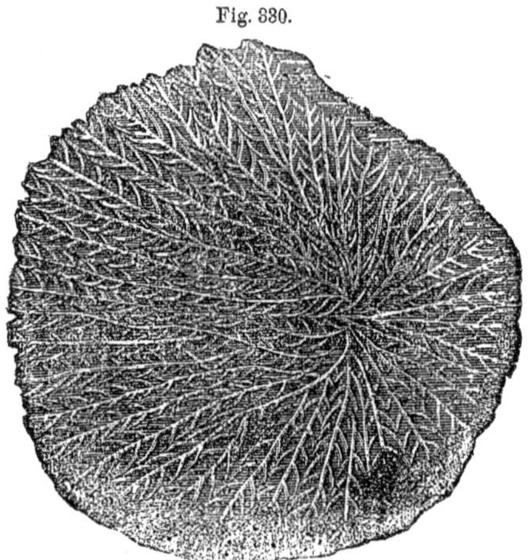

Reticulipora obliqua.

Fig. 331 gives an example of the Actinozoa, viz., the Diploctenium cordatum, and Fig. 332 another genus of the same class, the Anthophyllum atlanticum of this country.

Fig. 331.

Fig. 332.

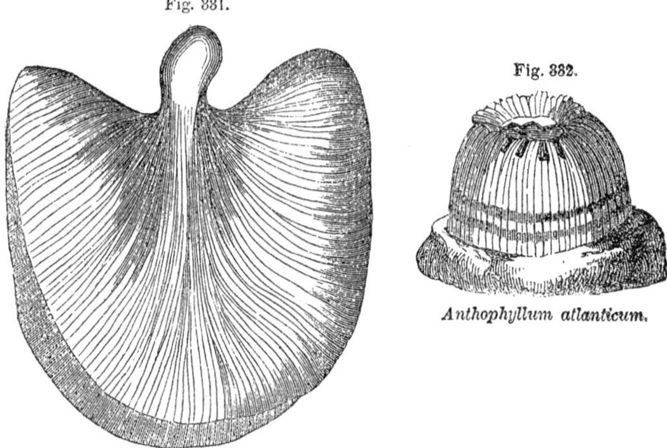

Anthophyllum atlanticum.

Diploctenium cordatum.

The Molluscs are abundant, and we give a few examples.

Among the Conchifera is Spondylus spinosus, shown on Fig. 333. Fig. 334 represents the Ostrea pectinata; Fig. 335 the Lyriodon scaber; Fig. 336

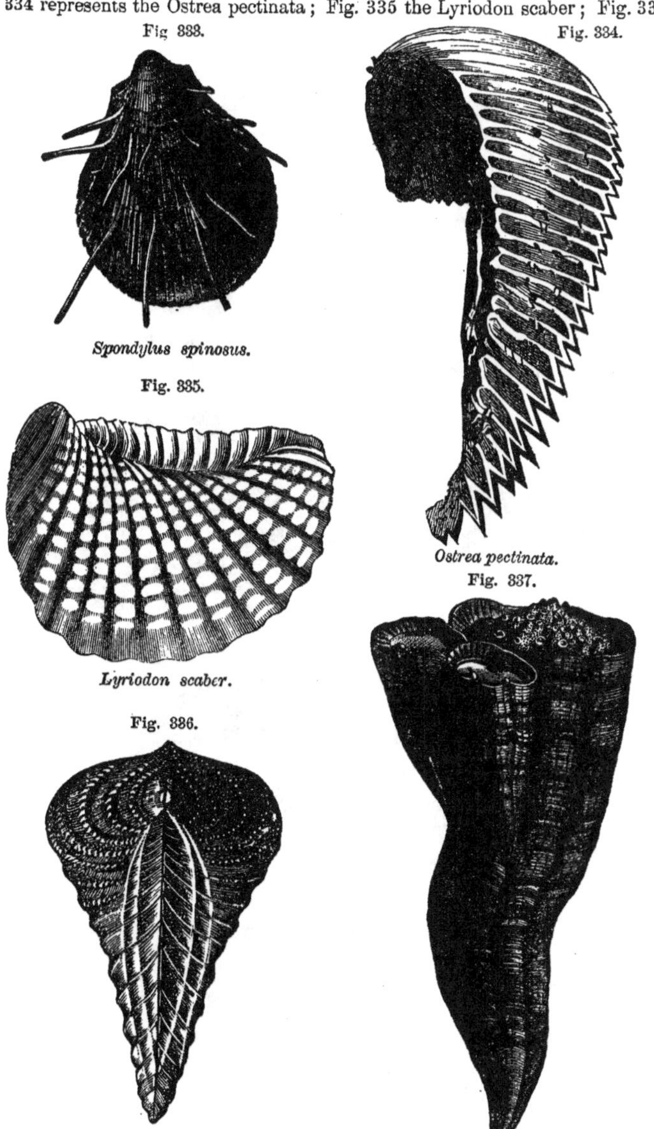

Fig. 333. *Spondylus spinosus.*

Fig. 334. *Ostrea pectinata.*

Fig. 335. *Lyriodon scaber.*

Fig. 336. *Trigonia caudata.*

Fig. 337. *Hippurites Toucasiana.*

the beautiful Trigonia caudata, and Fig. 337 the Hippurites toucasiana—a genus quite characteristic of the chalk. Figs. 338 and 339 show a side and a flat view of Inoceramus sulcatus.

Fig. 338.

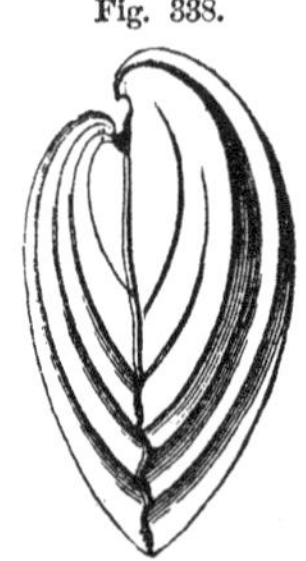

Inoceramus sulcatus

Fig. 339.

Inoceramus sulcatus.

Among the univalves, Fig. 340 shows the Turritella catenatus.

Fig. 340.

Turritella catenatus.

Fig. 341.

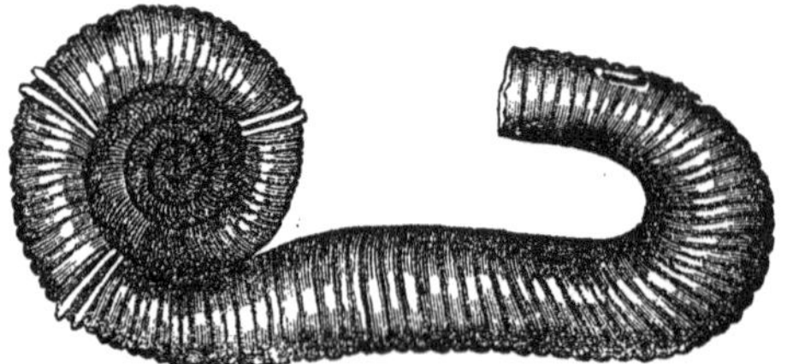

Scaphites Yvanni.

Fig. 342.

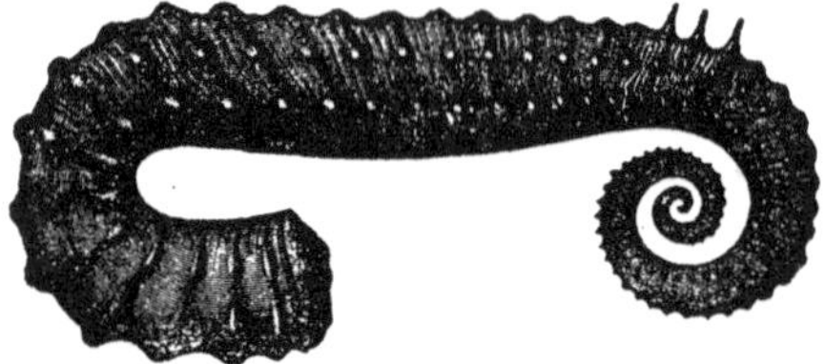

Ancyloceras Matheronianus

Of the Cephalopods, the Scaphites Yvanni is shown on Fig. 341, and the Ancylocerus Matheronianus on Fig. 342.

Of the Echinoderms, which are abundant and beautiful in this formation, we give only the star-fish, Palæocoma Fustenbergii, in Fig. 343.

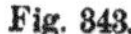

Fig. 343.

Palæocoma Fustenbergii.

Not less than seventy-eight genera of fishes have been described in the chalk, which we must pass by, because they present no important new phase not already noticed.

Fig. 344.

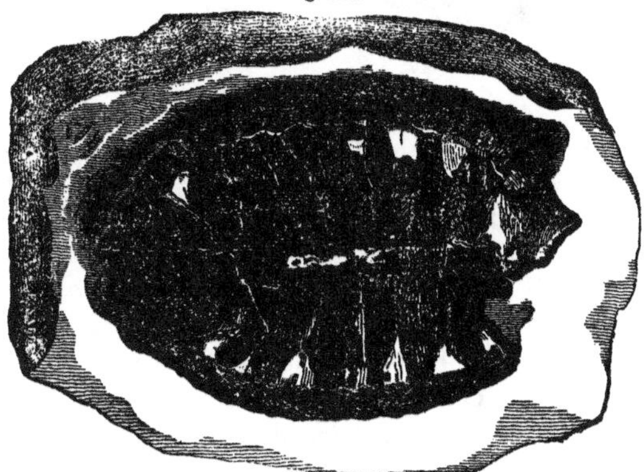

Chelonia Benstedi.

Fig. 344 shows a specimen of a Chelonian or tortoise, the Chelonia Ben-stedi, from the lower chalk of England.

Of reptiles, we give, on Fig. 345, a view of the head of the Mososaurus Hofmanni, as it appears in the Mæstricht limestone. Up to the time of the deposition of the chalk, the ichthyosaurus and plesiosaurus appear to have ruled in the ocean; but then they disappeared, and the mososaurus took their place, to keep the multiplication of the species of other animals within proper limits. It was most nearly related in its structure to the monitor, a species of lizard now living. While the head of the largest monitor does not exceed five inches in length, that of the mososaurus is four feet long; and the whole animal is twenty-five feet, while the monitor is only five feet in length. It had paddles instead of legs, and the number of its vertebræ was 133.

Fig. 345.

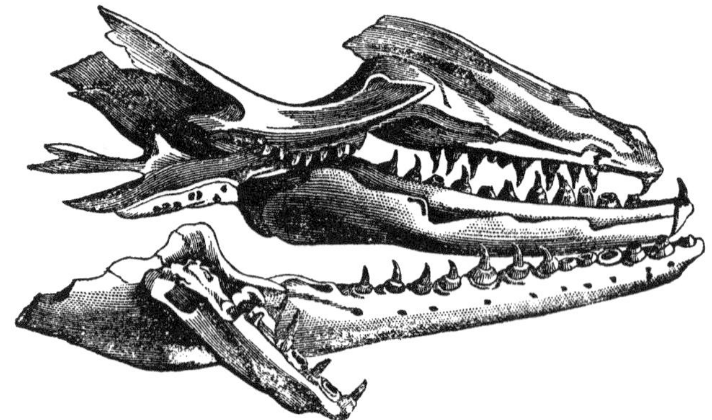

Mososaurus Hofmanni.

In 1858, Professor Leidy described a remarkable reptile from the cretaceous marl pits of New Jersey, to which he gave the name of Hadrosaurus Foulkii. It was a huge herbivorous saurian, closely allied to the Iguanodon, probably twenty-five feet long, whose thigh bone is nearly a third longer than that of a common mastodon. Its tail was three feet deep. Though dug out of a marine deposit, it was probably amphibious.

According to Prof. Owen (*Art. Palæontology in the Encyclopedia Britannica*) the "trifid metatarsal of a bird, about the size of a woodcock" has been found in the Cambridge Green sand of

England. This is the earliest example of the bones of birds in the rocks.

We have now reached the top of the Mesozoic or Secondary Period. Its peculiar characters are as follows:

1. In it Mammiferous Animals are found but rarely, and of small size, belonging to the Marsupial sub-class.

2. The Reptiles have an immense development in this period. Their great size and abundance have led some authors to call this the Palæosaurian Age, or the Reign of Reptiles. But such phrases are rather poetical than scientific.

3. The beautiful group of Ammonites and Belemnites with ramified divisions between the chambers, belongs exclusively to this period.

4. The Echinoderms are quite different from those of the Palæozoic Period. The Echinides and Stellerides have a great development.

5. The Polypi belong to peculiar groups not found scarcely in the palæozoic.

10. TERTIARY PERIOD.

As we pass from the secondary into the tertiary period, we find a decided change in the character of the organic remains, scarcely a species being common to the two divisions. Those which come in with the tertiary strata bear a strong resemblance to existing animals and plants, and numerous species are regarded by most zoologists as identical with those living now. But other eminent naturalists, among whom Agassiz stands at the head, are of the opinion that the fossil and living species are not in any case, perhaps, identical; but only closely related. At any rate, we find the fossil species becoming more and more like those alive, as we ascend in the tertiary series, so that it becomes more and more difficult to distinguish between them.

Sir Charles Lyell's well-known division of the tertiary strata into Eocene, (the lowest group), Miocene and Pliocene, is founded on the per cent. of living species in the different groups. It does not exceed five per cent. in the Eocene, twenty-five in the Miocene, and is between fifty and seventy in the Pliocene. In fact the per cent. varies all the way from nothing to seventy from the bottom to top, and hence it seems a merely arbitrary assumption to stop at any particular per cent. Moreover, if the opinion of other eminent zoologists is correct, that the species are all extinct, this classification falls to the ground. Yet it is generally adopted by English writers, and it would seem as if there must be something natural about it. Yet other divisions are made by continental writers. Pictet and D'Orbigny make six divisions. 1. The Inferior or Suesonian. 2. The *Calcaire grossier*, or Parisien in part. 3. The Eosene superior, or Parisien in part. 4. The Superior Sandstones, or Falunien in part. 5. The Miocene, properly so-called, or Falunien in part. 6. The Pliocene, or sub-appenine. We shall not in this work be able to go enough into details to render it necessary to use either of these classifications, but shall treat of the tertiary strata as a whole.

Plants.—The tertiary flora is characterized by the abundance of Angiospermous Dicotyledons, and Monocotyledons, especially Palms. These plants constitute more than three-fourths of the present vegetable productions of the globe. They began to appear in the chalk, but were not fully developed till the tertiary period. In the earlier part of the tertiary, marine and coniferous plants predominated. In the middle part, there was a mixture of tropical and temperate forms, and in the upper part, a great resemblance to the plants of the temperate regions of Europe, North America, and Japan.

We shall give illustrations only of certain remarkable fruits which occur in the upper tertiary, along with brown coal, at Brandon, in Vermont. They have not yet been referred to known genera.

They do not, however, correspond with any plants now growing in the northern parts of our country, and are doubtless of a tropical character. The figures below will give an idea of their size, shape and markings, and may serve as an example of tertiary fruits.

Figs. 346, 347 and 348 represent the most common species; the first two show the specimen flatwise; the other edgewise. Figs. 349, 350 and 351

Fig. 346. Fig. 347. Fig. 348.

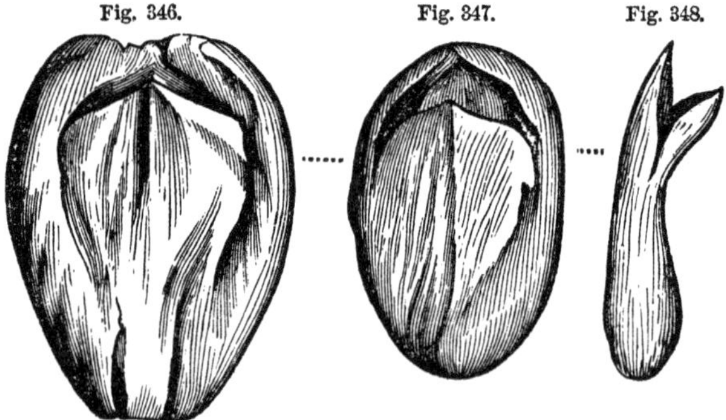

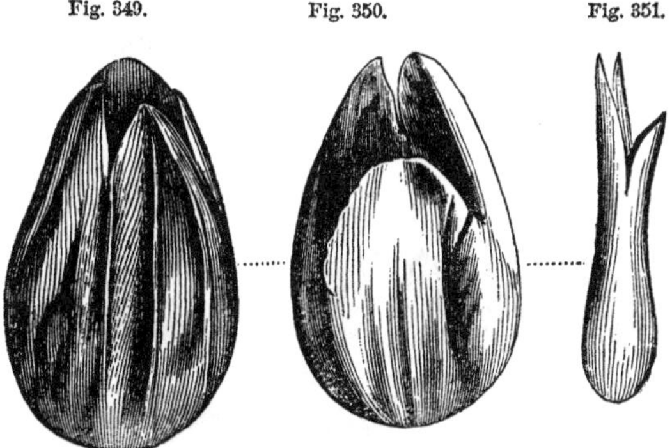
Fig. 349. Fig. 350. Fig. 351.

show the same as to a less common species. Figs. 352 and 353 show a somewhat different species. Figs. 354 and 355 represent a not unusual species, almost exactly spherical. Fig. 356 is similar but elongated. Fig. 357 shows a single carpel. Figs. 358 and 359 exhibit specimens with the apex quite aside from the geometrical axis. Figs. 360 and 361 have longitudinal ridges quite prominent. Figs. 362, 363 and 364 are more or less triquetrous. Figs. 365 and 366 are elongated, slightly striated, small fruits, with a rather thick epicarp. Figs. 367, 368 and 369 are leguminous seeds. Fig. 370 is an elegant and frail seed, with delicate waving striæ.

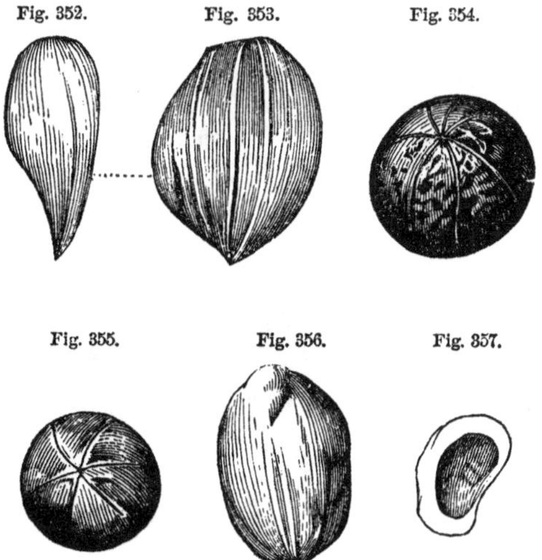
Fig. 352. Fig. 353. Fig. 354.

Fig. 355. Fig. 356. Fig. 357.

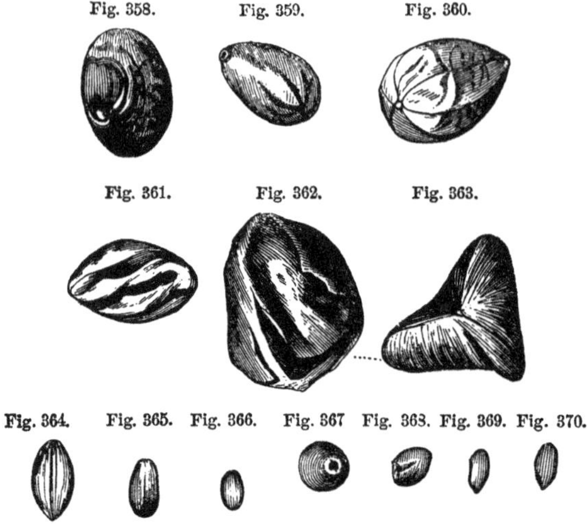

Fig. 358. Fig. 359. Fig. 360. Fig. 361. Fig. 362. Fig. 363. Fig. 364. Fig. 365. Fig. 366. Fig. 367 Fig. 368. Fig. 369. Fig. 370.

Animals.—Of the Protozoa we shall give only an example of the largest known species of Foraminifera. These occur in all the formations, 657 species and 73 genera having been already described. They increase in number and variety as we ascend through the strata, and reach their maximum development in the present seas. They are thought to be the oldest of all animals. Several species now living can not be distinguished from some in the chalk, and one or two go back as far as the lias. But on this subject, as we have seen, there are two opinions.

Fig. 371 exhibits a horizontal section of a Nummulite of the natural size (they are sometimes nearly two inches across), showing the division into chambers. Fig. 372 shows several of the shells in rock. This is the species which forms a large part of the pyramids of Eygpt and the Sphinx.

Fig. 372.

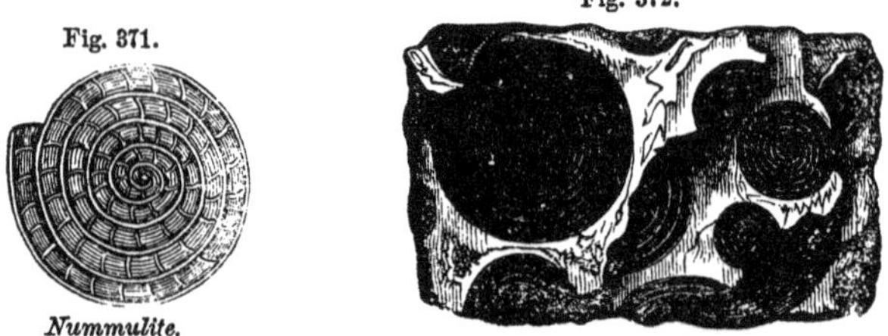

Fig. 371.

Nummulite.

Fig. 373 shows a delicate species of Zoantharian Polyp, the Turbinolia

Dixonii. A second, the Astrea semispherica, is given on Fig. 374. On Fig. 375 is given a Bryozoa, the Fasciculipora Marsillii.

Fig. 373.

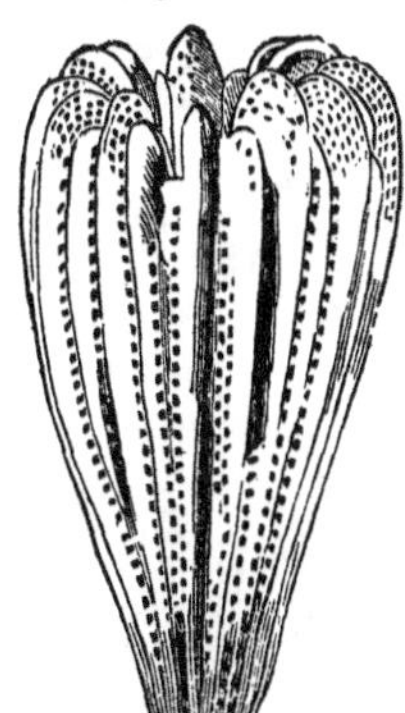

Turbinolia Dixonii.

Fig. 374.

Fig. 375.

Fasciculipora Marsillii.

Of the conchiferous molluscs, or ordinary bivalves, we give only two, and that mainly to show how nearly they resemble existing species. Fig. 376 shows the Pholadomya Mellevilli, and Fig. 377 the Corbula Gallica.

Fig. 376.

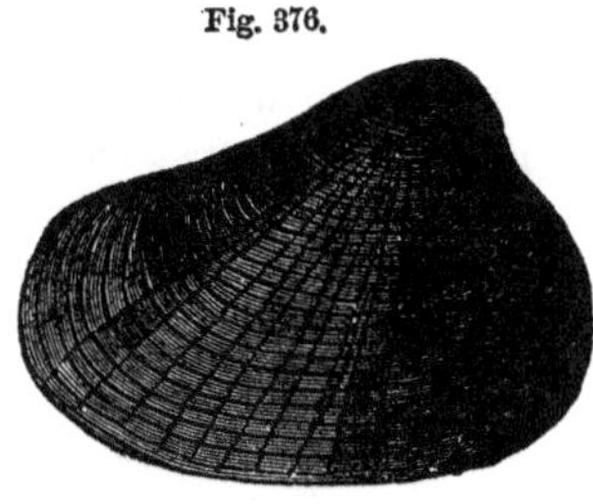

Pholadomya Mellevillei.

Fig. 377.

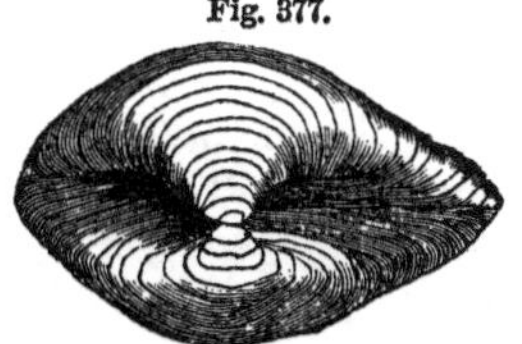

Corbula Gallica.

The fossil species of ordinary bivalves are nearly 6,000, while the recent species are little more than half that number. Yet as a group it attains its maximum in the present seas. There are seven times more genera in the newer tertiary than in the Silurian, which has yielded less than 100 species, while the chalk contains 500, and the miocene tertiary 800.

The four following figures will give some idea of the Gasteropod Molluscs in the tertiary. Fig. 378 shows the Fulgur canaliculatus from Maryland. Fig. 379 the Murex tricarinoides. Fig. 380 the Terebra fuscata, and Fig. 381 the Cypraea elegans.

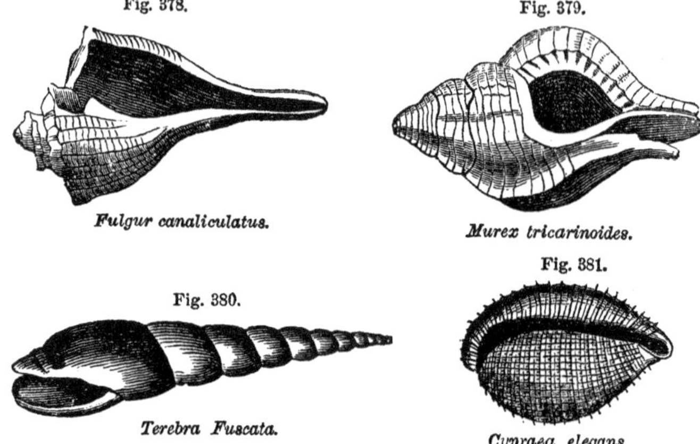

Fig. 378. Fulgur canaliculatus.

Fig. 379. Murex tricarinoides.

Fig. 380. Terebra Fuscata.

Fig. 381. Cypraea elegans.

The fossil univalve shells, which are less than 100 species in the Silurian, have increased upward to the newer tertiary, which has yielded twenty times as many species. All the fossil species are less than 6,000, while the recent species exceed 8,000. The air-breathing molluscs found fossil bear a still smaller proportion to those now alive. Only 300 species of land snails, and half as many other air-breathers occur fossil, but the living land snails exceed 4,000.

Passing by the other groups of the lower animals, we come to the vertebrates. The number of fish is greatly increased, amounting to 188 genera; but they approach existing forms so much that we give only a few examples, and those rather peculiar; for one-third of the genera of the lower tertiary have become extinct, and these are some of them.

Fig. 382 shows the Semiophorus velicans of Agassiz, from the famous locality at Monte Bolca, in Italy.

Fig. 383 shows a small fish called Lebias cephalotes, from the fresh water tertiary strata, in France. It gives a good idea of their crowded condition sometimes.

The Squalidæ or sharks have prevailed in all periods of the world's history since fishes first appeared. Many of those in the present seas are large and justly dreaded. But they are mere pigmies compared with those that swam in the seas that washed the shores of North and South Carolina during the eocene and micocene periods, as Fig. 384 will prove. It is copied from a

Fig. 382.

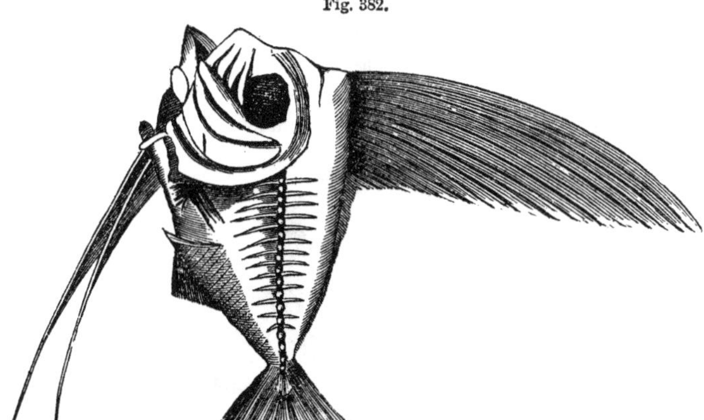

Semiophorus velicans.

Fig. 383.

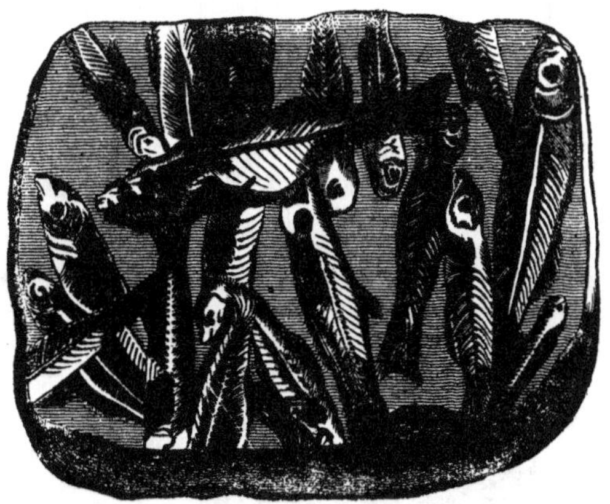

specimen found in North Carolina, and belonged to the Carcharodon megalodon. Prof. Owen describes a Carcharodon thirty-seven feet long, with teeth two inches long and nearly two broad. Yet this tooth is five inches long and four and a half broad, and Professor Gibbes, in his Monograph of the fossil Squalidæ of the United States, says that they are sometimes 6.5 inches high and five inches broad. "If," says Prof. Owen, "the proportions of these extinct Carcharodons correspond with those of the existing species, they must have equaled the great mammiferous whales in size; and combining with the organization of the shark its bold and insatiable character, they must have constituted the most terrific and irresistible of the predaceous monsters of the deep."

Fig. 384.

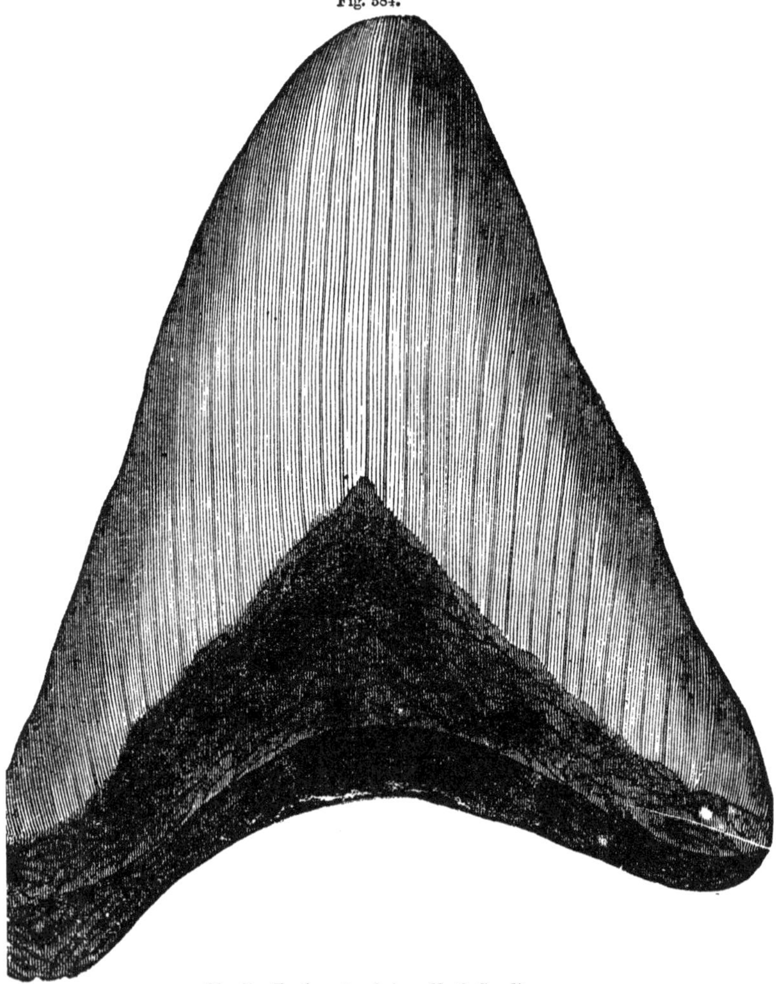

Shark's Tooth, natural size; North Carolina.

The reptiles of the tertiary, some of them at least, bear a strong resemblance to existing races. As an example, we give in Fig. 385 the Palæobatrachus Goldfussii, dug out of the *paper coal*, as it is called, on the Rhine.

Fig. 385.

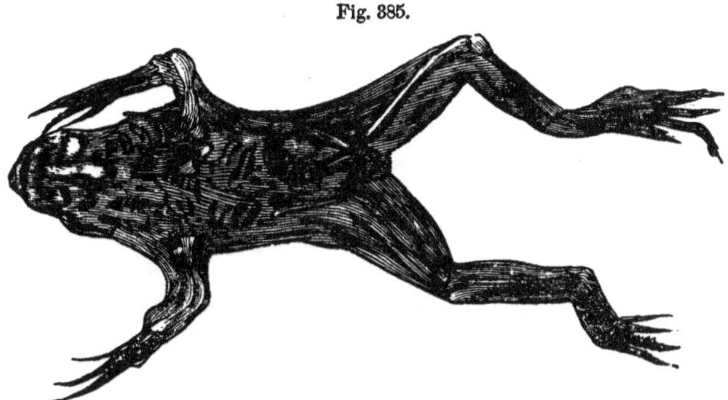

Rana Diluviana.

Not less than 104 species of Reptiles and Amphibia have been described from the tertiary. Among them we find seven species of crocodiles, embracing the alligator. Fig. 386 shows a part of the jaw of one of these latter animals from the tertiary of the Isle of Wight.

Fig. 386.

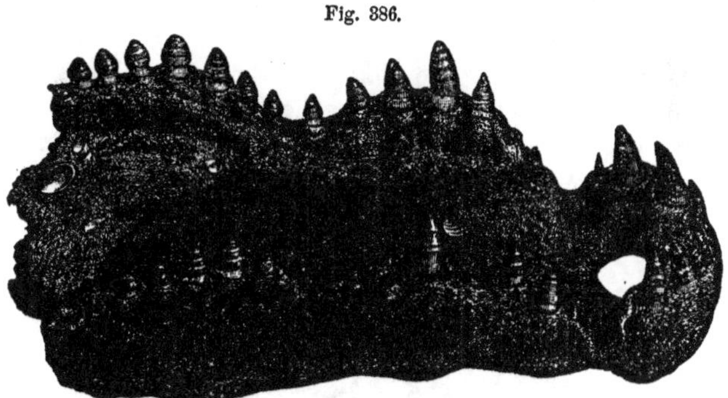

Jaw of Alligator, Isle of Wight.

Not less than eighteen species of crocodiles in a fossil state have been found in the tertiary; also nineteen species of land tortoises, seventeen species of pond tortoises, eighteen species of river tortoises, and sixteen species of sea tortoises, or turtles, as they are generally called; likewise seven species of lizards.

The Ophidia, or serpents, first appear in the tertiary, where at least ten species have been described.

The first certain remains of birds, with one exception in the green sand, are found in the tertiary, where twenty-three species, belonging to six known orders have been found. The most interesting is the Gastornis Parisiensis, described by Prof. Owen from the eocene tertiary of Paris. It was as large as an ostrich, and its affinities seem to place it between the Gallinaceæ, the Grallatores, and Cursores.

The influx of mammalia during the tertiary period is most remarkable. While only some ten or twelve species, and these of the most imperfect tribes, have been found in all the rocks below, already over 400 species have been described in the tertiary. Of these, ten species were monkeys, ninety-four carnivora, 109 Artiodactyla, or even-toed (two or four) animals (to adopt Owen's classification), fifty-nine Perissodactyla or odd-toed (one or three), eleven Proboscidea (elephants), three Toxodontia, ten of the Sirenia, twenty-seven of the Cetacea or whale tribe, three of the Chiroptera or bat tribe, twenty-six of the Insectivoraor insect-eaters, thirty-eight of the Rodentia or gnawers, and nine of the Marsupialia. We can give only a few examples from this great number.

Among the Carnivora, the dog, sometimes resembling the wolf, sometimes the fox, and sometimes the domestic dog, appeared in the eocene tertiary, as did also a species of hyena as large as a leopard. The bear, also, and the seal, came in somewhat later.

Fig. 337.

Anoplotherium Commune.

Among the Artiodactyla may be mentioned the hippopotamus and the hog; also several extinct allied animals, dug up in the vicinity of Paris, of which the Anoplotherium commune, shown on Fig. 387, will give an example. This animal was about the size of the wild boar, and could swim well.

Another of these animals from the Paris basin was the Palæotherium, of which there were several species, ranging in size from that of a sheep to a horse, and of which Fig. 388 will give an idea.

Fig. 388.

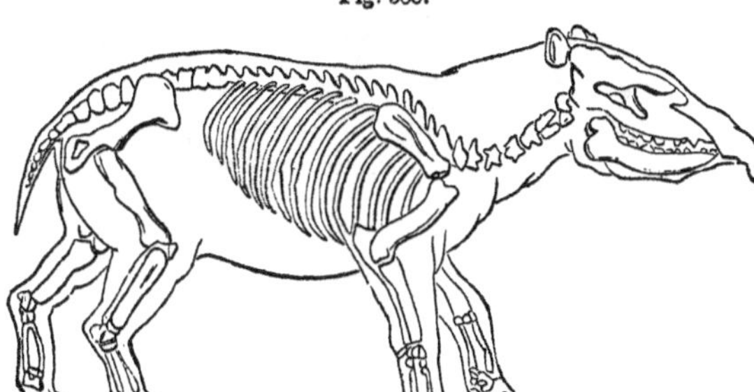

Palæotherium.

Among the ruminants of the Artiodactyla found fossil may be mentioned the camel, the giraffe, the musk, various kinds of deer, and the giant animals found in the miocene of India, called the Sivatherium and Bramatherium, which almost equaled the elephant in size.

Among the pachydermatus, or thick-skinned animals, reckoned by Owen among his Proboscidea, we find the living genera elephas, rhinoceros and tapir in a fossil state, as well as the extinct genus

Fig 389.

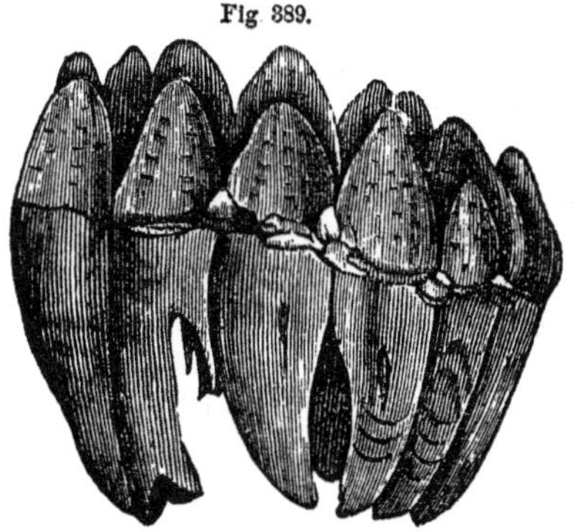

mastodon. This last animal appears to have been the elephant of tertiary days, and is distinguished from the elephant chiefly by the form of the teeth. Fig. 389 shows the tubercular character of the mastodon's tooth, and Fig. 390 the flat surface of the elephant's tooth.

Fig. 390.

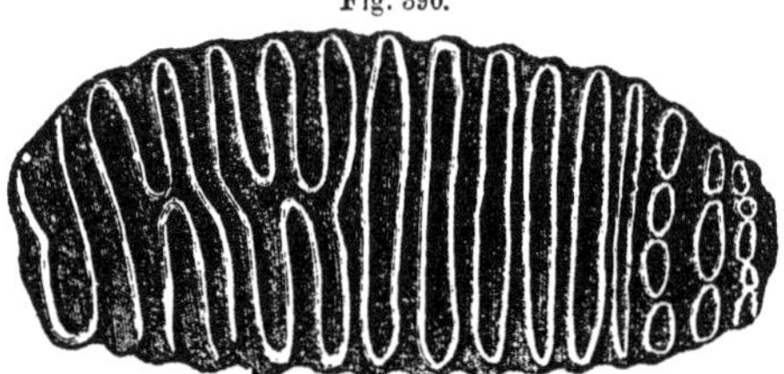

Three species of mastodon have been found in the Miocene tertiary, eight in the Pliocene, and three in Alluvium. The great mastodon of this country occurs in the latter, and we shall recur to it again. The elephant has been found in the tertiary of India, but only in alluvium or pleistocene in Europe and America.

The order Sirenia furnishes a remarkable and probably the largest of quadrupeds that have lived on the globe. The mammoth and mastodon have been supposed to be the most gigantic, but they must give place to the Dinotherium, described by Cuvier as a gigantic tapir, but by Professor Kaup as a new genus between the tapir and the mastodon; and adapted to that lacustrine condition of the earth which seems to have been so common during the deposition of the tertiary strata. Its remains have been found in tertiary strata, in the south of France, in Austria, Bavaria, India, and especially in Hesse Darmstadt. Its length must have been as much as eighteen feet. One of its most remarkable peculiarities consisted in two enormous tusks, at the anterior extremity of the lower jaw, which curved downwards, like those of the walrus. Its general structure seems to have been adapted to digging in the ground; and for this purpose its feet as well as tusks, projecting a foot or two beyond the jaws, which were four feet long, were intended. It lived principally in the water, like the hippopotamus; and it probably used its tusks for tearing up the roots of aquatic vegetables, which, as is shown by its teeth constituted its food. They might have been useful also to aid in dragging the body out of the water and for defense.

Fig. 391 is a sketch of the *Dinotherium giganteum* as restored by Professor Kaup. One or two other species have been found.

Fig. 391.

Dinotherium giganteum.

Fig. 392 represents the head of the same animal.

Fig. 392.

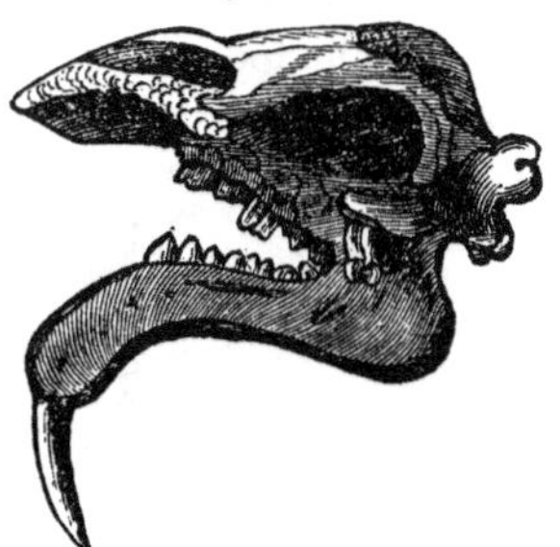

Dinotherium giganteum Head.

The tertiary strata furnish several species of dolphin and whale among the cetacea; but the most remarkable animal is the extinct genus Zeuglodon found in Alabama in the eocene tertiary, of which three or four species are recognized, but some doubt whether they are true cetaceans.

The annexed sketch, Fig. 393, may give some idea of the Zeuglodon macrospondylus, but we have no great confidence in its accuracy.

Many other kinds of mammiferous animals have been found in the tertiary. but we have not room to describe them. Among those best known are the bat, the hedgehog, the shrew, the mole, the squirrel, the jerboa, the rat and mouse, the beaver, the porcupine, the hare, the opossum, etc.

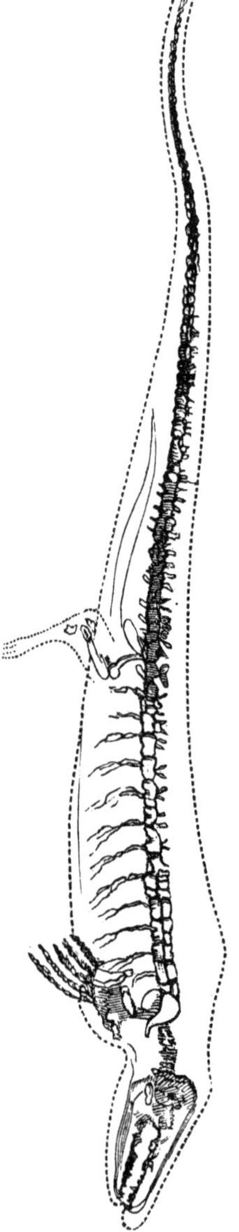

Fig. 393.

Zeuglodon macrospondylus.

The palæontological characteristics of the tertiary period, at which we have already hinted, are very marked. They are the following, principally.

1. The appearance and great development of mammiferous animals is the most important feature. With the exception of some ten or twelve species of marsupials in the rocks below, all the other mammalia, to the number of 400, open before us in the tertiary, and seem to be the precursors of the 2,000 species now inhabiting the globe.

2. The tertiary reptiles and fishes come near the living forms, and many correspond so closely that the best naturalists can not distinguish between them.

3. The Belemnites and Ammonites, which were so abundant to the top of the cretaceous period, suddenly disappear and have no representatives in the tertiary.

11. Alluvial or Pleistocene Period.

Under these names we include all the aqueous deposits above the tertiary. In countries, however, where drift is not fully developed, it is not easy to draw the line between the tertiary and the alluvial; but there is no deposit in the tertiary that would easily be confounded with the coarse, almost unstratified mass called drift. But when this drift has been comminuted, sorted, and redeposited by water, the layers are easily confounded with those of the tertiary period. Hence it is very probable that some of the organic remains which have been referred to pleistocene deposits really belong to the tertiary, and *vice versa;* especially as the fossils do not indicate any such great and decided change of life between these periods as there was at the close of the cretaceous period. The most we can say is, that more than three fourths of the fossils in alluvium correspond to existing species.

The most important feature of the alluvial formation was the introduction of man near the close of the period, and of numerous species, both of animals and plants, much better adapted to his wants than the analogous races of earlier times.

Another interesting fact is, that during the drift period, called the *glacial* period by some, when a lower temperature prevailed, the species both of molluscs and of mammals had a more arctic character than before, and that afterwards, as a warmer climate succeeded, the more southern species again moved northward, and the northern species retreated within their present limits.

The fossil birds and mammals of this period belong almost exclusively to extinct species, and often to extinct genera. The number of species of birds is fifty-four, or more than double those in the tertiary.

By far the most important of these extinct birds are those found in New Zealand by English missionaries, and fully described by Prof. Owen. He had at first only the fragment of a femur; but by applying to it the principles of comparative anatomy, he was able to construct the whole bird, and subsequent discoveries proved his conclusions to be true. It belonged to the Struthious or ostrich tribe, strongly resembling the Apteryx, a small wingless bird still living in the island. It had no wings, and its skeleton was extremely massive, its toe bones being almost equal to those of the elephant, and the leg bones quite as large as those of an ox. Prof. Owen has been able to describe eleven species of this bird from New Zealand, under the name of Dinornis; though to some of the species that had a short hind toe, he gives the name of Palapteryx. They varied in height from three to ten feet. The natives called them Moas, and there is evidence, from their occurrence with the half burnt bones of man on spots where cannibal feasts had once taken place, that they must have lived within a few hundred years, and possibly some may still be found alive. Their bones now occur in the banks of the rivers.

The species of the Moa described by Professor Owen are as follows: Dinornis giganteus, elephantoidea, ingens, struthioides, rheides, dromioides, casuarinus, robustus, crassus, geranoides, and curtus. Of these we give in Fig. 394 a restored sketch of Dinornis elephantoides.

Fig. 394.

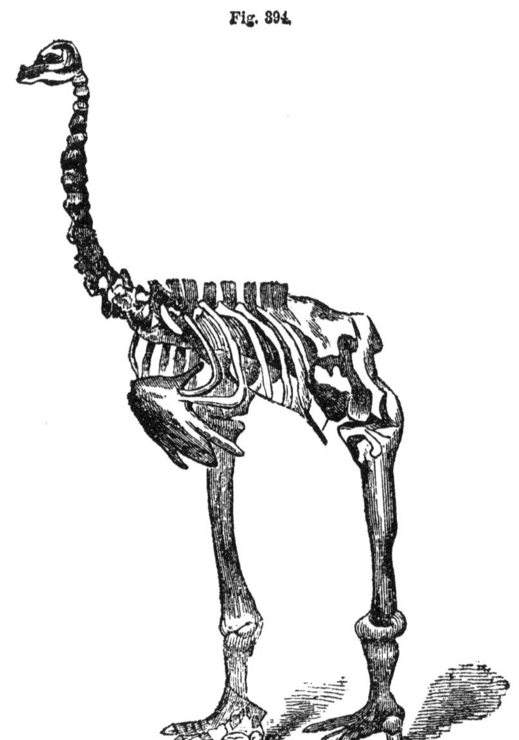

Dinornis elephantoides.

In Madagascar similar bones, quite as large as any of those of the Dinornis, along with some egg shells, are preserved in Paris. The bird has been called Æpiornis maximus. Its egg was over 13 inches in diameter, and over 33 inches in circumference, and equaled 148 hen's eggs, and 50,000 humming bird's eggs in size.

In New Zealand, along with those of the Dinornis, were found the bones of another extinct bird as large as the swan, called the Aptornis, and a large coot, the Notornis. So that at least thirteen species of remarkable birds have become extinct there at a comparatively recent date, and two species of Apteryx which were cotemporaneous with the Dinornis are nearly extinct.

In the island of Rodriguez, in the Indian Ocean, once lived another wingless bird, the Pezohaps solitarius, which has become extinct. In Fig. 395 we give a sketch of it.

Fig. 395.

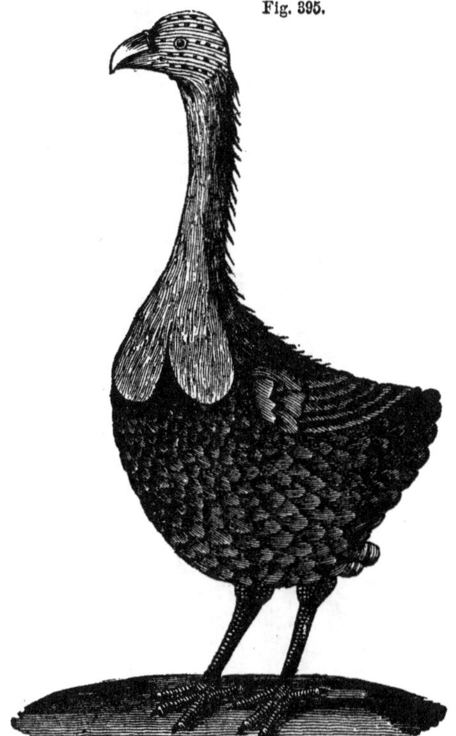

Pezohaps Solitarius. The Solitaire.

Other wingless or short winged birds existed about 200 years ago, in the islands of Mauritius and Bourbon. They belonged to the Columbidæ or pigeon tribe. The Dodo, which weighed fifty pounds, and inhabited Mauritius, is the most remarkable. It is represented in Fig. 396. One or two heads and feet of this bird are all that remain in the cabinets of Europe, although the earlier voyagers saw it alive and figured it.

One or two species of the Sturthio Rhea, or South American Ostrich, have been found fossil in that country.

We have seen that the earliest mammals that appeared on the globe were marsupials. Among existing animals, Australia is re-

Fig. 396.

The Dodo.

markable for the predominance of this tribe of animals. And there accordingly we find numerous fossil species in the more recent formations. We show in 397 the head of the gigantic thick-

Fig. 397.

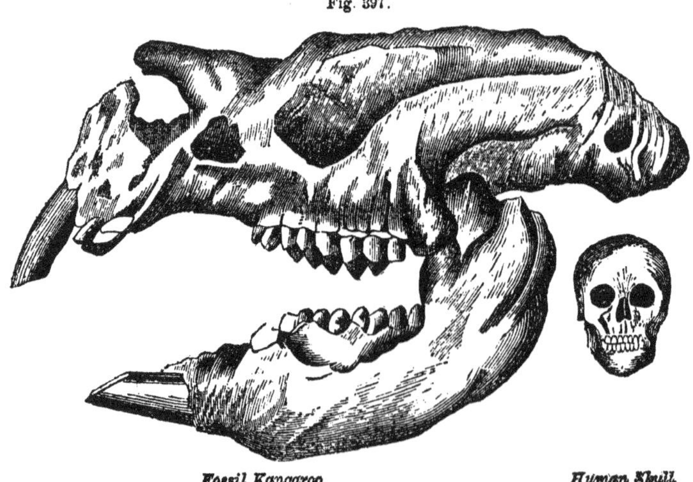

Fossil Kangaroo. *Human Skull.*

skinned kangaroo, called Diprotodon Australis. This head was three feet long, and its size, as compared with that of man, may be judged of by the human skull placed by its side.

The number of species of the different orders of mammalia found in the post-tertiary or alluvial strata, may be seen in the general table of fossils which we shall give at the close of this Section. Among them we notice only a few.

These remains occur, not merely in the common aqueous deposits of alluvium, but many of the most interesting have been obtained from caverns, where the bones are preserved by the deposition of stalagmite, which has dripped down from the cavern's roof to the floor, enveloping the bones. We give, in Fig. 398, a section of the cave of Gailenreuth, in Franconia, where the situation of the stalagmite, the bones, etc., is obvious to inspection.

Because they are so large, and found in Europe and America in regions too far north for the living elephant, the Mastodon and the Mammoth excite great interest. We have already indicated the difference between these two genera from the character of their teeth. The Mastodon appeared earliest; three species being found in the miocene tertiary, and eight in the pliocene. In still newer strata three species are described. They occur in Europe, North and South America, and in that famous locality of mammalian bones, the Sewalik Hills of India.

In this country the most remarkable locality of fossil mastodons, elephants, and other animals, is the Big Bone Lick, in Kentucky, about twenty miles southwest of Cincinnati. It is estimated that the bones of 100 mastodons, 20 elephants, two oxen, two deer, and one megalonyx, have been carried from this spot.

In general the bones of the mastodon in our country occur in superficial deposits; many of them in peat bogs, where the animal is sometimes found standing. The largest and most perfect skeleton ever found, we believe, occurred in such a situation, in Newburgh, Orange County, New York, from whence, many years before, another specimen had been obtained, and was put up in Peale's Museum in Philadelphia. That found in 1845 was from a peat bog, with marl beneath, and weighed 2,000 pounds. In the place where the stomach lay was found a quantity of broken twigs, perhaps of the white cedar. This was his last supper. A poor sketch of this mastodon is given in Fig. 399. It was purchased and fully described by the late Prof. John C. Warren of Boston, and by him placed, with many other splendid analogous fossils, in a fire-proof cabinet in Boston, where they now are, the property of his heirs.

Fig. 398.

Cavern of Gailenreuth.

Fig. 399.

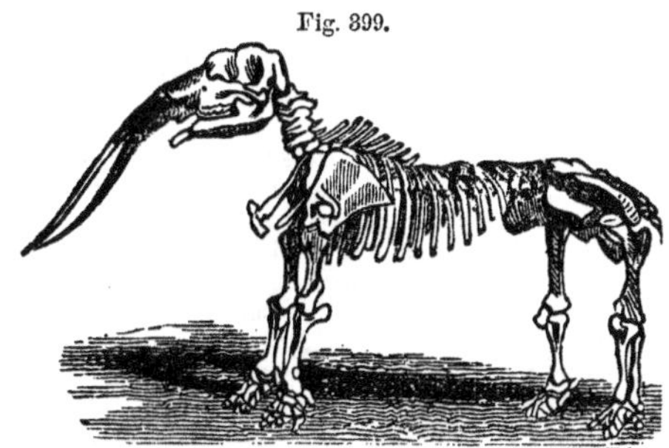

Newburg Mastodon.

The mastodons preceded, for the most part, the Mammoth (*Behemoth*, Arabic), which was a fossil elephant. "The transition," says Prof. Owen, "from the mastodontal to the elephantine type of dentition is very gradual." Two species of elephant preceded in Europe that which is called Mammoth. The most remarkable of this species was found in Siberia, encased in frozen mud at the mouth of the river Lena. Its flesh was not decayed, and it was covered with a reddish wool and long black hairs, indicating its existence in a colder climate than those countries where the elephant now lives. It is preserved in the Museum of Natural History in St. Petersburgh, and has a length of sixteen feet and a height of nine feet. We give a sketch of this animal in Fig. 400.

Fig. 400.

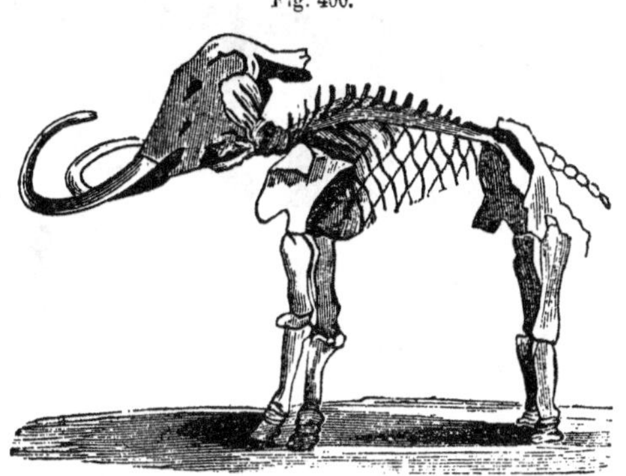

Mammoth.

There are two living species of elephant; the Asiatic or Indian, which extends only to the thirty-first degree of north latitude, and the African, which occurs as far south as the Cape of Good Hope. Ten species of fossil elephants have been described, all in post-tertiary strata.

Rhinoceros, Hippopotamus, Hyena, Horse, Ox, Deer, Sivatherium, Monkey, Camel, etc.—Most of these animals in their fossil state, differ so little from the existing species, that they need not be particularly described in this work. They are generally, however, of larger size than the living species. The rhinoceros found undecayed in the frozen gravel of Siberia, has already been noticed; and several other species of this animal occur in Europe and in India, associated with the bones of the elephant, also with several species of hippopotamus, and one or two of oxen, aurochs, and deer. The horns of the fossil ox are sometimes very large; in one example thirty-one inches long. Of deer some thirty or forty species have been found in the tertiary and post-tertiary. We give, in Fig. 401, a sketch of the Megacerus Hibernicus, or

Fig. 401.

Great Irish Elk.

great Irish Elk, that measured nearly eleven feet between the tips of its horns.

The most interesting remains of the hyena are those found in caverns. The *sivatherium* is an extinct animal recently found in India, in concretionary drift, larger than the rhinoceros, furnished with four horns and a proboscis, and forming an intermediate link between the ruminantia and pachydermata. In the same deposit were found the remains of a gigantic species of monkey and of a camel. Ten species of monkey have been discovered in tertiary deposits: so that the important fact seems now well established, that the animals approaching nearest to man in their structure, have been found in a fossil state.

Glyptodon, Megatherium, and Mylodon.—The armadillo, as is well known, is covered with a bony armor for defense against enemies, dust, etc. The few living species of this animal are small and confined chiefly to South America, where they burrow like the woodchuck.

But the ancient armadillo, called the Glyptodon by Prof. Owen, was a giant. Its carapace, which is preserved in the Museum of the Royal College of Surgeons in London, resembles a huge cask. Below is a sketch of the animal as restored. Fig. 402.

Fig. 402.

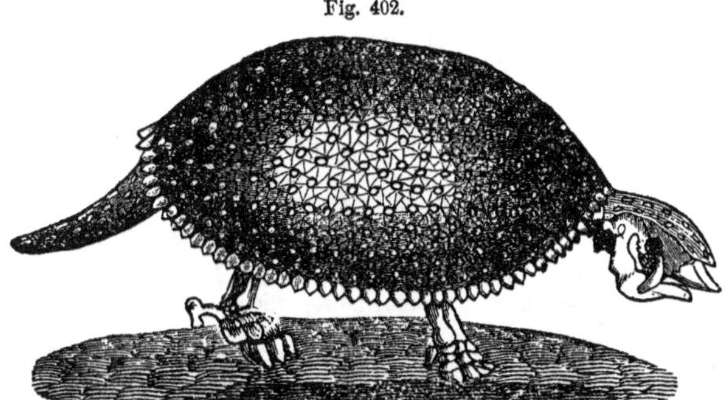

Glyptodon Clavipes.

The Megatherium is an enormous extinct animal, which was once abundant in the vast plains or pampas of the same continent. They have been found by Mr. Darwin over an extent of 600 miles, accompanied with bones and teeth of five other quadrupeds, some of them of a similar construction. Bones of this animal are found also on the island of Skiddaway, on the coast of Georgia. It was larger than the rhinoceros, and its proportions were perfectly colossal. With a head and neck like those of the sloth, its legs and feet exhibit the character of an armadillo, and the ant-eater. Its body was twelve feet long and eight feet high. Its forefeet were a yard in length and more than twelve inches wide, terminated by gigantic claws. Across its haunches it measured five feet, and its thigh bone was nearly three times as thick as that of the elephant. Its spinal marrow must have been a foot in diameter, and its tail, at the part nearest the body, twice as large, or six feet in circumference! Its teeth were admirably adapted for cutting vegetable substances, and its general structure and strength seem intended to fit it for digging in the ground for roots, on which it principally fed. Fig. 403 exhibits the entire skeleton of this animal, as seen in the Museum at Madrid, in Spain.

Fig. 403.

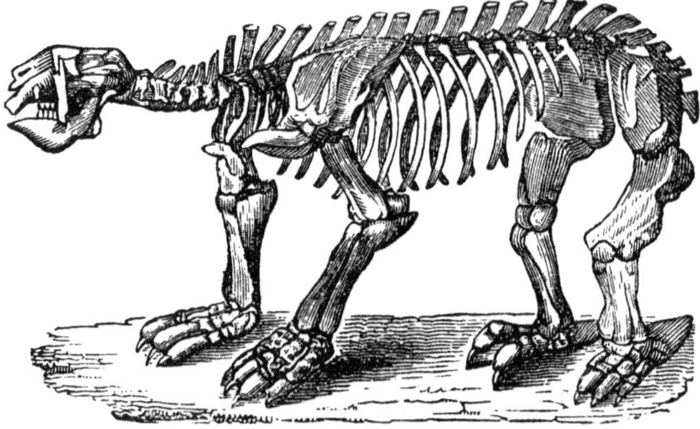

Megatherium.

In the superficial deposits of South America several other interesting extinct animals have been found, belonging mostly to the Pachydermata, or thick-skinned, and the Edentata. The Toxodon, which had a skull twenty-eight inches in length, approximates in its structure to several families of animals, viz., the Rodentia, the Ruminantia, and Cetacea; although, in fact, a Pachyderm. The Macrauchenia greatly resembled the llama, and had a neck almost as long as that of the giraffe, with a body nearly as large as that of the rhinoceros. This, also, was a Pachyderm. The Mylodon, an Edentate animal, was of massive and singular proportions. Its body was shorter than that of the hippopotamus, but was terminated by a pelvis as broad as that of the elephant and deeper, resting on two massive but short hind legs, with feet as long as the thigh bones. The tail, as long as the legs, and very thick and strong, was probably used like that of the kangaroo, to support the body when the animal raised up its anterior extremities. It is supposed by Mr. Owen that the peculiar structure of this animal adapted it, first for digging around trees, and then, resting upon the tripod base of its hind legs and tail, it seized the trunk with its fore legs, and rocked it to and fro until it was prostrated, and its leaves furnished food for several days, perhaps. The following sketch will give a good idea of this animal. (Fig. 404.) The Scelidotherium was an analogous animal not larger than some of the existing ant-eaters of South America, but with excessively large hind legs. These animals are all called Megatheroids, because they resemble the Megatherium.

Megalonyx.—This animal was first described by Thomas Jefferson. It was found in the nitre caverns of Virginia and Kentucky, and has since been discovered in other places. It was of the size of the ox, and appears to have been nearly related to the sloth.

As a contrast to the gigantic animals above described, we ought to mention those microscopic organisms that have gone by the name of Infusoria. The opinion seems to be gaining ground that the larger part of them are vegetables; but the astonishing facts as to their minuteness and rapid increase, as discovered by Ehrenberg, the great master of the microscope, still remain true. They occur most abundantly in very recent deposits, forming beds be-

Fig. 404.

Mylodon robustus.

neath the black mud and peat in swamps. But they form a part also of bog iron ochre, now forming from water, and also in flint and semi-opal. They make up most of the polishing slate of the tertiary strata, as at Bilin in Bohemia, and Richmond in Virginia. Of that from Bilin a single cubic inch contains 41,000 million skeletons, yet the deposit is fourteen feet thick, and thicker still at Richmond. A cubic inch from a deposit at Maidstone, Vermont, contains, according to Prof. Bailey, 15,000 million skeletons. A cubic inch of ochre sometimes contains a billion of skeletons.

The smallest animalcule is only the 24,000th part of an inch in size, and a single shield weighs only 187 millionth part of a grain. 500 millions of them could live in a drop of water. Their increase is prodigious. An individual of the Hydatina senta in ten days increased to 1,000,000; on the eleventh day to 4,000,000, and on the twelfth to 16,000,000. Another in four days increased to 170 billions.

Of eighty fossil species of these skeletons, Ehrenberg found half to belong to extinct species. They abound in the chalk and are all marine; but those in newer deposits are all fresh-water organisms. Fig. 406 represents a microscopic view of some of these organisms of the family Baccillaria, which are probably vegetables.

Fossil Man.—From the coast of Guadaloupe two specimens of human skeletons have been obtained in solid rock, one of which is in the British Museum and the other in the Royal Cabinet in Paris. The rock is a quite hard limestone, made up of minute fragments of shells and corals, ground down by the waves and

Fig. 406.

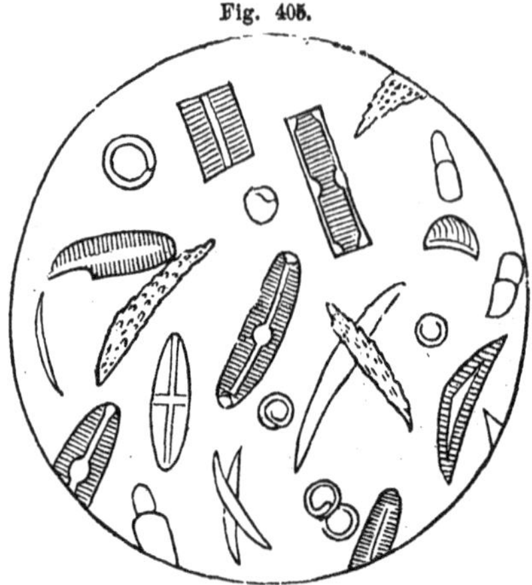

Fossil Infusoria.

cemented together in those places most frequently left dry by the waves. Such accumulations are common in the Antilles, and in them sometimes fragments of vessels and human workmanship are found at a depth of twenty feet. These bones still retain all their phosphate of lime and some of their animal matter. Some suppose them to be the remains of shipwrecked persons, others that they are the remains of Caribs and Gallibis, who had a battle on this spot in 1710.

Whether these be fossil men will depend upon the meaning which we give to the term fossil. According to our definition in the last Section (a body buried by natural causes in the earth) they are distinctly fossil. But those who suppose the body must have been buried in the earth "in a state different from the normal and actual conditions of existence," would exclude them. We give, in Fig. 405, a sketch of the specimen in the British Museum.

Many other examples of the bones or works of man have been described of late years, so deeply buried in the earth, or so connected with the relics of extinct animals, that some have concluded, not only that they are fossil, but of the same age as the extinct mastodons, rhinoceroses and hippopotami. Such examples often occur in caverns, buried beneath mud and stalagmite, as they are found in Greece, the south of France, Belgium, the Suabian Alps,

Fig. 406.

Fossil Man.

and in Brazil. In these cases the bones both of man and the animals are usually separated from one another, so that no whole skeletons exist, and often the human remains are found only in the upper part of the deposits.

We should call these relics fossils. But several difficult questions must be settled before we can say confidently that they were not introduced into these caverns subsequent to that of the extinct species. For often such caverns, in rude times and in days of persecution, were inhabited by men, who buried their dead there. Again, carnivorous animals often dragged in there for food the bones of other animals. Streams, also, have sometimes, especially in time of flood, drifted bones as well as other things into the caverns, and deposited them promiscuously, and sometimes earthquakes have changed the original levels and mixed together drift and alluvial deposits. It is a reasonable conclusion, then, as Sir Charles Lyell remarks, that "it is not on the evidence of such intermixtures that we ought readily to admit, either the high antiquity of the human race, or the recent date of certain lost species of quadrupeds."

Appeal has also been made to cases of human bones, arrowheads, pottery, etc., in alluvial deposits on the banks and at the mouths of rivers. These cases occur in the south of France, at the mouth of the Nile, at Natchez, on the Mississippi, etc. But here again we have many difficult questions to settle as to the rate at which river deposits are made, as to the changes in that rate, as to the power of heavy substances to sink through semi-plastic materials, etc., before we can be certain that man was a cotemporary of very ancient extinct animals.

The point of chief interest affected by these investigations, is the question whether any of the facts conflict with the common opinion that Adam was the earliest created human being. At all times great haste has been manifested by some to make the facts sustain the negative. But every geologist who fully understands the difficulties of the subject, and recollects how many analogous facts, confidently relied upon in years past to prove the great antiquity of man, have been given up as unsatisfactory, will be very cautious in respect to new facts. The following positions seem to us capable of satisfactory proof:

1. Man did not appear upon the globe till a very late epoch of the pleistocene or alluvial period.

2. It does not show his pre-Adamic existence to admit that he is found in a fossil state; since, according to our views, a fossil condition does not prove great age.

3. Nor is that implied if we admit that his remains occur in what some geologists call drift. The true drift occurs only in high latitudes, and when men say that human bones in Egypt, the south of France, or at Natchez, are found in drift, they must mean modified drift, or alluvium; since no true drift is found in those places, and the age of such drift remains to be proved.

4. Nor are human relics necessarily pre-Adamic because they occur with those of extinct animals. For we have shown that not a few animals, some sixteen species of birds, and some quadrupeds, have become extinct within historic times. The great Cetacean called Stelleria seems lately to have disappeared, and the arctic buffalo (Ovibus moschatus) is on the point of extinction. But, says Owen, "fossil remains of Ovibus and Stelleria show that they were cotemporaries of Elephas primigenius and Rhinoceros tichorrhinus." Are we sure that the mastodon has not lived within historic times? The Newburgh specimen seems certainly quite recent.

5. If it should turn out that fossil men exist in deposits decidedly older than Adam, they may belong to extinct species, and therefore not prove the pre-Adamic existence of the present race.

6. The creation of man, along with a vast number of cotemporary species of a higher grade than the earth had before seen, and forming the culmination of organic existence on the globe, is the most remarkable fact of geological history, and marks off the alluvial period from all others.

7. This last creation is distinguished from all that have pre-

ceded it on the globe in two respects: first, it presents by far the fullest and most perfect fauna and flora; secondly, it was not preceded immediately by any such violent catastrophes as in most other cases destroyed the existing races; but in this case many species lived on, or were recreated. In the more delicate organization and higher powers of the present races we can see why much previous preparation was necessary and catastrophes undesirable.

Lithichnozoa.—It may seem incongruous to denominate tracks in mud and clay lithichnozoa, since this term means *stony-track animals;* but it is not more improper than to call mud and marl rock. At any rate the tracks in mud and clay are a complete counterpart of those in consolidated rock; so that he who has seen the first, will no longer doubt as to the last. Dr. Buckland first described (in 1841) the tracks of deer and oxen upon mud, beneath a bed of peat in Pembrokeshire, England. Dr. A. A. Gould was the first to describe a famous locality on the Bay of Fundy, where the tracks of birds are preserved in great perfection. We have found, and one of us long ago described, a large variety of tracks with rain drops on the clay at Hadley, in Massachusetts, on the banks of Connecticut river. Fig. 407 shows the tracks of a snipe with rain drops on that Hadley clay.

Fig. 407.

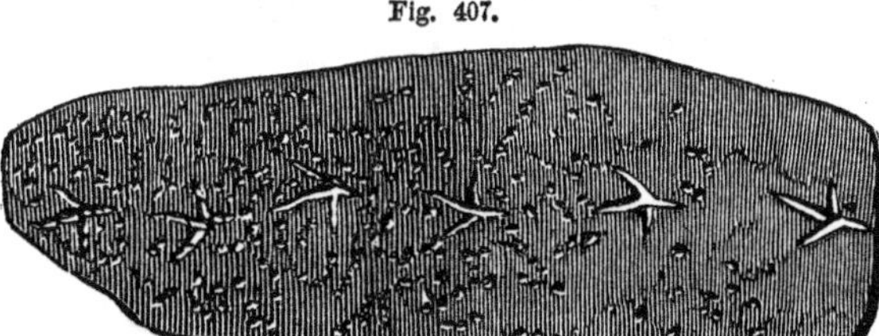

On Fig. 408 we have the track of an annelid, or myriapod, with the hairs

Fig. 408.

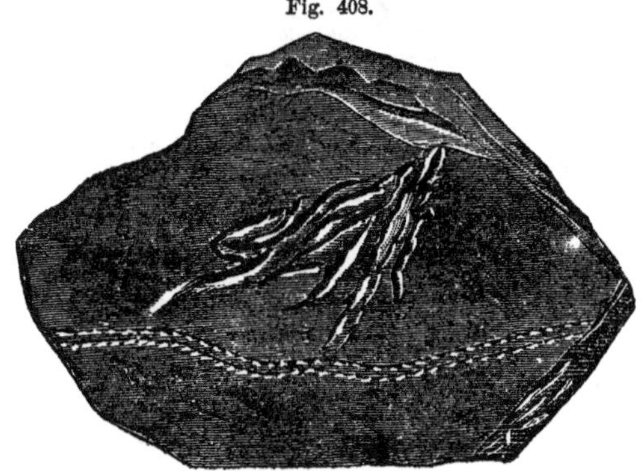

or legs on the side less delicate than in the original. It shows also a frog's track.

On Fig. 409 we have the addition of a man's tracks to those of a bird, probably of a crow, and rain drops. This shows us that had man lived in sandstone days, his tracks would probably be found with those of other animals.

Fig 409.

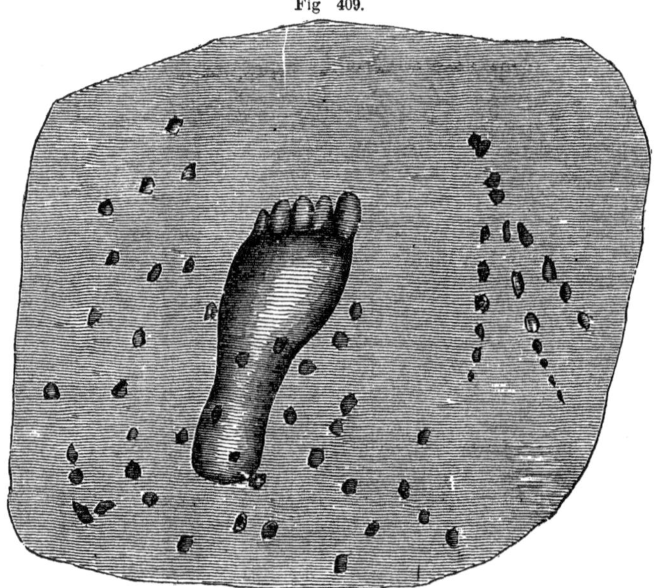

TABULAR VIEW OF FOSSIL ANIMALS.

The following Table is an abstract of the catalogue of fossil animals given in Juke's able Manual of Geology (London, 1857). He derived it almost entirely from Pictet's *Traite de Paleontologie*, etc. (Paris, 1853 to 1857). Pictet does not profess to give a complete list of the species, but notices only the most important. Hence this Table contains only 20,299, whereas the Table in the last edition of our work, taken from Bronn's Index Palæontologicus, gave 24,397, in 1849; and doubtless many thousands have been added since. But Pictet's enumeration gives a more recent and better view of the present distribution of the fossils, and therefore we use it with this explanation.

In Juke's catalogue we found quite a number of species referred to the Silurian formation without distinguishing between the Upper and Lower Silurian. In such a case we divided the number equally between these groups, as, for instance, 33 Trilobites, 14 Echinoderms, 66 Conchifera, 29 Cephalophora, 4 Rugosa, and 14 Zoantharia. So in the tertiary, several species are given, and the particular division of that formation is not specified. Thus, 67 species of Crustacea were divided equally among the Eocene, Miocene, and Pliocene groups; also 96 Cephalophora, 387 Conchifera, 19 Bryozoa,

	Cambrian.	Lower Silurian.	Upper Silurian.	Devonian.	Carboniferous.	Permian.	Triassic.	Oolitic.	Cretaceous.	Eocene.	Miocene.	Pliocene.	Alluvial.	Total Number.	Living Species.
MAMMALIA	..	..	..	..	..	..	2	9	..	107	210	104	219	650	2030
Bimana	..	..	..	..	..	..	..	..	..	..	..	..	1	1	
Quadrumana	..	..	..	..	..	..	..	..	..	3	..	..	8	11	
Carnivora	..	..	..	..	..	..	..	..	..	13	50	39	41	143	
Artiodactyla	..	..	..	..	..	..	..	1	..	34	65	31	37	168	
Perissodactyla	..	..	..	..	..	..	..	..	..	37	12	9	13	71	
Proboscidea	..	..	..	..	..	..	..	..	..	..	3	8	13	24	
Toxodontia	..	..	..	..	..	..	..	..	..	..	..	3	..	3	
Sirenia	..	..	..	..	..	..	..	..	..	3	5	1	..	9	
Cetacea	..	..	..	..	..	..	..	..	..	4	9	12	5	30	
Cheiroptera	..	..	..	..	..	..	..	..	..	1	2	1	7	11	
Insectivora	..	..	..	..	..	..	..	..	..	..	25	..	5	30	
Edentata	..	..	..	..	..	..	..	..	..	..	..	..	29	29	
Rodentia	..	..	..	..	..	..	..	1	..	6	35	10	41	83	
Mausupialia	..	..	..	..	..	..	2	7	..	6	4	..	19	38	
BIRDS	..	..	..	..	..	..	..	..	1	9	7	2	55	74	7000
Raptores	..	..	..	..	..	..	..	..	..	3	1	2	6	12	
Passeres	..	..	..	..	..	..	..	..	..	1	1	..	15	17	
Gallinaceæ	..	..	..	..	..	..	..	..	..	..	1	..	10	11	
Cursores	..	..	..	..	..	..	..	..	..	..	..	..	12	12	
Grallatores	..	..	..	..	..	..	..	..	..	3	3	..	5	11	
Palmipedes	..	..	..	..	..	..	..	..	..	2	1	..	7	10	
REPTILES	..	..	..	..	..	13	17	97	23	49	33	11	9	257	1055
Chelonia	..	..	..	..	..	..	..	25	4	32	26	7	5	99	120
Crocodilia	..	..	..	..	..	..	..	32	1	10	6	1	2	52	
Lacertilia	..	..	..	..	..	12	6	13	12	1	5	..	1	50	460
Dinosauria	..	..	..	..	..	..	1	1	..	..	..	..	..	2	
Enaliosauria	..	..	..	..	..	1	10	12	4	..	..	..	..	27	
Pterodactylia	..	..	..	..	..	..	..	14	2	..	..	..	..	16	
Ophidia	..	..	..	..	..	..	..	..	..	6	1	3	1	11	300
AMPHIBIA	..	..	..	1	4	2	9	..	..	..	13	..	..	29	
Labyrinthodonta	..	..	..	1	4	2	9	..	..	..	..	..	..	16	
Batrachia	..	..	..	..	..	..	..	..	..	..	8	..	..	8	125
Sauro-Batrachia	..	..	..	..	..	..	..	..	..	..	5	..	..	5	

FISHES	..	..	8	139	200	34	72	416	220	245	42	13	1	1385	8000
Teleostei	..	..	..	..	..	..	..	..	76	210	18	4	1	309	
Ganoidea	..	..	..	109	77	29	28	320	48	18	5	..	..	634	
Placoidea	..	..	3	80	123	5	44	96	96	17	19	9	..	442	
INSECTS	..	..	..	..	11	..	..	134	2	..	1551	..	..	1688	65.800
Myriopods	..	..	..	..	..	..	..	2	..	..	15	..	..	17	200
Arachnida	..	..	..	..	2	..	..	1	..	..	127	..	..	131	600
Insects proper	..	..	..	..	9	..	..	131	2	..	1409	..	..	1551	65.000
CRUSTACEA	1	213	198	58	13	10	8	46	64	89	26	25	1	702	791
Trilobites	1	213	198	58	13	..	..	..	..	..	..	..	..	483	
Other Orders	..	..	2	22	12	10	8	46	64	89	26	25	1	255	
ANNELIDA	2	7	8	9	..	2	4	25	23	21	23	21	..	144	
ECHINODERMATA	..	46	48	84	157	1	44	333	421	158	46	48	9	1325	498
MOLLUSCA	2	300	421	822	622	109	382	2174	2429	1811	1261	1325	126	11.804	11.482
Cephalophora	..	125	94	404	324	28	193	929	939	1222	964	933	1	6152	
Conchifera	..	..	36	190	120	34	154	1025	862	532	250	378	123	3704	
Brachiopoda	..	137	267	218	212	33	35	186	284	21	4	4	..	1356	
Bryozoa	2	38	24	10	26	9	..	34	394	36	43	10	2	628	980
ACTINOZOA	..	25	83	127	109	4	11	150	176	113	85	25	2	910	
Alcyonaria	..	..	..	..	..	1	..	..	3	4	2	..	..	10	
Rugosa	..	5	51	86	84	1	..	..	..	..	..	..	..	227	
Zoantharia	..	20	42	41	25	2	11	150	173	108	83	25	2	677	
Hydrozoa	..	..	..	..	..	..	..	..	..	1	..	..	1	2	
ASTOMATA	..	2	8	8	4	8	35	166	464	213	300	128	..	1331	
Spongiadae	..	2	8	8	1	4	35	102	209	3	1	..	..	368	
Foraminifera	..	..	..	..	3	4	..	64	255	211	299	128	..	964	1000
LITHICHNOZOA	2	11	10	5	5	9	4	115	..	..	..	..	10	171	
Mammalia	..	..	..	..	..	..	..	..	..	..	..	..	4	..	
Marsupialoids	..	..	..	..	..	..	1	5	..	..	..	..	..	..	
Birds	..	..	..	..	..	..	..	31	..	..	..	..	3	..	
Lizards	..	..	1	2	3	2	1	23	..	..	..	..	..	..	
Chelonians	..	..	..	1	..	5	..	8	..	..	..	..	..	..	
Batrachians	..	..	..	1	1	2	1	16	..	..	..	..	1	..	
Fishes	..	..	2	..	1	..	..	4	..	..	..	..	..	..	
Crustaceans	..	11	..	..	..	..	1	1	..	..	..	..	..	..	
Crustaceans and Insects	..	..	..	..	..	..	..	19	..	..	..	..	..	..	
Annelids	2	..	4	..	..	..	..	8	..	..	..	..	1	..	
Molluscs	..	..	3	1	..	..	..	..	..	..	..	..	1	..	

5 Zoantharia, and 4 Foraminifera. This is very unsatisfactory; but these numbers can easily be deducted from those given in the Table, if any one pleases.

We have made no changes in the number given by Jukes, save in a very few cases, where some interesting species have been quite recently discovered. We have, however, annexed to the Table a group of animals, not yet thought to be determined with sufficient certainty to be placed in the divisions to which they belong, if there is no mistake as to their nature. These are the *Lithichnozoa*, or animals known only by their tracks. We believe that in many instances a track furnishes quite as good a means of determining the character of an extinct animal, as the imperfect fragments of their skeletons from which their nature has been inferred. But palæontologists are reasonably slow in admitting any new principles, and therefore let this group stand by itself, and pass for as much as it is worth. The enumeration which we give must of course be very imperfect; yet it is interesting to see that there is scarcely a formation that has not already its Ichnology. Professor Owen has fully installed this branch of Palæontology into its proper place in his admirable work on Palæontology, from the Encyclopedia Britannica, *Eighth Edition*, and there is the best account of Ichnology as a whole which we have seen.

FOSSIL PLANTS.

Unger in his work on Palæophytology has presented us with the following estimate of the genera and species of fossil plants arranged under the three divisions of Dicotyledons, Monocotyledons, and Acotyledons.

	Genera.	*Species.*
Dicotyledons.		
Thalamifloræ,	24	84
Calycifloræ,	56	182
Corollifloræ,	23	60
Monochlamydeæ Angiospermæ,	48	221
——— Gymnospermæ,	56	363
Monocotyledons.		
Dictyogenæ,	2	5
Petaloideæ,	36	125
Glumiferæ,	5	12
Acotyledons.		
Thallogenæ,	31	203
Acrogenæ,	121	969
Doubtful,	35	197
	437	2421

These are distributed through the rocks as follows:

	Species.
Cambrian, Silurian and Devonian,	73
Carboniferous,	683
Permian,	76
Magnesian Limestone,	21
Trias or Upper New Red Sandstone,	38
Trias, Shell Limestone,	7
Trias, Variegated Marls,	70
Lias,	126
Upper Middle and Lower Oolite,	168
	1262
	Forward.

	Species.
Brought forward	1262
Wealden,	61
Green Sand and Chalk,	122
Eocene,	414
Miocene,	496
Pliocene,	35
Pleistocene,	31
Fossil Species,	2421

SECTION III.

LAWS BY WHICH ORGANIC REMAINS HAVE BEEN DISTRIBUTED.

We have seen in the last Section that the animals and plants have experienced great changes, as we have briefly reviewed them in their several formations. We now proceed to point out the laws by which these changes have been regulated. For, however irregular and capricious the operations of nature may seem to superficial observation, we find that wise and harmonious laws are always concerned.

First Law.—Species of animals and plants have had a limited duration, rarely extending from one formation into another.

We apply the word species in fossils just as we do in living animals and plants. A number of species closely related constitute a genus; a number of genera, having certain common characters, form an Order: several orders a Class, and several classes a Province, or Sub-kingdom, or Kingdom.

Now it is of the species only that we speak under this law. The larger divisions, genera, orders and classes, do extend through more or less of the formations; but in nearly all cases the species become extinct at the close of the great periods pointed out in previous pages of this work. Some distinguished naturalists, if we understand them, as Agassiz and D'Orbigny, are of opinion that there are no exceptions. Even the tertiary species they regard as extinct; but how far they would extend this view into the post-tertiary, we do not know. "The number of species still considered identical in several successive periods," says Agassiz, "is growing smaller and smaller, in proportion as they are more closely compared." Hence he reasonably infers that probably all will be found unlike. Professor Bronn thinks that species sometimes pass not only into a second but into a third formation, and others state, as we have mentioned, that some living species of foraminifera began their existence as low down as the oolite. According to Bronn, out of 2,055 species of plants, 12 pass into other formations; and of 24,366 animals, 3,322 pass out of the rocks where they are most abundant. So that each species had an average duration of 1.12 of a formation. In respect to the rocks below the tertiary, all would agree that the law has scarcely an exception, and still higher, as stated above, they will doubtless diminish upon more careful examination.

Second Law.—With the exception above named, the fossil species have all perished.

This is merely an inference from the first law; bnt it is so common to suppose recent species identical with the fossil, that we make the inference a distinct law.

Third Law.—" The duration of types and species as a general rule, is usually proportioned to rank and intelligence. The most highly organized fossils have the smallest range."—(Owen.)

The great lizards of the Jurassic series, the mammals of the tertiary, and especially man, are examples of this law.

By a type we mean a set of characters by which a genus, or family, or group is distinguished from all others. It is the model or pattern on which such groups are formed. Thus the horse family, the cat family, the ostrich family, the pigeon family, have certain characteristics by which we know them, though sometimes difficult to describe; and it is found that many of these types have gradually changed. This law declares that these types have the shortest duration among the higher tribes.

Fourth Law.—Each type of organism has had but one term of uninterrupted existence, and sometimes has extended only through part of a formation.

There are some seeming exceptions to this law; as for instance the appearance of the marsupial animals in the trias and oolite, and then their failure in the chalk and tertiary, and reappearance in the alluvial. But the probability is that they existed during these intermediate periods, since they are found in the pleistocene, of Australia.

Among the animals extending through a part of a formation, we may mention such us the Mastodon, Elephant, Dinotherium, Zeuglodcn, and Man, which are found only in parts of the tertiary and alluvial.

Fifth Law.—Most of the great Sub-Kingdoms of animals and plants, two thirds of the classes and nearly half the orders, and a few of the genera extend through all the formations.

The only exception in respect to the sub-kingdoms, is, that vertebrate animals are not found in the Lower Silurian, and no deeper in the Upper Silurian than the lower Ludlow Rock; and flowering plants are not found lower than the Devonian, where Hugh Miller has detected coniferous trees in the lower Old Red Sandstone of Scotland. But perhaps in considering this subject we ought to have reference to a palæontological classification, rather than one founded partly on lithological characters, and this would bring all the sub-kingdoms into the lowest life period, which reaches as high as the top of the Permian.

It will be seen by referring to the Table of Organic Remains, which we have presented at the end of the last Section, that while many of the classes and orders of the less perfect animals and plants extend through all the formations, those of the higher vertebrate type rarely reach through the whole series. The number of orders has more than doubled since the earliest times, so that more than half do not reach through all the strata.

In the Palæozoic formations there were	31
In the Triassic Period	21
In the Jurassic	41
In the Cretaceous	41
In the Tertiary	71

Much fewer is the number of genera that have survived all the changes which the globe has undergone. The following statement by D'Orbigny shows strikingly how great have been the changes of the organic world. It is confined to animals:

Number of living genera of animals,	1324
Number of fossil genera,	1457
Of these there yet live,	539
Have become extinct	933
Have survived all changes, only	16

All of these surviving *venerable* genera belong to the different families of Molluscs, while of all the other animals not a genus has been extended through all past periods.

Sixth Law.—Complexity and perfection of organization as well as intelligence increase as we ascend in the rocks.

This is true as a general fact; but in particular tribes we find the reverse, viz., retrogradation from a lower to higher condition. "All our most ancient fossil fishes," says Professor Sedgwick, "belong to a high organic type; and the very oldest species that are well determined, fall naturally into an order of fishes which Owen and Miller place, not at the bottom, but at the top of the whole class." Says Hugh Miller, "in the imposing programme of creation, it was arranged as a general rule, that in each of the great divisions of the procession, the magnates should walk first. We recognize yet further the fact of degradation specially exemplified in the fish and the reptile." "The Cephalopods, the most perfect of the molluscs, which lived in the early period of the world," says D'Orbigny, "show a progress of degradation in their generic forms. The molluscs as to their classes have certainly retrograded from the compound to the simple, or from the more to the less simple."

Such statements are not inconsistent with the law we have stated above; for there may be upward progress by the introduction of higher and higher forms of life, while some of the groups may suffer deterioration, as seems to have been the case. A simple inspection of the tabular view we have given of organic remains will show how strong is the evidence of progress. The only way to escape the inference is to say that higher forms may yet be discovered in the lower rocks. But this is a point of so much importance in its bearings upon certain hypotheses that we shall recur to it again in the next Section.

More impressively to exhibit these facts and to show to the eye the periods when the most important races came upon the globe, we copy Fig. 409 from Professor Owen. We shall have occasion to refer to it again in another connection.

Seventh Law. Particular classes, orders, and genera, as well as whole faunas and floras, have had their periods of expansion, culmination, diminution, and sometimes extinction.

Fig. 410, prepared by Prof. Owen, shows these facts in respect to the orders of Reptiles. The shaded lenses and triangles indicate the periods of their

Fig. 400.

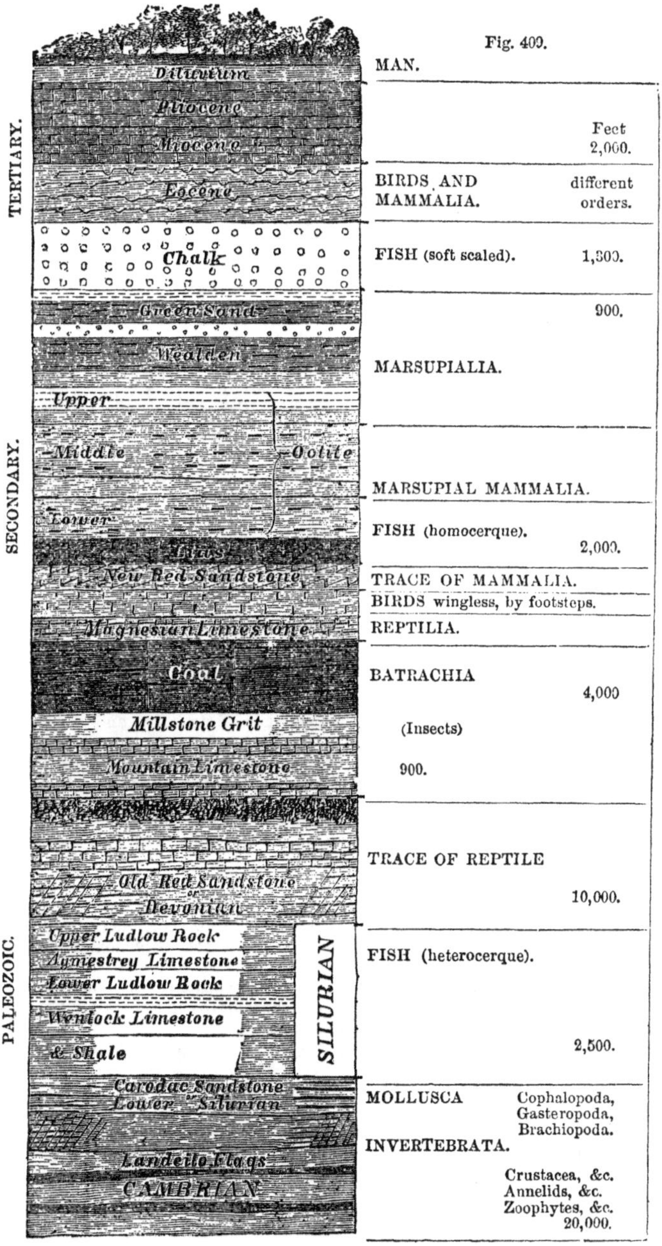

TERTIARY.
Diluvium
Pliocene
Miocene
Eocene
Chalk
Green Sand
Wealden
SECONDARY.
Upper
Middle
Lower
Oolite
Lias
New Red Sandstone
Magnesian Limestone
Coal
Millstone Grit
Mountain Limestone
Old Red Sandstone or Devonian
PALEOZOIC.
SILURIAN
Upper Ludlow Rock
Aymestrey Limestone
Lower Ludlow Rock
Wenlock Limestone & Shale
Carodac Sandstone
Lower Silurian
Landeilo Flags
CAMBRIAN

	Feet
MAN.	
	2,000.
BIRDS AND MAMMALIA.	different orders.
FISH (soft scaled).	1,300.
	900.
MARSUPIALIA.	
MARSUPIAL MAMMALIA.	
FISH (homocerque).	2,000.
TRACE OF MAMMALIA.	
BIRDS wingless, by footsteps.	
REPTILIA.	
BATRACHIA	4,000
(Insects)	
900.	
TRACE OF REPTILE	10,000.
FISH (heterocerque).	2,500.
MOLLUSCA	Cophalopoda, Gasteropoda, Brachiopoda.
INVERTEBRATA.	Crustacea, &c. Annelids, &c. Zoophytes, &c. 20,000.

commencement, expansion, diminution, and extinction. The Chelonia, Lacertilia, Ophidia, and some of the Batrachia, are shown as on the increase at the commencement of the alluvial period; while the Crocodiles and some of the Batrachia then nearly died out. Several of the orders became extinct at the close of the cretaceous period.

Fig. 410.

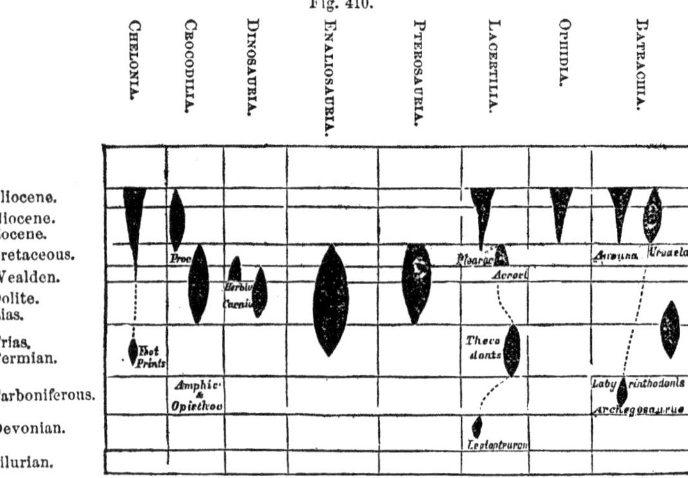

Distribution of Reptiles.

Indeed, Prof. Owen says that "the class of reptiles, unlike that of fishes, is now on the wane; and that the period when Reptilia flourished under the greatest diversity of forms, with the highest grade of structure, and of the most colossal size, is the mesozoic."

D'Orbigny finds that of seventy-seven orders of fossil animals fourteen have decreased in the number of their genera since their first appearance, and sixty-four have increased. These are distributed as follows:

	Decreasing.	Increasing.
Radiated animals	4	12
Molluscs	4	10
Annelids	1	18
Vertebral kingdom	5	23

Of these decreasing orders six are found in the palæozoic rocks, viz., the Placoid and Ganoid Fishes, the Trilobites, a part of the Cephalopod and Brachiopod Molluscs, and the fixed Crinoids. In the Jurassic series occur the Saurian Reptiles and the free Crinoids. In the Cretaceous series, two families of Molluscs, one of Foraminiferæ and one of Amorphozoa. In the tertiary series are the Edentate and Pachydermatous Mammals.

The greatest expansion of particular and peculiar Faunas and Floras has been employed to characterize certain periods. Thus the Palæozoic Period has been called by the botanists the Reign of Acrogens, because that tribe of plants then predominated; the Mesozoic Period, the Reign of Gymnosperms; and the Tertiary Period, embracing also the living plants, the Reign of Angiosperms. In respect to animals, the Palæozoic Period has been called the Reign of Fishes, the Mesozoic the Reign of Reptiles, and the Tertiary the Reign of Mammals.

Eighth Law. The older the rock the more unlike the existing fauna and flora are the fossil animals and plants.

If we compare the plants and shells of the tertiary with those now living, a casual observer would see but little difference. But let the successive groups in the lower rocks be brought into comparison, and the naturalist would be obliged to form new genera and orders for their reception. Still more rapidly do the forms of the higher animals and plants deviate from existing types as we descend.

There are some exceptions to this statement; for some forms are wonderfully persistent. Take, for example, the ammonite and nautilus; how much like the living nautilus! So the terebratulidæ, living and extinct, are closely related. So in the tertiary, although the Dinotherium, the Palæotherium, the Zeuglodon, etc., are quite unlike living forms, yet in the same formation certain small mammifers can hardly be distinguished from those living.

Ninth Law. The fossil faunas and floras were, for the most part, of a tropical character, whatever be the present climate where they are found.

Even the tertiary plants and animals agree, for the most part, with those of intertropical regions better than those of temperate regions; and it was essentially the same even in post-tertiary days, when Europe and the United States were filled with elephants, rhinoceroses, lions, tigers, hyenas, etc., though the hair and wool of the Siberian fossil elephant indicate a colder region than the intertropical; but unless warmer than that at present along the shores of the Arctic Ocean, so many of these huge animals could not have subsisted as are found buried there.

As we go deeper into the rocks the evidences of a former tropical, or even ultra tropical climate, multiply. The coal formation especially, which has been traced beyond Melville Island in N. latitude 75°, is decidedly and strikingly tropical everywhere. The old fossil corals found over equally wide arctic regions—at Melville Island, for instance—tell the same story. And so do the numerous and sometimes gigantic chambered shells so widely diffused.

Some facts seem to indicate an occasional alternation of a colder with the tropical climate, at an earlier date than drift, when we know that in northern regions there was a glacial period. Similar temporary reductions of the temperature may have taken place earlier. But these cases do not invalidate the general law of the prevalence of a tropical climate.

Even in the Pleistocene Period, "Grand, indeed," says an English naturalist, "was the fauna of the British Islands. Tigers as large again as the biggest Asiatic species, lurked in the ancient thickets; elephants of nearly twice the bulk of the largest individuals that now exist in Africa or Ceylon roamed in herds; at least two species of rhinoceros forced their way through the primeval forest, and the lakes and rivers were tenanted by hippopotami as bulky and with as great tusks as those of Africa." To these he might have added the great cave bear and cave hyena, two species of huge oxen, and an elk ten feet and four inches high.

Tenth Law. In the distribution of species in the ancient faunas and floras, they had a much greater range than at present, while in the newer rocks their limits differed but little from existing zoological and botanical provinces.

In the palæozoic strata animals and plants have a striking resemblance over almost the whole globe. As we ascend, diversity increases when we compare species together from widely separated localities; and of the mammalia in the tertiary and post-tertiary, Prof. Owen says, "particular forms were assigned to particular provinces, and the same forms were restricted to the same provinces at a former geological period as they are at the present day."

If we only admit the high temperature of the globe in palæozoic days, and that it has since gradually decreased, the facts above stated are just what we should expect. When a tropical climate existed over the whole earth, the same animals and plants essentially would be placed on every part. But as the temperature fell and the diversities of climate now existing came on, the animals and plants would be gathered more and more into provinces, which would gradually approach to, and finally culminate in those now existing.

Eleventh Law. The fossil animals and plants had the same general structure as those now on the earth, and their modes of living in both classes have been the same.

Comparative anatomy has not found it necessary to frame any new law to embrace the relations of the extinct to the living races. Physiology, also, finds that these extinct races, although greatly differing in form from existing nature, were sustained by the same kinds of food which was digested by analogous organs. They had the same senses; they breathed in the same modes; they were reproduced in the same manner; they were carnivorous and herbivorous; they suffered and enjoyed, and were subject, like the living species, to accident, disease, and death.

Twelfth Law.—"The phases of development of all living animals correspond to the order of succession of their extinct representatives in past geological times." (*Agassiz.*)

This law represents the extinct adult animal as corresponding more nearly with the embryonic than the adult state of its living representative. In the ancient world the individual, though an adult, did not pass beyond the present embryo state; but among living species the analogous animal passes on to a higher state, or more complete development.

Pictet thinks that this law is not applicable to the whole animal kingdom, but to certain groups. Agassiz, however, regards it "as a general fact, very likely to be more fully illustrated as investigations cover a wider ground." To name a few examples, he regards the Trilobites embryonic types of Entomostracea (a tribe of living crustaceans); the Oolitic Decapods embryonic Crabs; the Zeuglodonts embryonic Sirenidæ; and the Mastodonts embryonic Elephants.

Thirteenth Law.—Many of the fossil animals had a combination of characters which among living animals are found only in several different types or classes.

Agassiz very appropriately calls such types *Prophetic Types.* For they form the pattern of animals that were to appear afterward. It is found that almost all the existing animals were thus typified by some characters that existed in the fossil animals. The facts show how completely the whole plan of creation lay in the Divine Mind. We give a few examples:

The Sauroid Fishes were true fishes, yet they had some strongly marked reptilian characters. "The Plesiosaurus," says Buckland, "to the head of a lizard united the teeth of a crocodile, a neck of enormous length resembling the body of a serpent; a trunk and tail having the proportions of an ordinary quadruped; the ribs of a chameleon, and the paddles of a whale." The Ichthyosaurus, as its name denotes, had a close affinity to fishes. "Its general external figure," says Owen, "must have been that of a huge predatory abdominal fish, with a longer tail and a smaller tail-fin; scaleless, moreover, and covered by a smooth, or finely wrinkled skin, analogous to that of the whale tribe." The Archegosaurus seems to have been "a transitional type between the fish-like Bratrachia and the lizards and crocodiles." The Labyrinthodonts were "reptiles having the essential bony characters of the Batrachia, but combining these with other bony characters of crocodiles, lizards, and ganoid fishes." (*Owen.*) The Rhynchosaurus had a "lacertine structure leading towards Chelonia and birds, which before were unknown." (*Owen.*) The Dicynodontia were a race of "reptilian animals once living in South Africa, presenting in the construction of their skull characters of the crocodile, the tortoise, and the lizard, coupled with the presence of a pair of huge sharp-pointed tusks, growing downwards, one from each side of the upper jaw, like the tusks of the mammalian morse or walrus." (*Owen.*) The Pterodactyle, the most anomalous of ancient forms, had the head and neck of a bird, the mouth of a reptile, the wings of a bat, and the body and tail of a quadruped.

If more examples were wanted, ichnology would furnish them abundantly in such remarkable animals as the Otozoum, Anomœpus, Plesiornis, and Gigantitherium.

Fourteenth Law.—The fossil far exceeded the living species in number.

We should expect this if there have been several distinct creations; and in respect to quite a number of classes it is proved by the facts in a most satisfactory manner; though we can not suppose that half the fossil species have yet been found, and many sorts of animals and plants are too soft and frail to be preserved. As to plants, so small is the number found fossil compared to those now living, that we may perhaps doubt whether the single flora now living is not more numerous than all those which have ever lived.

The following table will show the proportion between the fossil and living species in Great Britain:

	Living Species.	Fossil Species.	Proportion of Living to Fossil.
Plants	1600 flowering 2800 flowerless	655	6.7 to 1
Zoophytes	70	435	1 to 6.2
Polyzoa	70	258	1 to 3.7
Testacea (Molluscs, etc.)	513	4580	1 to 8.9
Echinodermata	70	492	1 to 7.0
Crustacea	225	298	1 to 1.3
Fishes	162	741	1 to 4.6
Reptiles	18	180	1 to 10.0
Birds	332	11	30 to 1.0
Mammals	70	110	1 to 1.5

Here we find that six times more zoophytes, nine times more mulluscs, seven times more echinoderms, five times more fishes, and ten times more reptiles have lived in Great Britain during geological times than now exist there. The argument is irresistible to show that many distinct creations have occupied the surface successively and passed away; corroborating the same conclusion drawn from other facts.

Fifteenth Law.—Contemporaneous species in any one locality, or in localities not distant from one another, have appeared and disappeared together.

Some have maintained that the formations pass insensibly into one another, so that near the limits, the fossils of the two adjoining formations are mixed together; and that as individual species have died out, others have taken their place. And it is sometimes true, that there is no trenchant division between adjacent formations. Moreover species do sometimes become extinct, as we have shown elsewhere in respect to existing nature, though there is not the slightest evidence that these species, as they drop out, are replaced by new ones. But in the rocks the group of species that characterize a formation in almost all cases, show themselves together at the bottom, and continue to live together till the close of the period, when all disappear, and the new formation that follows contains an entirely distinct group. So few are the exceptions to this distribution of the species, that it must be considered as the general law, and the exceptions the result of local and unusual causes.

Sixteenth Law.—Numerous and successive systems of life, all different from one another, have occupied the globe since it became habitable.

Long ago Deshayes, a distinguished naturalist, declared that "in surveying the entire series of fossil animal remains, he had discovered five great groups so completely independent that no species whatever is found in more than one of them." Adding the existing group, it makes six entire changes of inhabitants, which accords with the palæontological classification which we have given, viz., the first reaching to the top of the Permian; the second embracing the Trias; the third the Oolite; the fourth the Chalk, and the fifth the Tertiary.

But the ablest palæontologists of the present day feel as if this were a very inadequate view of the subject, falling far short of the number of changes in inhabitants which the earth has experienced. Says the late eminent palæontologist, M. Alcide D'Orbigny, "A first creation took place in the Silurian stage. After that was annihilated by some geological cause, and after a considerable time, a second creation took place in the Devonian stage, and successively *twenty-seven times have distinct creations repeopled all the earth with plants and animals*, following each time some geological disturbance, which had totally destroyed living nature. Such is the certain but incomprehensible fact, which we are bound to state, without trying to pierce the superhuman mystery that envelops it."

Seventeenth Law.—All the diversities of organic life that have appeared on the globe were only wise and necessary adaptations to its changing condition.

There is abundant evidence that changes of climate, food, etc., have been great and numerous, and had there not been a corresponding change in the nature and habits of animals and plants, suffering and death must have been the consequence, as the history of existing races proves.

But there is not the slightest evidence that any such effect followed the modification of forms. Peculiar as they often were, they seem to have been wisely prepared to subserve the wants and happiness of the species, nor was life thereby shortened.

Eighteenth Law.—All the minor systems of life that have appeared, were but harmonious parts of one all-comprehending system of organization, whose culmination we witness in existing nature.

Diverse as the different floras and faunas are in the different creations, they are all embraced in the same system of classification, which groups together existing organisms. They have all had similar organs and similar senses, have been both carnivorous and herbivorous, have had the same relations to light and heat as at present. Nowhere do we find different and antagonistic systems, but all the wide diversities of structure and habit coalesce into one harmonious whole; showing that the complicated and numberless details, stretching over almost interminable ages, were but the development of the vast plan of creation in the Divine Mind.

SECTION IV.

INFERENCES FROM PALÆONTOLOGY, IN CONNECTION WITH DYNAMICAL GEOLOGY.

Inference 1. *The present continents of the globe (except, perhaps, some high mountains) have been for long periods beneath the ocean, and have been subsequently elevated.*

Proof 1. Two thirds at least of these continents are covered with rocks, often several thousand feet thick, abounding in marine organic remains; which must have been quietly deposited, along with the sand, mud, and calcareous or ferruginous matter in which they are enveloped, and which could have accumulated but slowly. 2. Some very high mountains contain marine fossils at or near their summits. For example, there are marine shells of cretaceous age upon the tops of the Pyrenees; cretaceous and tertiary fossils upon the summits of the Rocky Mountains, and foraminifera of cretaceous age high up on the flanks of Mt. Lebanon.

The amount of land above the ocean has varied in every period of the earth's history, and it may be that large tracts, now submerged, once were important theatres of terrestrial life.

Inference 2. *The periods of repose between catastrophes have been long.*

Proof 1. Catastrophes are indicated by unconformability of the strata, or a great change in the character of the deposits. 2. Catastrophes have been comparatively infrequent, while deposition has always continued slowly to build up formations. For example, several thousand feet of strata were deposited during the Lower Silurian period, between two catastrophes. The periods of disturbance must have been very short, and the interval of repose very long. 3. The deposits appear generally not to have been disturbed by any elevating force while in a state of formation, as this would have changed the character of the organic ramains.

There are instances where there seems to have been a quiet, gradual elevation for an immense period, without catastrophes. But often this elevation has been sudden and very great. Some single local dislocations are of enormous size, amounting to 3,000 or 4,000 feet; as in the Penine region of the north of England; and it is difficult to conceive how such faults could have resulted from a succession of minor forces acting through long intervals.

Inference 3. *Catastrophes have generally corresponded to changes in fossils.*

Elie de Beaumont has long maintained that the changes in the zoological and botanical characters of the formations correspond in general to the epochs of elevation; that is, the period of elevation seems to have been the time for the destruction of one group of organic races and the introduction of new species. The progress of Palæontology tends greatly to increase the number of distinct systems of life, and it may not be possible in all cases to find evidence of any great geological disturbance at the close of all the life periods. Yet D'Orbigny, who contends for the largest number of these, still maintains, by a course of strong arguments, that the faunas and floras have all been destroyed by catastrophes, such as the sudden elevation of mountains, though they may have taken place at a distance, and the destruction may have resulted from the great inundating waves that spread far and wide from the center of disturbance. He believes, also, in the existence generally of a long interval between the destruction of one group and the creation of a new one. "We can not then explain," says he, "the annihilation of all the faunas which have succeeded each other twenty-seven times, but by powerful geological disturbances. We have seen that whenever in past ages a dislocation of the crust has taken place, capable of effecting a displacement of the seas, the existing fauna has been annihilated by the movement of the waters at the points dislocated, and even in other points not dislocated."

These decided views may need some modification when the whole subject of the disappearance of species has been more fully studied. At present we know not how to resist the evidence adduced by D'Orbigny in his *Cours Elementaire de Paleòntologie et de Geologie.*

Inference 4. *The whole period since life began on the globe has been immensely long.*

Proof 1. There must have been time enough for water to make depositions more than ten miles in thickness, by materials worn from previous rocks, and more or less comminuted. 2. Time enough, also, to allow of hundreds of changes in the materials deposited: such changes as now require a long period for the production of one of them. 3. Time enough to allow of the growth and dissolution of animals and plants, often of microscopic littleness, sufficient to constitute almost entire mountains by their remains. 4. Time enough to produce, by an extremely slow change of climate, the destruction of several nearly entire groups of organic beings. For although sudden catastrophes may have sometimes been the the immediate cause of their extinction, there is reason to believe that those catastrophes did not usually happen, till such a change had taken place in the physical condition of the globe, as to render it no longer a comfortable habitation for beings of their organization. 5. Time enough for erosions to have taken place in the rocks, in an extremely slow manner, by aqueous and atmospheric agencies, on so vast a scale that the deep cut through which Niagara River runs, between Niagara Falls and Lake Ontario, is but a moderate example of them. We must judge of the time requisite for these deposits by similar operations now in progress; and these are in general extremely slow. The lakes of Scotland, for instance, do not shoal at the rate of more than six inches in a century.

Obj. 1. The rapid manner in which some deposits are formed at the present day; *e. g.*, in the lake of Geneva, where, within the last 800 years, the Rhone has formed a delta two miles long and 600 feet in thickness.

Ans. Such examples are merely exceptions to the general law, that rivers, lakes, and the ocean are filling up with extreme slowness. Hence such cases show only that in ancient times rocks might have been deposited over limited areas in a rapid manner; but they do not show that such was generally the case.

Obj. 2. Large trunks of trees, from twenty to sixty feet long, have sometimes been found in the rocks, penetrating the strata perpendicularly or obliquely; and standing apparently where they originally grew. Now we know that wood can not resist decomposition for a great length of time, and therefore the strata around these trunks must have accumulated very rapidly; and hence the strata generally may have been rapidly formed.

Ans. Admitting that the strata enclosing these trunks were rapidly deposited, it might have been only such a case as is described in the first objection. But sometimes these trunks may have been drifted into a lake or pond, where a deep deposit of mud had been slowly accumulating, which remained so soft, that the heaviest part of the trunks, that is, their lower extremity, sunk to the bottom by their gravity, and thus brought the trunks into an erect position. Or suppose a forest sunk by some convulsion, how rapidly might deposits be accumulated around them, were the river a turbulent one, proceeding from a mountainous region.

Obj. 3. All the causes producing rocks may have operated in ancient times with vastly more intensity than at present.

Ans. This, if admitted, might explain the mere accumulation of materials to form rocks. But it would not account for the vast number of changes which took place in their mineral and organic characters; which could have taken place, without a miracle, only during vast periods of time.

Obj. 4. The fossiliferous rocks might have been created, just as we find them, by the fiat of the Almighty, in a moment of time.

Ans. The possibility of such an event is admitted; but the probability is denied. If we admit that organic remains from the unchanged elephants and rhinoceroses, of Siberia, to the perfectly petrified trilobites and terebratulæ of the Palæozoic strata, were never living animals, we give up the whole groundwork of analogical reasoning; and the whole of physical science falls to the ground. But itis useless formally to answer an objection which would never be advanced by any man, who had ever examined even a cabinet collection of organic remains.

Inference 5.—The period before life appeared, was also immensely long.

Proof 1. We can trace indications of life into the upper part of the Cambrian series. Below this horizon there are at least 30,000 feet of stratified rocks, which must have required an immense period for their formation. 2. Previously to the production of the stratified rocks, the globe had cooled from an incandescent state, at an inconceivably slow rate. It is not unlikely that this period of time was greater than the whole of the fossiliferous era. 3. If we admit the truth of the hypothesis that the world was condensed from a gaseous to the liquid state, we have another period previous to the existence of life immensely protracted, to cool the surface sufficiently to allow of the presence of water.

Inference 6.—The changes which the earth has experienced, and the different species of organic beings that have appeared, were not the result of any power inherent in the laws of nature, but of special Divine creating power.

The opposite hypothesis, when fully stated, embraces three distinct branches. The first supposes the present universe to have been developed by the power of natural law from nebulous matter, without any special Divine interposition, according to the views of the eminent mathematician, La Place. This has been called the cosmogony of the subject. The second supposition is, that certain laws, inherent in matter, are able of themselves to produce the lowest forms of life without special creating power. This forms the Zoogony of the subject. The third supposition is, that in the lowest forms of organization thus produced, called monads, there exists an inherent tendency to improvement. And thus from a mere mass of jelly vitalized, higher and more complicated organic forms have been eliminated, until man at last was the result. This called the Zoonomy of the subject.

The supposed proof of this hypothesis is derived from astronomy, physiology, galvanism, botany, zoology, and geology. But it is only the argument from the latter subject that can receive any attention in this work. When this hypothesis is fully carried out, it is intended and adopted to vindicate atheism. When advocated by a professed believer in the Deity and even in revelation, it is made to assume a much more attractive aspect.

In favor of this hypothesis of *creation by laws*, it has been argued, 1. That in the oldest fossiliferous rocks we find chiefly the more simple invertebrate animals and flowerless plants, and the more perfect ones came in gradually, increasing in numbers and complexity of organization to the present time. The lowest vertebrate animals were fish; then reptiles succeeded, then birds, then mammals, then man. Here we see the series gradually expanding, just as this theory requires. 2. There was probably a distinct *stirps*, or root, for each of the great classes of animals and plants, with which it started, from which the development proceeded along as many great lines as there are classes. This supposition shows why we find representatives of all the classes in the lowest rocks.

In answer to these arguments, and as proofs of the sixth inference, we remark 1. That in all the more than 30,000 species of organic remains dug from the rocks, they are just as distinct from one another as existing species, nor is there the slightest evidence of some having been developed from others. 2. The gradual introduction of higher races is perfectly explained by the changing condition of the earth, which being adapted for more perfect races, Divine Wisdom introduced them. 3. For the most part the new races were introduced by groups, as the old ones died out in the same manner. The new groups were introduced at once; pointing clearly to creation rather than development. 4. If anywhere, we ought to find evidence of development and metamorphosis in the human species. But so immeasurably is man raised by his moral and intellectual faculties above the animals next below him in rank, that the idea of his gradual evolution from them is absurd. Man's moral powers, for instance, which are his noblest distinction, do not exist at all in the lower animals. Nothing but miraculous creation can explain the existence of man. 5. The admission of a distinct stirps for each of the classes, is a virtual abandonment of the whole hypothesis; for it admits, for example, that a flowering plant and a vertebral animal commenced two of these series, although to reach such a height or organization, requires, by the same hypothesis, a transmutation through all the flowerless plants and invertebrate animals. 6. There is decisive evidence that in many cases during the geological periods, animals, instead of ascending, descended on the scale of organization from the more to the less perfect. 7. Geology shows us that there was a time when organic life first appeared on the globe, and an indefinitely long period when no animals or plants existed. What gave the laws of nature the power, all at once, to start the new races? Why was not that power put forth earlier, or even from eternity, if the world existed from eternity? In short, of all the sciences, geology affords the fewest facts to sustain this hypothesis. No other science presents us such repeated examples of special miraculous intervention in nature.

Inference 7. The changes which have occurred on the globe, both organic and inorganic, have shown progress from the less to the more perfect.

Proof 1. As the temperature of the interior of the earth is much higher than that of surrounding space, by the laws of heat there must be a constant radiation of heat into space, and unless this can be proved to have proceeded in a cycle, or without end,—which can not be done,—the earth must have been constantly undergoing physical changes. If this process of refrigeration has been going on long enough, there must have been a time when the surface was too hot for any kind of organic beings to exist upon it. And when it became possible for some sorts to be placed upon it, it was still unadapted for those of complicated organization. 2. Accordingly, we find but a few of the flowering plants, or of vertebral animals, in the lowest rocks, and their number and perfection have for the most part increased from the first, while the lower classes have made but little progress, and perhaps in some instances have retrograded. 3. The surface has been rendered capable of sustaining beings of a higher organization in three modes; first, by the operation of aqueous and atmospheric agencies the quantity of soil has been increased; secondly, animals and plants have eliminated lime from its more hidden combinations, and converted it into carbonate and sulphate; thirdly, the surface has reached a statical condition, and the climate is more congenial to such natures.

Obj. Almost every year brings to light in the rocks evidence of the existence of more perfect animals and plants at an earlier date than had been known, and since the greater part of the earlier fossils are marine, perhaps the number of air-breathing vertebrate animals and of flowering plants found among them, is almost as great as we ought to expect, even if the present condition of things has existed from the earliest Silurian periods.

Ans. It is true that one or two examples of Batrachians and Chelonians have been found as low as the Devonian series, but not one in the vast formations below, nor a single example of mammals till we rise to the trias; whereas in the tertiary we find 392 species of mammals, and in the alluvial 358 species; and among existing animals 2,030 species; and a similar prodigious increase of more perfect forms exists in almost all other vertebral tribes and vascular plants. While, therefore, the discovery of now and then a species of higher organization shows that their existence was possible at the earlier periods, yet it will require a vast number of such discoveries to prove the proportion between the more and the less perfect to have been then as now. And until that be proved, the evidence of progression remains unaffected.

Inference 8. *The causes of geological change have varied in intensity. There are two theories upon this subject.*

In his address before the London Geological Society in 1851, Sir Charles Lyell states what is called the uniformitarian hypothesis, as follows:—"That the ancient changes of the animate and inanimate world, of which we find memorials in the earth's crust, may be similar, both in kind and degree, to to those which are now in progress."

Proof 1. It is agreed on all hands that the nature of geological causes has been the same in all ages; although even as late as the time of Cuvier, he says that "none of the agents nature now employs were sufficient for the production of her ancient works."

2. An indefinite repetition of an agency on a limited scale, can produce the same effects as a paroxysmal effort of the same agency, however powerful; provided the former is able to produce any effect, as, for instance, in the accumulation of detritus, the elevation of continents, the dislocation of strata, etc. Now it is unphilosophical to call in the aid of extraordinary agency, when its ordinary operation is sufficient to explain the phenomena.

3. Nearly every variety of rock found in the crust of the globe has been shown to be in the course of formation by existing aqueous and igneous agencies; and if a few have not yet been detected in the process of formation, it is probably because they are produced in places inaccessible to observation.

The opposite hypothesis admits that no causes of geological change different in their nature from those now in action, have ever operated on the globe; in other words, that the geological processes now going on, are in all cases the antitypes of those which were formerly in operation; but it maintains that the existing causes operate now, in many cases, with less intensity than formerly.

Proof 1. The spheroidal figure of the earth, and other facts already detailed, seem to render almost certain the former fluidity of the globe. Now, whether that fluidity was aqueous or igneous, or both in part, it is certain that the agencies which produced it must have operated in early times with vastly greater intensity than at this day, and that their energy has been constantly decreasing from that time to the present.

2. Still more direct is the evidence from the character of organic remains in high latitudes, of the prevalence of a temperature in early times hotter than tropical; too warm, indeed, to be explained by any supposed change of levels in the dry land. And if this be admitted, heat must have been more powerful in its operation than at present; and this would increase the aqueous, atmospheric, and organic agencies of those times.

3. No agency at present in operation, without a vast increase of energy, is adequate to the elevation, several thousand feet, of vast chains of mountains and continents, such as we know to have taken place in early times. A succession of elevations by earthquakes, repeated through an indefinite number of ages, the vertical movements being only a few feet at each recurrence, is a cause inadequate to the effect, if we admit that earthquakes have exhibited their maximum energy within historic times.

4. In a majority of cases, the periods of disturbance on the globe appear to have been short compared with the periods of repose that have intervened; as is obvious from the fact that particular formations have the same strike and dip throughout their whole extent; unless some portions have been acted upon by more than one elevatory force; and then we find a sudden change of strike and dip in the formations above and below. Whereas, had any of the causes of elevation now in operation lifted up these formations by a repetition of their present comparatively minute effects, there ought to be a gradual decrease in the dip from the bottom of the formation upwards, and no sudden change of dip between any two consecutive formations, unless some strata are wanting. At the periods of these elevatory movements, therefore, the force must have been greater than any that is now exerted, to produce analogous effects.

5. The sudden and remarkable changes in the organic contents of the strata, as we pass from one formation to another, even when none of the regular strata are wanting, coincides exactly with the supposition of long periods of repose, succeeded by destructive catastrophes. Nor is the supposition that species of animals and plants have become gradually extinct, and have been replaced by new species, by a law of nature during periods of repose, sustained by any facts that have occurred within the historic period: no example having been discovered of the creation of a new species by such a law; and only a few examples of the extinction of a species.

6. Upon the whole, were we to confine our attention to the tertiary and alluvial strata, it might be possible to explain their phenomena by existing causes, operating with their present intensity. But when we examine the secondary, palæozoic, and hypozoic rocks, we are forced to the conclusion that this hypothesis is inadequate; and that we must admit a far greater intensity in geological agencies in early times than at present.

8. But the question here arises, how long a period shall we assume as a measure of the intensity of existing agencies? The most strenuous advocates of the doctrine of uniformity will admit of some oscillation in the intensity of these agencies; because a single year shows it. How, then, shall we determine how wide that oscillation may be? In order to obtain the average intensity, how can we say but that all geological cycles must be included? To make any particular portion of time the measure of all the rest, must be an arbitrary assumption. And, therefore, we can not ascertain what is the standard or the average of intensity; and until this can be done, is the subject considered under this head any thing more than a controversy about words?

PART III.

CONNECTION BETWEEN GEOLOGY AND NATURAL AND REVEALED RELIGION.

1. ILLUSTRATIONS OF NATURAL RELIGION FROM GEOLOGY.

1. *Geology shows us that the existing system of things upon the globe had a beginning.*

Proof 1. Existing continents have been raised from the bottom of the sea, where most of their surface was formed by depositions. 2. With a few exceptions, the existing races of animals and plants must have been created since the deposition of all the rocks except the alluvial, because their remains do not occur in the older rocks. Hence it appears that not only the present races of organic beings, but the land which they inhabit, are of comparatively modern production.

Inf. 1. Hence it is inferred that the existing races of animals and plants must have resulted from the creative agency of the Supreme Being; for even if we admit that existing continents might have been brought into their present state by natural causes, the creation of an almost entirely new system of organic beings, could have resulted only from an exertion of an infinitely wise and powerful Being. Indeed, the bestowment of life must be regarded as the highest act of omnipotence.

Inf. 2. Hence the doctrine which maintains that the operations of nature have proceeded eternally as they now do, and that it is unnecessary to call in the agency of the Deity to explain natural phenomena, is shown to be erroneous.

Inf. 3. The preceding inferences being admitted, natural theology need not labor to disprove the eternity of matter, since its eternal duration might be admitted, without affecting any important doctrine.

2. *In all the conditions of the globe from the earliest times, and in the structure of all the organic beings that have successively*

peopled it, we find the same marks of wise and benevolent adaptation, as in existing races, and a perfect unity of design extending through every period of the world's history.

Proof 1. The anatomical structure of animals and plants was very different at different epochs; but in all cases the change was fitted to adapt the species more perfectly to its peculiar condition. 2. To communicate the greatest aggregate amount of happiness, is a leading object in the arrangements of the present system of nature; and it is clear from geology, that this was the leading object in all previous systems. 3. The existence of carnivorous races among existing tribes of animals tends to increase the aggregate of enjoyment, first, by the happiness which those races themselves enjoy; secondly, by the great reduction of the suffering which disease and gradual decay would produce, were they not prevented by sudden death; and thirdly, by preventing any of the races from such an excessive multiplication as would exhaust their supply of food, and thus produce great suffering. Now, we find that carnivorous races always existed on the globe, showing a perfect unity of design in this respect. Thus, when the chambered shells, so abundant in the secondary rocks, and which were carnivorous, became extinct at the commencement of the tertiary epoch, numerous univalve molluscs were created, which were carnivorous; although till that time these races had been herbivorous.

Inf. From these statements we infer the absolute perfection, and especially the immutable wisdom of the Divine character. A minute examination of the works of creation as they now exist, discloses the infinite perfection of its Author, when they were brought into existence; and geology proves Him to have been unchangeably the same, through the vast periods of past duration, which that science shows to have elapsed since the original formation of the matter of our earth.

3. *Geology furnishes many peculiar proofs of the Divine benevolence, so peculiar that they have sometimes been quoted in proof of penal inflictions.*

Most of these proofs are derived from agencies whose immediate effects are destructive and desolating. Thus soils, which are little else than comminuted rocks, can not be prepared and spread over the valleys without long and powerful erosions by ice and water, storms and inundations, glaciers and icebergs. But though sometimes involving men and animals in destruction, yet who will doubt the benevolence of the operation? So the processes by which the various ores have been put into the earth's crust, have been accompanied by violent fracture and dislocation, and a semifusion of most of the strata. How little like benevolence, also, to have seen the crust bent, crumpled and fractured, here ridged into mountains, and there sunk into valleys. Yet without all this man never could have got access to many of the useful minerals and rocks, water would have stagnated on the level surface, and the beautiful scenery of the globe would never have been seen. In the fearful history of volcanoes and earthquakes, though full of scenes of appalling suffering, yet who knows how essential they may be to preserve the balance of nature, and prevent the great furnace of heat within the earth from rending it to atoms?

If any inquire why God could not have secured the good without the evil? it can only be said, this is a fallen world, where man requires the discipline of evil, and therefore it is mixed with all sublunary things.

4. *Geology furnishes interesting examples of what may be called prospective benevolence.*

By this is meant a special benevolent provision for the happiness of animals, made long before their existence. The following are examples:

1. The vast amount of coal found in the earth is the result of long and slow processes in the ages far back towards the beginning. Vast forests, almost untenanted by animals, and seemingly of no use then, were buried beneath the soil and waters, and gradually changed into peat, brown coal, bituminous coal, and some of it into anthracite. What if this storehouse of fuel had not been laid up? Human society could not have advanced much beyond barbarism, nor have multiplied as it has done.

2. Gold seems not to have been introduced into the rocks till just long enough before man's appearance to allow erosive agencies to collect it in the low spots, where man could obtain it. Before man no animal needed it, but how great a blessing to man! It does seem as if the time and manner of its introduction into

the earth's crust pointed most unmistakably to man as an act of prospective benevolence.

3. It looks like the same benevolence that prepared by slow processes a richer soil to greet man than had ever before existed, and afford him nourishment.

4. So, too, there is reason to suppose that certain miasms, such as an excess of carbonic acid, were gradually removed from the atmosphere to adapt it to his health and happiness.

5. *Geology proves repeated special divine interpositions, or miracles, in nature as well as special providences.*

A miracle is an event that can not be explained by the laws of nature, but takes place in opposition to those laws or by their agency intensified or diminished.

A special Providence is an event brought about apparently by second causes, but those causes have been so arranged or modified by Divine agency out of sight, that some specific object is accomplished, which would not otherwise be effected.

Geology abounds with examples of miracles and special providences as thus defined. We know that the time was when no animal or plant lived on the globe, because it was a molten world. What but a miracle could have filled it with inhabitants? We know that in after ages whole races died out and new ones came in, so that numerous entire changes of population occurred. A miracle certainly was essential at each change—to create the new ones, if not to destroy the old races. Or if we set aside all these cases, we know that man was introduced among the latest of animals; and if his creation was not a miracle, no event could be.

So the various circumstances mentioned under the last head as examples of prospective benevolence, all pointed through long ages so significantly to man, that true philosophy must regard them as arranged with special reference to him by the Deity, and are therefore indicative of special providence.

Thus may we with confidence put down miracles and special providences as articles in the creed of natural religion, where they have not till lately been found. They of course take away all presumption against analogous doctrines in revelation.

6. *In spite of these evidences of Divine benevolence, geology unites with all other sciences, and with experience, in showing the world*

to be in a fallen condition, and that this condition was foreseen and provided for, long before man's existence, so that he might find a world well adapted to a state of probation.

Proof 1. It appears that the laws and operations of nature, have been the same, essentially, as at present in all ages. 2. That the same systems of sustenance, reproduction, and death, have always prevailed.

Inf. 1. Hence it must always have been impossible, in this world, to have avoided severe suffering; *e. g.*, pain and death.

Inf. 2. Hence it has never been a such a world as perfect benevolence would have prepared for perfectly holy and happy beings; though benevolence has always so decidedly predominated in it, as to show it to be a world of probation and mercy, not of retribution.

7. *Geology enlarges our conceptions of the plans of the Deity.*

1. The prevailing opinion, until recently, limits the duration of the globe to man's brief existence, which extends backward and forward only a few thousand years. But geology teaches us that this is only one of the units of a long series in its history. It develops a plan of the Deity respecting its preparation and use, grand in its outlines, and beautiful in its execution; reaching far back into past eternity, and looking forward, perhaps indefinitely, into the future.

2. Each successive change in the condition of the earth thus far, appears to have been an improved condition; that is, better adapted for natures more and more perfect and complicated. In its earliest habitable state, its soil must have been scanty and sterile, and almost destitute of calcareous matter, except in the state of silicates, which plants decompose with difficulty. The surface, also, was but little elevated above the waters; and of course the atmosphere must have been very damp; though the temperature was very high. Every subsequent change appears to have increased the quantity and fertility of the soil, the amount of the salts of lime and humus, and the dryness of the atmosphere. Should another change occur, similar to those through which it has already passed, we might expect the continents to be more fertile, and capable of supporting a denser population.

3. It appears that one of the grand means by which the plans of the Deity in respect to the material world are accomplished, is

constant change; partly mechanical, but chiefly chemical. In every part of our globe, on its surface, in its crust, and we have reason to suppose, even in its deep interior, these changes are in constant progress; and were they not, universal stagnation and death would be the result. We have reason to suspect, also, that changes analogous to those which the earth has undergone, or is now undergoing, are taking place in other worlds; in the comets, the sun, the fixed stars and the planets. In short, geology has given us a glimpse of a great principle of *instability*, by which the *stability* of the universe is secured; and at the same time, all these movements and revolutions in the forms of matter essential to the existence of organic nature, are produced. Formerly the examples of decay so common everywhere, were regarded as defects in nature; but they now appear to be an indication of wise and benevolent design;—a part of the vast plans of the Deity for securing the stability and happiness of the universe.

2. BEARINGS OF GEOLOGY UPON REVEALED RELIGION.

Since many truths are common to natural and revealed religion, it is not easy to draw the line exactly between the bearings of geology upon these two departments of theology.

There are, too, some erroneous notions widely prevalent on the subject, which need to be corrected before a person can look at it in its true light.

One is, that geologists in their writings have arrayed the facts of their science against revelation. But the fact is, that the whole range of geological literature scarcely furnishes an example of this sort from any geologist of distinction. Such attacks, when made, have come from mere sciolists in the science, or from men learned in other departments, but no geologists.

Another is, that the bearings of geology upon religion are those of conflict rather than of illustration and corroboration. The fact is, that most cases of supposed collision have turned out already to be mere illustration: just as modern astronomy has shown us how to understand certain passages of the Bible relating to the rising of the sun and immobility of the earth, so has geology cast similar light upon passages relating to the age of the world and the introduction of evil. And although some few points may still have an aspect of collision, the reverse is almost universally true; and we may now say that geology illustrates rather than opposes revelation.

A third false notion is, that the principles of geology are unsettled and constantly changing, and that in fact it is chiefly made up of vague and conflicting hypotheses. That there are in geology, as in other physical sciences, unsettled points and doubtful hypotheses, is admitted. But its leading principles are as well settled nearly as those of chemistry, astronomy, and physiology. Especially is it true that those principles which bear upon religion are rarely modified by new discoveries, but rather established more firmly.

Hence we see how false is the position some professed friends of religion take, who say that the time has not yet come to attempt a reconciliation of geology and religion, and therefore they will believe the latter on the principle of faith, because the Church does, and wait for further developments. Such a sort of belief, with philosophic minds, is usually little else but covert infidelity, and instead of honoring, it dishonors religion, by admitting that as yet it can not be defended against the attacks of science.

Hence, too, we see the error of maintaining, as some do, that geology ought not to be allowed to modify at all our views of the meaning of Scripture, or any of its truths. For astronomy, chemistry, and physiology, as well as civil history, have been allowed to make such modifications; why should a like power be denied to geology, if its leading principles are settled?

Different stand-points from which to judge of the Religious Bearings of Geology.

Three classes of men have written concerning the connection between geology and religion. The first are professed believers in revelation; but they do not suppose the Mosaic record to be inspired and infallible as to history or science; and hence they are not surprised to find discrepancies and absurdities in what they regard as a myth or fable of the creation got up by Moses to accomplish some important purpose, but not inspired.

The second class are firm believers in the Bible, but not in geology, which they consider so unreliable that it ought not to be taken into account at all in the interpretation of Scripture; nay, they consider the science, as well as its teachers as really hostile to Scripture, and therefore to be met by the most determined resistance.

The third class believe in the divine inspiration and authority of every part of the Bible; but they admit also the great principles of geology, and think the two records not only reconcilable, but that they cast mutual light upon each other, and that geology lends important aid to some of the most important truths of revelation.

With this last class our views coincide entirely, and we regard it as useless in this work to describe the theories by which the other classes attempt to sustain their views, since the authority of the Bible is destroyed by the first, and the settled principles of science ignored by the second. The third is the stand-point which we shall occupy in enumerating the most important illustrations and corroborations derived by revelation from geology.

We think it is an error committed by some of the ablest writers on this subject, that they have attempted to draw out a complete system of reconciliation and illustration between Genesis and geology. For it is obvious that the Mosaic account is fragmentary; or, as an able writer has expressed it, it gives us only the *memorabilia* of creation, but not a full and detailed account. Hence if we expect to find in the Scriptures something corresponding to all the details of geology, and in the same order, we shall be disappointed; because it was not the object of Moses to give us a full account of the creation, and in a scientific dress. Let us now enumerate some of the points in revelation that derive support or illustration from geology, and also show the harmony of the two records.

1. *The Scriptures and geology agree in not fixing the time of the creation of the world.* The Bible says it was made "in the beginning," and language is scarcely capable of more indefiniteness as to time; nor is there any necessary connection between this general proposition and the facts which follow.

Geology is alike indefinite. We see, indeed, on its records a great number of distinct facts, but no clue is given as to their chronology; and in fact no hint as to the first act, the production of matter.

We might stop here, and with good reason take the ground, that having proved the preceding proposition, nothing further is necessary entirely to answer all objections against revelation on the ground that its chronology does not agree with the records of geology. No matter how old geology makes the world; it is not older than the "beginning" of Scripture.

2. *They do fix the time when man appeared.* The Bible represents him as the last of the animals created, and from him a series of chronological dates is carried forward to the present time. His remains, too, are found only in alluvium, the most recent of the formations. This is a most interesting coincidence.

3. *They agree in representing creation as the work of God.* This is very marked in the Bible, and geology presents numerous exigencies in which no law of nature, no transmuting process will answer,—nothing but the special creating power of the Deity.

4. *They agree in representing instrumentalities as employed in the work of creation.* God commanded the earth to bring forth grass, and herb yielding seed, on the third day, and the waters every living thing that moveth, on the fifth. Divine energy was of course concerned, but these were the instruments. So from geology we learn that immense periods were consumed in preparing by natural operations for the introduction of animals and plants.

5. *They both represent creation to be a progressive work, completed by successive exhibitions of Divine power, with intervals of repose.* How long the intervals were, according to the scriptures, will depend on the meaning which we attach to the word day. But if they were only common days, the acts of creation would still be successive and the work progressive.

Geology, too, teaches us most distinctly that the various animals and plants were not introduced at once, but at intervals widely separated. This is an interesting coincidence between the two records; because we should beforehand presume that all the races would be introduced by one creative act.

6. *They agree in representing the continents as covered an indefinite period by the ocean, and subsequently elevated above it.* Geology testifies to several vertical movements of this kind; the scriptures mention but one, which perhaps was intended to stand as a representative of all.

7. *They agree in giving to the earth a very early revolution on its present axis.* The very first day in the Bible, while yet the ocean covered the continents, is represented as having its evening

and morning, just like all the rest. This was before the existence of animals and plants. But geology shows that this evening and morning commenced still earlier, even while yet the earth was in a molten state; for we find the earth flattened at the poles exactly so much as it would be by a revolution on its axis in twenty-four hours. After its consolidation, such a revolution would not have thus flattened the poles; and while fluid, if it had turned faster than it now does, the poles would have been more flattened; or if slower, they would have been less flattened. The proof is conclusive, therefore, that it revolved as it now does as early as when it was in a molten state. This fact is fatal to several fine theories, which have been based on the supposition that before the fourth day, when the sun and moon was created, the earth's revolution was much slower than afterward; and, therefore, Moses did not intend us to understand the days as periods of twenty-four hours. Science now shows that such has always been their length.

8. *The Mosaic account of creation allows us to suppose an indefinite interval between the beginning and the first day, which may correspond to the vast periods of geological history.* After the first production of matter, it is said to have been covered by water and darkness, and to be without form and void, that is, invisible, or waste, and unfinished. Now how long it may have remained in such a condition, who can tell? It may have been long enough to pass through the changes which geology discloses, except that which prepared the way for the introduction of the present races. All this may be admitted, whatever views we take of the nature of the six days.

If all will admit this, as nearly all do, why may we not rest here, and say that it is unnecessary to go farther in order to show the harmony between geology and scripture. For here we have an admitted interval in the Mosaic account, sufficient to stretch over all the geological periods, and why need we trouble ourselves to inquire into the nature of the six days, whether they be natural days or longer periods. We fully vindicate the scriptures from collision with science, by planting ourselves on this admitted interval. And this is the second resting-place of this kind which we have already found. But inquisitive minds are not satisfied without an attempt to enucleate the meaning of the term day in Genesis, and therefore we take up that subject.

9. *The six days of creation, in the view of eminent writers, may be used figuratively for indefinite periods.* This opinion found advocates as early as the times of the Christian fathers, Augustine, Origen, etc., and in more modern times has been ably defended by Hahn, De Luc, Professors Lee and Wait of England, and by Professors Silliman and Guyot of this country. They maintain that the word day is used thus figuratively in all languages; that it is so used in Gen. 2–4; that the seventh day, or God's Sabbath, has not yet terminated, and, therefore, the previous days may have been equally long, and that such an interpretation corresponds remarkably with the traditions and cosmogonies of many heathen nations. Yet others object that such a meaning is forced and unnatural in a passage where everything else seems literal, and that the sacred writers have shown what meaning they attached to this word in the fourth commandment, where it is impossible to doubt that the six days in the first part are literal days, because they are days of labor; and so must also be the six days referred to in the latter part, in which the Lord made heaven and earth.

But though it is difficult to believe that Moses had any other than natural days in mind, most reflecting persons who read the whole chapter, will feel that in reality they must be different, and perhaps they will say, like St. Augustine, "it is very difficult to conceive, much less to explain, what sort of days those were." Another view has been proposed which excites unusual interest at the present time. It is the following:

10. *We may understand the days as symbolically representing indefinite periods.* A symbol is the representative of something else. The word is taken in all respects in its literal signification, yet it has a higher meaning. Moses probably understood, and meant his readers should understand, the days of creation as literal days; but they actually symbolize higher periods; just as days, weeks, and times are used in prophecy (which often has a symbolical form) for years.

The great advantage of this view of the subject over that which makes the days a figurative representation of long periods, is, that hereby we can take the scriptural statement in its plain, literal sense, yet those literal days may be stretched by symbolization over the widest periods which geology shows to have separated the Divine creative acts. It is no error, if a man chooses to understand the six days of creation as literal days; nor any error for the geologist to make them symbolize vast periods.

11. *The biblical account of creation may be regarded as a succession of pictures with existing nature on the foreground.* Ever since this, the pictorial method, was suggested by Dr. Knapp, in 1789, it has been a favorite mode of representation among authors; the most brilliant exhibition of which was by Hugh Miller. But three errors have generally pervaded these representations. The first is, that the six pictures in Genesis embrace every geological change the earth has undergone; secondly, that they are given in true chronological order; and thirdly, that in the life pictures the plants and animals now found fossil, not the existing species, occupy the foreground. Inextricable confusion and discrepancy have resulted from the mixture of such elements. But only admit that the sacred writer intended to give only certain prominent scenes in creation (its most important memorabilia), and not always in true chronological order, and that existing animals and plants were the models before him, the fossil species coming in on the background only by implication, and all the pictures become luminous, beautiful, and harmonious.

12. *By such a mode of description the sacred writer was not bound to give, and indeed could not give, always the true chronological order of creation.* To make this evident we subjoin in parallel columns the principal events as they are revealed by the sacred penman and by geology.

The right hand column gives as fair a view as we can of the order of creation as developed by geology; the names of the several classes being given when they first appear, and their greatest development by small capitals. The left hand column gives the principal results of the six days' work according to Scripture; and where there seems to be no doubt of parallelism, they are placed opposite to events in the geological record. An examination of this table leads to several important conclusions.

1. We learn that some events found in one column do not occur in the other. The igneous fluidity of the globe is one of the best established conclusions of geology; but it is not named in the Bible. The introduction of numerous groups of animals and plants at different periods is another settled fact in geology; but

Day	Scriptures	Geological record
6 Day	MAN, Mammals, and Reptiles.	MAN: FULL FAUNA AND FLORA. Alluvium.
		MOLLUSCA, ARTICULATA. MAMMALIA; DICOTYLEDONS. Tertiary.
5 Day.	Birds and Sea Animals.	RADIATA; MOLLUSCA. Chalk.
		BIRDS; REPTILES. Oolite.
		Reptiles. Trias.
4 Day.	Sun, Moon, and Stars Created.	Saurian Reptiles. Permian.
		Dicotyledons; ACROGENS. Carboniferous.
		Batrachians. FISHES. Coniferæ? Devonian.
3 Day.	Plants of all sorts. Land emerges.	Fishes.
2 Day.	Atmosphere Created.	Articulata. Radiata. Mollusca. Algæ. Silurian and Cambrian.
1 Day.	Light, Darkness and Ocean.	Mostly ocean. Azoic.
		Igneous Fluidity.

the Scriptures name only one creation of the great classes. On the other hand, the creation of the atmosphere on the second day, and of the sun, moon, and stars on the fourth, have no counterpart in the geological record.

2. There are several rather striking coincidences between the two records as to the order of events and the kinds of organisms introduced. Both show us, in early times, the continents beneath the ocean, and subsequently lifted out of it. Birds and sea animals are introduced on the fifth day, which may reasonably correspond to oolitic times, when birds and reptiles appeared in large numbers, if we may depend upon the tracks of the former as proof. Land reptiles and mammals, or quadrupeds, come in not till the sixth day, which may well be regarded as synchronous with the tertiary formation, when, according to geology, they were first fully developed. Man, too, on both records is represented as the last animal created: a coincidence of great interest.

3. There exist also several diversities on the two records as to the nature and order of events. We do not call them discrepancies; for they are so different in nature as to be incapable of being compared. Thus, the creation of the atmosphere is repre-

sented as occupying the whole of the second demiurgic day. But geology has no record of such an event, and therefore no comparison can be instituted. The same is true of the creation of the sun and moon on the fourth day. It does seem remarkable, however, that these luminaries should be represented as created not until after the vegetable world on the third day, if the writer had intended to preserve the true chronological order of events. No impostor would have been so short-sighted as to commit such a blunder; hence there must be some other reason for such an arrangement. Alike strange is it to find the creation of the atmosphere placed so much before that of the heavenly bodies, when these, as things now are, seem indispensable to atmospheric phenomena.

4. The most important conclusion drawn from this table is, that the sacred writer did not and *could not* give the true chronological order of events. The different classes of animals and plants, according to the geological record, appeared at different periods; the same class often several times repeated, and with different degrees of development. Thus, plants began with the lowest class, the Algæ, and not numerous, in the Cambrian slates, the oldest of fossiliferous rocks. In the Devonian a few acrogens and coniferous plants appeared. In the Carboniferous there was an immense development of acrogens, or flowerless trees, and some dicotyledons. The latter, however, the most perfect of plants, were not fully developed till the tertiary, and still more fully in alluvium. Yet plants are all represented as created on the third day. How was it possible, then, to give the chronological date or order of their creation unless the sacred writer had gone into the scientific details above hinted at? The same is true of the groups of animals, which in the Bible are more comprehensive and indefinite than those of science, because they are such as are in popular use. By the plan of the inspired writer, the time and order of their appearance could not be given, and, therefore, the discovery of any diversity in this respect between revelation and science is no objection to the former, because it is not responsible for the time and order of events, but only for their truth. And if this is so in regard to the organic world, why may it not be so in regard to the other events described? Moses wished to give a pictorial representation of some of the principal events in the work of

creation, and, therefore, he conformed to a chronological order only so far as his leading object required. It would be natural for him to begin his pictures with the world in a chaotic state, buried by darkness and water, with the light just breaking in. According to ancient ideas there was an ocean above as well as below, and this might have suggested the formation of the firmament on the second picture. It was natural next to bring up the submerged land and adorn it with vegetation. This might awaken the thought of introducing the heavenly bodies. And now it might occur that everything was ready for the introduction of animals into the atmosphere and the waters; and last of all to let the most perfect of animals come in with man.

These may not, and probably were not the reasons why, as we suppose, Moses departed from a chronological arrangement of his six pictures; but they show that there might be reasons for doing this. It has been and still is almost universally assumed, that Moses gives a connected and chronological history of creation; and then ingenuity has been taxed to the utmost to accommodate the facts to such a supposition. But if we may reasonably suppose that he meant *only to give certain leading and selected facts, conformed to a chronological order only so far as suited his purpose*, just as one might select certain facts from the early history of the country, and show them by pictures arranged so as to produce the best effect, without reference to dates, it relieves the sacred writer from all responsibility as to chronological order and scientific arrangement, and really does more to bring out the beauty of the Mosaic history of creation, and to bring it into harmony with science, than almost all other principles.

13. *Geology and the Bible agree in representing physical evil as in the world before man.* Geology shows that the same mixed system of suffering and enjoyment, of liability to painful accident and inevitable death, has always prevailed as they now do. The Bible, too, intimates that death and other evils preceded man. Of what use was the threatening of death if no example of it existed among animals? Again, plants were created with seeds in them, and animals made male and female for the production of a succession of races, and such a system implies a correspondent system of death. The human family might have been specially preserved by the fruit of the tree of life, perhaps, from the com-

mon lot, till they had sinned, when they too must die. Again, the selection and fitting up of a spot eastward as the Garden of Eden, as a place for man while holy, and his expulsion from it after he had sinned, implies that the world generally was, as now, a world of evil and suffering. It was made so from the beginning, because it would ultimately become a world of sin, and sin and death are inseparable. If animal existence is, on the whole, a blessing in such a world as the present, or if animals may live hereafter, and receive some compensation for their sufferings here, the time when they suffer, be it before or after man's apostasy, makes no difference.

14. *Zoology and geology throw doubt over the literal universality of the deluge of Noah.* The many vertical movements of continents taught by geology afford a presumption in favor of the Noahian deluge. But the science also shows the absurdity of a wide-spread opinion, that the numerous marine shells and plants found fossil in the rocks were deposited by the deluge. For they extend through more than ten miles in thickness of rocks, and are arranged in systematic order, and most of them are changed into stone by a slow process; and to impute all this to a transient deluge of less than a year, is to impute effects to a totally inadequate cause.

The doubts about the flood's universality result, first from the difficulty of covering the whole earth for so long a time with water; secondly, to find a place in an ark 450 feet long, 75 feet broad, and 45 feet high, for 1,658 species of quadrupeds, 6,000 species of birds, 642 species of reptiles and tortoises, and 120,000 species of insects—all of which have been shown by naturalists to exist. But the grand difficulty is to collect them all in one spot, and then to disperse them again, without a special miracle; and if a miracle be introduced, all reasoning is nonsense. Moreover, if the regions inhabited by man, then probably quite limited, were covered, what was the use of drowning the rest of the world? The language of Scripture, though at first view seeming strongly to teach a literal universality, is in many other cases quite as strong, although we know that it does not imply universality; but is an example where universal terms are employed to designate only a great many. See Genesis xli. 57, Exodus ix. 25 and x. 15, Acts ii. 5, Colossians i. 23, etc.

15. *The Bible teaches that the earth will be, and geology that it may be, destroyed by fire and its surface renovated.* The Bible declares that the earth will be burnt up and its elements melted, which would reduce it to a molten globe. Geology shows that the globe contains all the elements necessary to bring about such a result. At the rate the internal heat increases, melted matter would be reached in less than 100 miles. How vast the amount of melted matter below, on that supposition, Fig. 125 will show. It is clear that if from any cause, natural or supernatural, such a crust in one part should be broken through and sink into the molten ocean below, all the rest might founder and disappear, and a melted globe alone remain. Then would begin anew the formation of another crust, on which another economy of life might be established, and this might be the new heaven and new earth described in the Scriptures as the future residence of man glorified.

Conclusions.

First, in order to show that there is no discrepancy between revelation and geology, we can take any one of three positions, each of which is sufficient. We may show that Moses does not fix the time of the material creation; or, secondly, that his account admits an indefinite period between the beginning and the first day; or, thirdly, that the days stand symbolically for long periods, and that on the plan of description adopted by the sacred writer he could not give, in all cases, the chronological order of creation. Either of these positions, in the view of any unprejudiced mind, completely vindicates the Mosaic account from any collision with geology.

Secondly, geology furnishes very important illustrations of the Mosaic account, and corroborates several truths of revelation.

Thirdly, still more remarkably does geology illustrate the principles of natural religion, and add to its creed several doctrines generally regarded as exclusively revealed.

Hence it is high time for believers in revelation to cease fearing injury to its claims or doctrines from geology, and to be thankful to Providence for providing in this science so powerful an auxiliary of religion, both natural and revealed.

PART IV.

ECONOMICAL GEOLOGY.

Economical Geology is an account of rocks with reference to their pecuniary value, or immediate application to the wants of society. These practical applications may be included under the three general heads of mining, engineering and architecture, and agriculture.

1. MINING.

Mining is usually understood to relate chiefly to the means employed for extracting metallic ores from veins. We shall apply it to the extraction of ores from all metalliferous deposits. Previously, then, to the details of the process, we must describe the different modes in which the metals are found, and their origin.

Metalliferous Deposits.

Metallic Veins.—The metallic matter, called *ore*, rarely occupies the whole of the vein; but is disseminated more or less abundantly through the quartz, sulphate of baryta, wacke, granite, etc., which constitute the greater part of the vein, and are called the *gangue*, *matrix*, or *veinstone*. Often the ore and the gangue form alternating layers. Sometimes there are cavities lined with crystals, which cavities are called *druses*.

Metallic, like other veins, vary very much in extent, both in a vertical and a horizontal direction. They are of unknown depth; for scarcely ever have they been exhausted downward. The deepest mine that has been worked, is that at Truttenburg in Bohemia, which has been explored to the depth of 3,000 feet.

In all cases metallic, like other mineral veins, are filled with matter different from the rocks which they traverse. In some instances they are obviously of the same age with the containing rock, but in a majority of cases they are fissures that have been subsequently filled. They exhibit almost every variety of dip and strike, and yet it has been thought that they very often affect an east and west direction, though frequently they run north and south, and their dip usually approaches the perpendicular. These veins often ramify and diminish until they finally disappear. Their width is very various; from a mere line up to some hundreds of feet. The metallic veins of Cornwall vary from an inch to thirty feet in width. The contents are sometimes arranged in successive and often corresponding layers on each side.

The contents of metalliferous veins often vary in the same vein, in different rocks through which it passes, both perpendicularly and in the direction of the vein. Its width also varies in the same manner.

It is often found that all the veins of the same neighborhood have essentially the same direction; and if there should be two distinct systems of veins in the same locality, one system, if they are both metalliferous, will contain a metal not found in the other.

The rock in which metallic veins are found is called the *country;* the veins themselves are *lodes;* unproductive veins, intersecting the lodes, are called *cross courses;* the dip or inclination of the vein is its *hade*, *slope*, or *underlie;*

strings and *threads* are small filaments into which the vein sometimes ramifies. The two sides of the sheet or lode are called its *walls;* and if the dip of the vein is considerable, the upper one is termed the *hanging wall,* and the lower the *foot wall.*

Metallic veins are most numerous in hypozoic and palæozoic rocks. No vein is worked in Great Britain above the new red sandstone. Nor are any explored of much importance, above the carboniferous limestone. In the Pyrenees, however, hematitic and spathic iron occurs in palæozoic strata, in the lias, and the chalk. In the Cordilleras of Chile, also, tertiary strata, which have become crystalline by the proximity of granite, are traversed by true metallic veins of iron, copper, arsenic, silver, and gold, which proceed from the underlying granite.

Fig. 412.

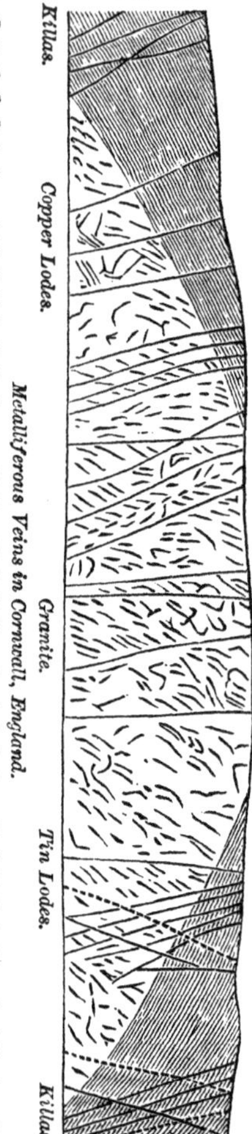

Metalliferous Veins in Cornwall, England.

As a general fact, metallic veins are most productive near the junction of stratified and unstratified rocks. Their productiveness depends also on their relative direction. If one lode is rich, another lode near it, with nearly the same direction and in nearly the same country, will probably be found rich in that part opposite the rich part of the first lode. It is also considered a favorable indication of rich metallic veins, to find at the surface decomposed masses of the ore called *gossan.*

The latest writers upon metallic veins argue that the ores are richer near the surface than at great depths.

Metallic, like common veins, have been produced at different epochs. Mr. Carne finds evidence in Cornwall of the existence of metallic veins of no less than six or eight different ages; a case analogous to the one exhibited in Fig. 31, in Section I.

Fig. 412 is a section of tin and copper veins near Redruth, in Cornwall. They generally pass from the killas, or slate, into the granite beneath. The section reaches to the depth of 1,200 feet. The dotted lines represent the tin lodes (veins), and the continuous lines the copper lodes.

Lead Veins of the Upper Mississippi.—The most extensive deposits of galena in this country are in the valley of the Upper Mississippi, in rocks of the Hudson River group. The simplest form of the lode is a vertical sheet, from the thickness of a knife-blade to several feet; or a number of these sheets may be grouped together.

Sometimes the sheets terminate downwards in a large horizontal bed of ore, not usually less than four nor more than fifteen feet thick. The sheets connected with these beds or *openings,* are called chimneys, as may be seen in Fig. 413.

Very frequently these openings are not filled with ore, but are merely cavities in the rock, and often contain bones of extinct species of animals—as the wolf, peccary, etc.

Or these openings may be partially occupied by ore. Fig. 414 represents

Fig. 413.

Fig. 414.

the Marsden's lode, showing both an opening, *e*, partially filled with galena, and the composite structure of some of the lodes. *a* is the cap rock, *b* a layer of blende, *c* of pyrites, *d* of blende, and the galena in the opening, *e*, is twelve inches thick. This lode is what is called a flat sheet.

These deposits are generally completely isolated, having no connection with each other laterally or vertically. Nor do they extend definitely downward.

Ores in the form of Beds.—The ores of iron sometimes occur in lodes, but more frequently as beds interstratified with sandstones, schists, etc. Tin, lead, and copper are rarely found associated with these beds of iron, but not in large quantities.

These beds are undoubtedly of sedimentary origin.

Alluvial Deposits.—Gold, platinum, and tin are often found in gravel and sand. The same forces that removed the gravel and sand from the ledges also washed away the ores from the veins, and deposited them as a part of alluvium. It is much more profitable, in general, to obtain gold and platinum from alluvium than from the original veins.

ORIGIN OF METALLIC VEINS.

1. Werner supposed that metallic veins were fissures filled by aqueous infiltration from above.

2. Hutton supposed that metallic veins were filled by melted matter injected from beneath. It is probable that many metallic veins were thus produced.

3. Professor Sedgwick supposes some metallic veins to have been produced by chemical segregation from the rock in which they occur, while that was in a yielding state; just as nodules of flint were segregated from chalk, or crystals of simple minerals from the rocks in which they are now found imbedded.

4. Mr. Fox and M. Becquerel refer the origin of many metallic veins to electro-chemical agencies which are operating at the present day, to transfer the contents of veins even from the solid rocks, in which they are disseminated, into fissures in the same. The former of these gentlemen has shown conclusively that the materials of metallic veins, arranged as they are in the earth, are capable of exerting a feeble electro-magnetic influence; that is, they constitute galvanic circuits, whereby numerous decompositions and recompositions, and a transfer of elements to a considerable distance may be effected. He was induced to commence experiments on this subject, by the analogy which he perceived between the arrangements of mineral veins and voltaic combinations. And he thinks if such an agency be admitted in the earth, it shows why metallic veins, having a nearly east and west direction, are richer in ore than others; since electro-magnetic currents would more readily pass in an east and west than in a north and south direction, in consequence of the magnetism of the earth. M. Becquerel has shown that even insoluble metallic compounds may be produced by the slow and long-continued reaction and transference of the elements of soluble compounds by galvanic action. He has also made an important practical application of these prin-

ciples, which is said to be in successful operation in France, whereby the ores of silver, lead, and copper are reduced without the use of mercury. This ingenious theory bids fair to solve many perplexing enigmas relating to metallic veins, and to prove that some of them may even now be in a course of formation.

5. M. Neckar and Dr. Buckland suggest that some mineral veins may have been filled by the sublimation of their contents into fissures and cavities of the superincumbent rocks, by means of intensely-heated mineral matter beneath. Thus it has been shown that by heating galena in a tube, and causing its vapor to unite with that of water, a new deposition of that mineral was produced in the upper part of the tube; and in a similar manner boracic acid, which by itself does not sublime, may be carried upward and deposited anew.

Probably it will be necessary to call in the aid of all the preceding hypotheses to explain the complicated phenomena of mineral veins. The third and fifth of these hypotheses meet with the greatest favor with geologists at the present day.

MINING.

Preliminary Operations.—Valuable veins may be discovered by attentively observing the fragments of rock strewed over the surface. Their sources will be either upon the sides of the valleys in which they are scattered, or in the direction of the drift current. Ravines and steep hill-sides in the neighborhood should be carefully explored for traces of veins, which are usually prominently marked, either by elevation above the enclosing rock, depression below it, or by peculiar products of decomposition.

When these means are not available, *shoading* or *costeaning* must be tried. This consists in digging a series of narrow pits, a few feet deep, at right angles to the supposed course of the lodes. If the course of the lodes can not be satisfactorily conjectured, there should be two series of pits at right angles to each other; and these should be connected by underground galleries, that no traces of ore may be overlooked.

If a productive lode has been discovered, the first operation, if the situation requires it, is to drive an *adit level.* This is a gallery intersecting the lode as far as possible below the surface, and yet secure the draining of the mine.

The second operation is to sink a pit or shaft intersecting the lode at a

Fig. 415.

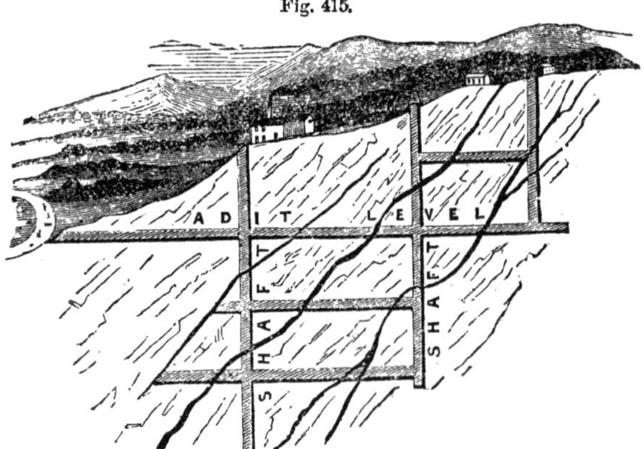

proper distance from the surface. It is designed as the thoroughfare through which the ores are brought to the surface, and ingress and egress are afforded to the miners. The simplest mode of conveyance is by means of a windlas mounted on the shaft, to which two buckets are attached by a long rope and made alternately to ascend and descend in the pit.

Fig. 415 represents three lodes traversed by an adit level. Three shafts are also represented. Fig. 416 shows the interior of a shaft.

As the work progresses, horizontal galleries are excavated at different levels, striking the lode at different points, and connecting with the principal shaft. These are called *cross cuts*; and by means of railways the ores are conveyed to the shaft, where they are drawn to the surface by the simple windlass, or a *whim*—a contrivance employing horse power—or by a steam engine.

Fig. 416.

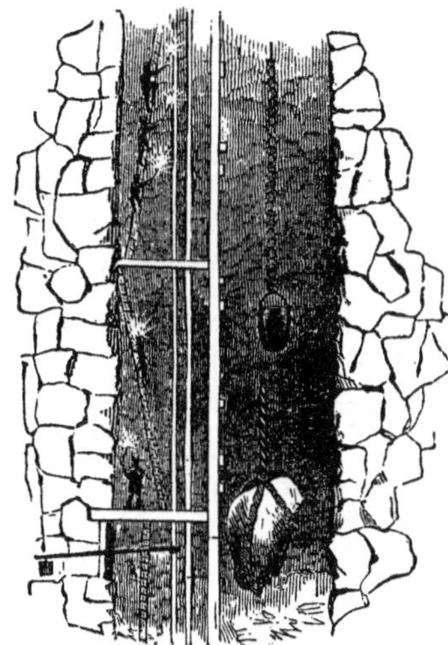

Sometimes, for purposes of ventilation and the readier working of the mine, pits are sunk from one level to another, without being directly connected with the surface. These pits are called *winzes.*

Methods of attacking the Rock.—These vary with the nature of the rock. If it be soft, pick-axes and shovels may be used. If it be hard, but traversed by seams, steel wedges or *gads* may be used to split off large fragments of ore. Most usually, however, the rock must be excavated by blasting with gunpowder. When the rocks are soft, or there is danger of sliding, walls of stone or frames of timber must be constructed, to preserve the openings and galleries.

Extraction of Ores.—The materials excavated to form the galleries may be largely composed of ores, but these form a very small part of the valuable fragments which must be preserved. After the ore has been removed, the worthless rubbish may occupy its place, and thus the valuable portions be easily obtained at the different levels. When the progress is from a higher to a lower level, the operation is called *stoping;* but if the progress is upward, the opening is called a *rise.*

Fig. 417 exhibits the excavations in the Huel Crofty Copper Mine in Cornwall. The perpendicular excavations represent the shafts and winzes, and the portions shaded black represent those parts of the lode which have been removed by stoping. The levels in this mine, like most of those in Cornwall, are ten fathoms apart.

Mechanical Preparation of Ores.—When the ores reach the surface, the valuable portions are picked out and prepared by various mechanical operations for metallurgic processes.

Crushing.—Many ores are prepared for smelting by *crushing.* The fragments are brought beneath two large cast iron cylinders, revolving in contrary directions, and kept in place by a heavy weight. After being crushed, the ores pass over a sieve, which separates the fragments of different sizes;

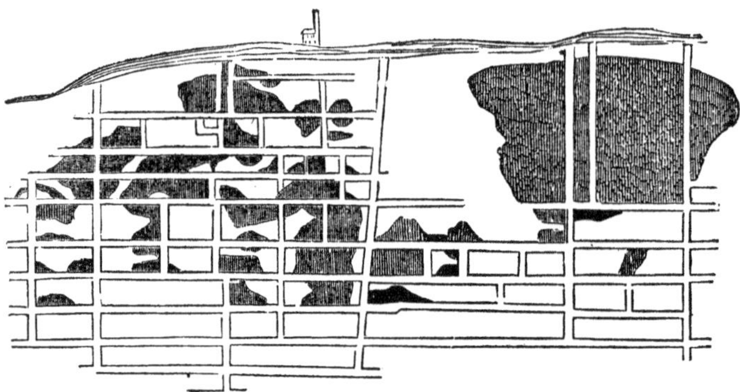

passing the larger pieces beneath the rollers again, until they are sufficiently reduced.

Stamping.—Many ores, instead of being crushed by rollers, are pounded into small fragments by huge pestles moved by water or steam power. The pestles usually weigh from 300 to 400 pounds.

Washing.—Though the ores have been thoroughly crushed or stamped, they are not yet quite ready for the furnace. There may be foreign substances mixed with them. These are commonly separated from the valuable parts by washing. The principle of the separation is, that in consequence of the different specific gravities of the ores and refuse matters, the two classes of fragments, if made to fall in water, will settle in different layers; and the most valuable layer, after the water has been poured off, may easily be separated from the others.

The simplest apparatus for the washing of ores is the hand-sieve. It may be compared to a large tub having a sieve at the bottom. The tub is partly filled with the crushed fragments, then it is placed in a large tank filled with water. The tub is speedily filled with water, and by giving it a sort of undulatory motion with the hands, the heavier particles will settle at the bottom, and thus be separated for the metallurgist.

There are other methods of washing the ores by machinery, which are in more general use than the hand-sieves; but they all involve essentially the same principle.

The native metals, such as gold and platinum, which are worked in alluvium, need only to pass through this process of washing to be prepared for use in the arts. But most ores, when they have been carried through the processes already described, must be reduced in a furnace to the metallic state. It is the province of METALLURGY to describe the methods of reduction.

Amount of Metals Mined.

It may be of interest to some to learn the amount of metals that are annually mined in the world. We add, therefore, two tables, the first giving the estimated value of metals obtained by mining in 1854, and the second giving their amount. For these tables we are indebted to *The Metallic Wealth of the United States*, by J. D. Whitney.

Estimated Value of Mining Products in 1854.

	Gold.	Silver.	Mercury.	Tin.	Copper.	Zinc.	Lead.	Iron.
Russian Empire	$14.880.000	$928.000	....	....	$3.900.000	$440.000	$92.000	$5.000.000
Sweden	496	56.000	....	....	900.000	4.400	23.000	3.750.000
Norway	....	272.000	....	....	330.000	....	....	125.000
Great Britain	24.800	1.120.000	....	$4.200.000	8.700.000	110.000	7.015.000	75.000.000
Belgium	....	....	....	....	....	1.760.000	115.000	7.500.000
Prussia	....	480.000	....	....	900.000	3.680.000	920.000	3.750.000
Harz	1.488	480.000	....	....	90.000	1.100	575.000	....
Saxony	....	960.000	....	60.000	30.000	....	230.000	175.000
Rest of Germany	....	48.000	....	....	....	....	115.000	2.500.000
Austrian Empire	1.413.000	1.440.000	$250,000	30.000	1.980.000	165.000	805.000	5.625.000
Switzerland	....	....	....	....	....	....	....	375.000
France	....	80.000	....	....	....	....	172.500	15.000.000
Spain	10.416	2.000.000	1.250.000	6.000	300.000	....	3.450.000	1.000.000
Italy	....	....	....	....	150.000	....	57.500	625.000
Africa	992.000	....	....	....	360.000	....	....	....
S. Asia and E. Indies	6.200.000	....	....	3.000.000	1.800.000	....	....	....
Australia and Oceanica	37.200.000	128.000	....	....	2.100.000	....	....	....
Chile	744.000	4.000.000	....	....	8.400.000	....	....	....
Bolivia	297.600	2.080.000	....	} 900.000	}	....	....	....
Peru	471.200	4.800.000	100.000	}	} 900.000	....	....	....
Equador, New Granada, etc.	3.720.000	208.000	....	....	}	....	....	....
Brazil	1.488.000	11.200	....	....	}	....	....	....
Mexico	2.480.000	28.000.000	....	....	....	....	....	....
Cuba	....	....	....	....	1.200.000	....	....	....
United States	49.600.000	352.000	500.000	....	2.100.000	550.000	1.725.000	25.000.000
Total	$119.523.600	$47.443.200	$2.100.000	$8.196.000	$34.140.000	$6.660.500	$15.295.000	$145.425.000

Amount of Metals obtained by Mining in 1854.

	Gold.	Silver.	Mercury.	Tin.	Copper	Zinc.	Lead.	Iron.
	lbs. Troy.	lbs. Troy.	lbs. Av.	Tons.	Tons.	Tons.	Tons.	Tons.
Russian Empire.......	60.000	58.000	....	......	6.500	4.000	800	200.000
Sweden..............	2	3.500	....	....	1.500	40	200	150.000
Norway............ ..	....	17.000	....	....	550	......	.... ..	5.000
Great Britain..........	100	70.000	....	7.000	14.500	1.000	61.000	3.000.000
Belgium...............	....	....	....	..... .	....	16.000	1.000	300.000
Prussia...............	....	30.000	....	......	1.500	33.000	8.000	150.000
Harz................	6	30.000	....	....	150	16	5.000	....
Saxony................	....	60.000	....	100	50	......	2.000	7.000
Rest of Germany......	....	3.000	....	....	....	......	1.000	100.000
Austrian Empire......	5.700	90.000	500.000	50	3.300	1.500	7.000	225.000
Switzerland..........	....	... .	....	....	....	......	....	115.000
France................	.. ....	5.000	....	....	....	......	1.500	600.000
Spain.................	4.	125.000	2.500.000	10	500	......	30.000	40.000
Italy......	....	....	....	....	250	......	500	25.000
Africa................	4.000	....	....	....	600	......	....	....
S. Asia and E. Indies..	25.00	....	....	5.000	3.000	......	....	....
Australia and Oceanica	150.000	8.000	....	....	3.500	......	....	....
Chile................	3.00	250.000	....	....	14.000	......	....	....
Bolivia...............	1.200	130.000	....	} 1.500	}	......	....	....
Peru......	1.900	800.000	200.000	}	} 1.500	......	....	....
Equador, N.Granada,&c	15.000	13.000	....	....	}	......	.. ....	....
Brazil................	6.000	700	....	....	}	......	....	....
Mexico................	10.000	1.750.000	....	....	....	......	....	... .
Cuba..................	....	....	....	....	2.000	......	....	....
United States..........	200.000	22.000	1.000.000	....	8.500	5.000	15.000	1.000.000
Total...........	481.950	2.965.200	4.200.000	13.660	56.900	60.550	133.000	5.819.000

2. ENGINEERING AND ARCHITECTURE.

The spheres of the engineer and the architect are so similar that we may conveniently bring under one head what we have to say of the uses to which they may apply geology. The engineer has to locate railroads, common roads, and canals, to tunnel mountains, to construct embankments, harbors, breakwaters, quays, and bridges. The architect selects the sites of public and private buildings; and both must select materials for their works. Their applications of geology, then, will fall under two heads—1. Location of their works. 2. The materials to be used in construction.

1. *Location.*

In the location of railroads, as well as of carriage roads, an engineer familiar with geology will be able often to prevent great losses and failures by a judicious selection of routes. The greatest danger lies in the loose or imperfectly consolidated materials at the surface. Where there is an alternation of sand, gravel, and clay, especially if the strata are at all inclined, and deep cuts are made through them, slides will be apt to occur subsequently in very wet or very dry weather. He who knows all this beforehand can, by a variety of expedients, guard against such accidents, to which he who has never studied surface geology will be liable.

The same difficulties meet the architect in selecting the site of large buildings. If he can find a little below the surface what is called hard pan, that is, gravel and sand more or less consolidated, he could not obtain a better foundation. But this hard pan may have but little thickness and be underlaid by loose materials, even by quicksand. Its thickness, therefore, should be ascertained, and no building, bridge, or embankment placed upon it which will be liable, by its weight, to break through the stratum.

Clay, especially such as occurs in alluvial formations, is one of the foundations most to be suspected. For although very solid when dry, powerful rains will convert it into mortar; or it may be underlaid by that most slippery of all foundations, quicksand, which, if a stream of water should find access to it beneath the clay, will be swept away with astonishing rapidity, undermining, of course, the superincumbent structure. A case of this kind occurred in East Hampton, Massachusetts, in the summer of 1860; when a factory, just erected by Hon. Samuel Williston, was injured in one night to the amount of some $50,000.

But the engineer and architect should be acquainted with the solid strata beneath alluvium; not only with their nature, but their position, whether horizontal or inclined. For if inclined, the loose materials above will be very liable to slide down, and therefore without due precaution no structures of great weight and importance, whether embankments, quays, or houses, should be placed upon them.

The dip and strike of the strata, as well as their nature, should also be known to the engineer, in laying out railroads and canals, on other accounts. To locate them on the line of strike is the most unfavorable of all directions, while the most favorable is to cross them at right angles. Still more important are the dip and strike in tunneling. To carry a tunnel through a hill on the line of strike, or with the rock dipping from the workman, is most unfavorable, because the work must be done on the edges of the strata. The most favorable is where the drilling can be made on the broad face of the strata. The nature of the rock, too, is very important. Some formations have so little coherence, that if a tunnel be made through them, the roof will fall in. Others are so hard, that it is almost as easy to drill and blast iron. This is especially true of the compact trap rocks and some varieties of porphyry. Cuttings through them are so costly, that, if possible, they should be avoided; though the tufaceous traps are not difficult to blast. These trappean rocks are apt to occur when least expected, and the engineer, before he decides upon an extensive cutting or tunnel, ought to be confident that he shall not unexpectedly encounter these hard materials. He ought to find out, if possible, also, where faults exist, what strata are pervious or impervious to water, and where springs may be expected.

The question as to the probable success of boring Artesian wells has become, at this day, one of great interest and importance, and also one of great difficulty, concerning which the most practiced geologist may be mistaken. Certain principles, however, are true in respect to such explorations. One is, that we can not expect success if the underlying rock in the region is all unstratified, nor unless some stratum can be reached whose outcrop rises higher than the surface of the well; that is, although water may be found, it will not rise to the surface. So if all the strata are equally pervious to water, no hydrostatic pressure will force it upward.

2. *Materials.*

For most common purposes of construction men are obliged to use such materials as are easily accessible, though perhaps not the best. The most valuable are often remote and costly. Some kinds of rocks, however, the world over, are always highly prized. Such are the marbles, granites, porphyries, serpentines, alabasters, soapstones, etc. The most valuable monuments of antiquity, that have survived the ravages of time, were of this description, and they are now used more extensively than ever. Analogous materials, however, of coarser kinds, answer well for common purposes; and the engineer and architect must make the best selection they can out of materials at hand, taking into account their accessibility, cheapness, and durability.

The following are the rocks generally employed for the purposes of construction, roofing, and flagging, as well as for macadamizing roads: 1. Limestone. 2. Sandstone. 3. Clay slate. 4. Micaceous and talcose schists. 5. Gneiss. 6. Soapstone. 7. Granite, syenite, and trap.

Limestone is, upon the whole, the most important. Sometimes it is simply carbonate of lime, or a double carbonate of lime and magnesia, called dolomite; in both which states it forms beautiful white marble, and is very enduring. In Philadelphia especially, and more or less in other Atlantic cities, this white marble forms the fronts of houses; and in the City Hall in New York, the entire edifice is made of it. The large pillars around Girard College in Philadelphia are of this stone, obtained from Sheffield, in Massachusetts. It is less common in European cities, though the new Houses of Parliament in London are built of dolomite; but it is scarcely crystalline, and comes from the permian formation. The oolite furnishes most of the best building stone in Great Britain, especially from the famous quarries in the Isle of Portland and near Bath. But this rock is entirely wanting in our country. Yet vast beds of the white and gray, or variegated limestones of the azoic rocks, run along the whole length of the Appalachian chain of mountains from Canada to Alabama. Farther west the limestones take argillaceous matter into their composition and form admirable building stones, as may be seen in our western cities.

The great amount of steatite, or soapstone, in the Appalachian chain of mountains, especially in New England, has led of late to its employment for the fronts of houses in New York and elsewhere. It has the advantages of being worked with great ease and of keeping free from mosses and lichens, but it is not handsome, and is easily marred.

Sandstones of various colors, hardness, and of different degrees of fineness from the tertiary to the azoic rocks, are widely employed in most countries. From the well-known quarries of Portland, in Connecticut, and near Newark, New Jersey, large quantities of this rock are carried to almost every portion of the country accessible by water. This rock is red or gray, and belongs to the oolitic or triassic group. Other sandstones, from the palæozoic rocks, are extensively used in many parts of the country. Most of these sandstones are very enduring and beautiful.

In this country clay slate is used almost exclusively for roofing. But in the vicinity of some of the European slate quarries, as in Wales, it is employed for the floors of houses, for doors, fences, troughs, coffins, and almost every thing, in fact, for which boards are used in other countries.

For flagging stones the most usual rocks employed in our country are devonian gritstone and mica and hornblende schists. The beautiful mica schist of Bolton, in Connecticut, and the gritstone of the Hamilton group of rocks along Hudson River, furnish flagstones for a large part of the cities of this country. The first is perhaps the most beautiful, but the latter the most enduring.

In Great Britain gneiss is hardly spoken of as a stone for construction, and hence we conclude that it is not used. But in this country, especially in New England, it is one of the most valuable of all our rocks. Composed of the same materials as granite, it is equally enduring, and it has the advantage of splitting easily in the direction of the stratification, though there is some difficulty in hewing it smoothly across the layers. The quarries of this rock in New England are very numerous, and some of them furnish most beautiful stone, much to be preferred to sandstone.

The unstratified rocks are also described by English writers as "not very often employed in the construction of public edifices." Very different is the case with us. Trap and porphyry are not, indeed, much used on account of the difficulty of bringing the blocks into a regular shape, as they can only be broken, but not hewn. But granite and syenite are used almost everywhere, if obtainable, and form the most solid and enduring of structures. The syenitic quarries at Quincy and Cape Ann, in Massachusetts, as well as those of pure white granite at Hallowell, in Maine, at Barre, in Vermont, at Chelmsford and Fitchburg, in Massachusetts, and many other places, furnish inexhaustible quantities, and in the northern cities form a large part of the most imposing public as well as private buildings. Enormous blocks are sometimes got out at these quarries to form solid columns of great size and length, as may be seen in several public edifices in Boston and elsewhere.

It is an important and difficult point to ascertain whether an entire rock will endure long exposure without disintegration. In Europe, where buildings from the quarries have stood for several centuries, this point can generally be determined. But in this country we must resort to other means. The mineral composition will give us some information, and in general the more perfectly crystalline the rock, the less liable it is to disintegration, though there are some exceptions. A better test is to examine ledges that have been for ages exposed to atmospheric agencies, and observe the amount of erosion. A method of testing the influence of dampness and frost by the use of a boiling solution of Glauber's salts is said to afford good results in a short time. The details, which we have not room to give, may be found in Ansted's Geology, vol. ii., p. 458.

3. AGRICULTURAL GEOLOGY.

The first inquiry in Agricultural Geology is, what is the composition of good soils?

The matter in all soils capable of sustaining vegetation exists in two forms, inorganic and organic. The first contains twelve chemical elements, viz., oxygen, sulphur, phosphorus, carbon, silicon, and the metals potassium, sodium, calcium, aluminum, magnesium, iron, and manganese. In the organic part the elements are four: oxygen, hydrogen, carbon, and nitrogen. The inorganic elements are derived from the rocks; the organic elements from decaying animal and vegetable matter. So that it is with the earthy constituents of soils that geology has to do. The above-named ought all to be present. They do not indeed occur in their simple state, but as water, sulphates, phosphates, carbonic acid, silicates of potassa, soda, lime, magnesia, alumina, iron, etc. The average amount of silicates or sand in soil is 89 in 100 parts.

The second inquiry is, whether these elements of the soil are found in the rocks. In the table of their analysis given on page 93, it will be seen that they are all present except phosphorus, which, however, is not unfrequently found in them in the condition of phosphates. Moreover, the proportion of the ingredients in the rocks does not differ much from that of the soils. Hence the conclusion is, that the latter are only the former comminuted, with the addition of from three to ten per cent. of organic matter.

Since the rocks differ considerably in composition, we should expect a corresponding difference in the soils derived from them. And such is the fact to a considerable extent where the soil is simply the result of the disintegration of the rock beneath it. It is enough so in many districts to form characteristic soils. Thus over quartz rocks and some sandstones we find a very sandy and barren soil, though it is said that in nearly all soils enough silicates of lime and magnesia are present to answer all the purposes of vegetation. But the alkalies and phosphates may be absent. When the rock is limestone, the soil is sometimes quite barren for the want of other ingredients, and in consequence of the difficulty of decomposition. Clay, also, may form a soil too tenacious and cold. The sandstones that contain marly beds, and some of the tertiary rocks of analogous character, form excellent soils. So does clay slate, and especially calciferous mica schist. The amount of potash and soda in gneiss and granite often makes a rich soil from those rocks, and the trap rocks form a fertile though scanty soil.

But, in the third place, in most countries aqueous and glacial agencies have so mixed the soils together that their original peculiarities are lost, and new and compound characters are given them. This is particularly the case in northern countries, where the drift agency has swept over the surface and torn off and mixed together the disintegrated portions of the several formations. Subsequently rains and streams have carried the finer portions of the drift into the lowest places, and there formed alluvial meadows; and although these are usually the best of soils, they are often derived from many different rocks. The drift left upon the higher grounds is generally quite barren, chiefly because of its coarseness.

A fourth service which the geologist renders to agriculture is by the discovery of fertilizers. Sometimes he can point out deposits of the phosphates either in a crystalline state, or as coprolites or guano. He can also show what rocks contain carbonate of lime, or discover sulphate of lime, or marl beds, or green sand, or decomposing fossil shells, or deposits of carbonaceous matter. He can also find what rocks contain enough of potash or soda to be of service when pulverized.

The subject of drainage, as well as the discovery of springs of water and the best means of bringing it to the surface, belong to Agricultural Geology; but our limits do not allow us to enter upon the details.

PART V.

NORTH AMERICAN GEOLOGY.

THE history of American Geology commences with the present century. A few collections of minerals and rocks served as the nucleus around which the interest of the public gradually accumulated. The first attempt at exploration was commenced by William Maclure in 1807, who published a geological map of the States then in the Union, giving the old Wernerian classification of the rocks. Great service has been rendered to American geology by the *American Journal of Science and Art*, commenced by Professor Silliman, Senior, in 1818, and continued to this day as the ablest American scientific journal.

An important feature in the history of American geology is the numerous geological surveys that have been executed, or are still in progress, under the patronage and direction of the different State authorities, as well as the United States government. The leading object of these surveys is to develop those mineral resources of the country that are of economical value. But, with a commendable liberality, the legislatures have encouraged accurate researches into the scientific geology, and sometimes also into the botany and zoology of their several States.

The first survey authorized by the government of a State was that of North Carolina, which was committed to Professor Denison Olmsted in 1824. Two small pamphlets embodied its results. A year or two later, Professor Vanuxem was commissioned to explore the geology of South Carolina, but its results were published only in the newspapers.

Massachusetts, in 1830, commenced a geological survey of its territory upon a more extensive scale, under the direction of the senior author of this work. The first report was made in 1832, a pamphlet of seventy pages. In 1833 a full report was made, of 702 pages, with an atlas of plates and a geological map; and in 1841 a final Report of 831 quarto pages, with fifty-five plates. Within ten years the example of Massachusetts was followed by fifteen other States. Nearly every State and Territory in the Union, at the present date (1860), has been more or less explored, or is now conducting a survey.

The survey of New York was commenced in 1836, and has been conducted upon the most liberal principles. Nearly twenty large quarto volumes have been published by her legislative authority upon all the branches of Natural History, including agriculture, at an expense of half a million dollars. In consequence of these accurate researches, the rocks of New York are classic ground for American geologists; and the names employed by the New York geologists, though derived from localities within the State, are applied to contemporaneous deposits over the whole continent.

These Reports relate chiefly to the Silurian and Devonian Systems. The magnificent Report of Professor H. D. Rogers upon the Carboniferous System of Pennsylvania, has laid a foundation for describing all North American coal fields. The New England and Canada Reports describe the azoic rocks more particularly. Morton has given a system to the cretaceous, and Conrad to the tertiary deposits of the country.

Besides the State surveys, scientific societies and associations in the principal cities have done much toward the development of our Natural History.

The Academy of Natural Sciences at Philadelphia, the Lyceum of Natural History of New York, the American Academy of Arts and Sciences, and the Society of Natural History in Boston, are prominent among them. The American Association for the Advancement of Science is an organization including members from all parts of the country, and meets annually in different places. It was originally an *Association of American Geologists*. Then it included all the Naturalists, and ultimately, in 1847, was enlarged so as to admit all sciences, and received its present name.

Nor should we neglect to mention those Cabinets of Geology and Natural History which begin to compare favorably with those of Europe. The largest collections may be found in the Academy of Natural Sciences at Philadelphia, the State Collection at Albany, N. Y., that of the Boston Natural History Society, the collection of the Canada Survey at Montreal, the Cabinet of the Smithsonian Institution at Washington, and that of the New York Lyceum of Natural History. A magnificent museum of Palæontology and Zoology is commenced at Cambridge. Among the Colleges, the most extensive Cabinets are those at Amherst and Yale. These museums are thronged with visitors. For example, the register of the Cabinet at Amherst shows that the collections are visited by 15,000 people annually.

GEOLOGICAL MAP OF NORTH AMERICA.

Accompanying this section, we present a small map of the geology of North America, compiled from the most reliable sources. Owing to its small size, only the more general classes of rocks can be represented. There are six distinctions upon it: 1, Azoic rocks; 2, Palæozoic rocks, including all the formations between the Cambrian and Permian series, except a part of the Carboniferous series; 3, that part of the Carboniferous series which is underlaid by valuable beds of coal; 4, Mesozoic rocks; 5, Cainozoic rocks; and 6, Igneous rocks, such as have been erupted since the commencement of the Triassic period.

A general division of the geology of North America is into three great fossiliferous basins resting upon azoic rocks. The first is the *Arctic basin*, occupying the greater part of the islands and peninsulas within the Arctic circle. This may be connected with the other basins. The second may be called the *Hudson's Bay basin*, because it is chiefly developed about Hudson's Bay. The third is the great *Continental basin* of the interior. The last is the one best known.

The Arctic basin has been explored by Arctic voyagers. An excellent map of it is given in McClintock's Narrative of the Expedition in search of Sir John Franklin. Silurian, carboniferous, and mesozoic rocks are found there. The Hudson's Bay basin is composed entirely of palæozoic rocks, so far as any thing is known concerning it. The Continental basin embraces fossiliferous rocks of every age, from the Cambrian to the latest Cainozoic.

The form of the continent is that of a great triangular basin, as described in Section V., Part I. The mountainous regions correspond very nearly with the areas occupied by the azoic rocks, except that along the Pacific coast they are mostly covered by cretaceous and tertiary strata—the latter constituting most of the summits of the Rocky Mountains. It corresponds also with the views already stated, to find that the igneous rocks are generally located near the oceans.

The Rocky Mountains belong to the longest chain of mountains upon the globe, and, with one exception, the highest. Commencing in the extreme southern part of South America, it extends through the whole length of that continental area, under the name of Andes or Cordilleras. On the Isthmus

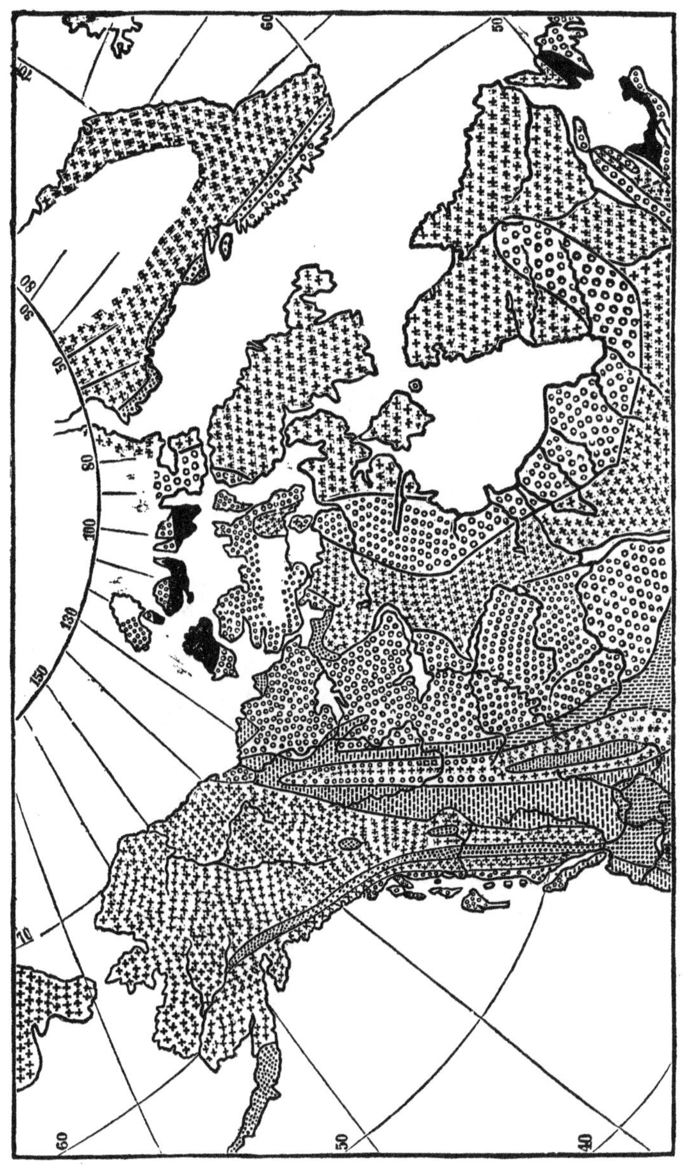
60
50
40

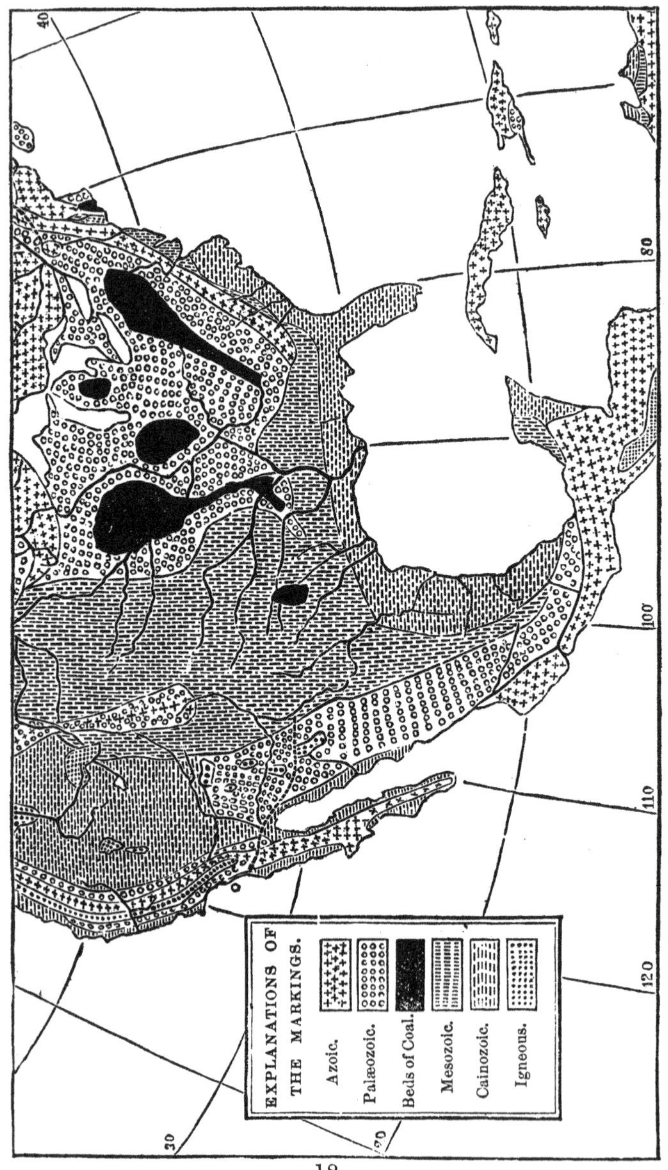
EXPLANATIONS OF
THE MARKINGS.
Azoic.
Palæozoic.
Beds of Coal.
Mesozoic.
Cainozoic.
Igneous.
40
80
100
110
120
30

of Panama it sinks into a comparatively low ridge, but rises into the table lands of Guatemala, Mexico, and the ridges and plateaux of the western United States and British America, quite to the Northern Ocean.

The whole of the interior of North America may be regarded as a vast plateau, gradually passing into the Appalachian ranges upon the east, and the elevated table lands upon the west; but gradually descending to the Northern Ocean and the Mexican Gulf, from a water shed in the middle of the continent.

AZOIC ROCKS.

The azoic rocks are represented on the map by those areas which are covered by small crosses. They embrace all the crystalline and unfossiliferous rocks of every age, but chiefly the hypozoic or Laurentian groups. They are gneiss, mica schist, talcose schist, quartz rock, clay slate, saccharoid azoic limestone, granite, syenite, and the ancient porphyries.

Laurentian System.—On the north shore of the St. Lawrence there are two ranges of mountains running parallel to the river; one at the distance of fifteen or twenty miles, and the other 200 miles distant. Here are the Laurentine Mountains, from whence the name was derived. The system extends from the shores of the Arctic Ocean, passing round the Hudson's Bay palæozoic rocks, including the Laurentines, and occupying the eastern shores of the continent, to the north of the St. Lawrence. Greenland, Grinnell Land, and other islands to the north, are supposed to be of the same age. The space thus occupied has the form of the letter V.

The other deposits of this age, east of the Rocky Mountain range, are mostly insulated. In the northern part of New York the Adirondacs belong to this system, and are hardly separated from the Laurentines north of the St. Lawrence. Another isolated area of these azoic rocks is in Northern Michigan. In New York they are composed chiefly of Labradorite and hypersthene rock. Of the same age are the Ozark Mountains in Missouri, the Washita Hills, south of the Arkansas River, the Whitchita Mountains in Northern Texas, and other eminences in Central Texas.

The azoic rocks of Russian America, which extend uninterruptedly as far as Mexico, are supposed to be Laurentian, although analogy would lead us to suppose that many of them are of palæozoic age. Two or three interrupted ranges, with a few isolated patches of these azoic rocks, are represented along this region, extending into Mexico. The same rocks occupy the southern part of Mexico and most of the larger islands of the West Indies.

The Laurentian rocks contain large masses of magnetic iron ore. Those in Missouri are among the largest on the globe. They are connected with porphyry, and are separated from metalliferous limestones by a deposit of granite with trap dykes, six miles in width. Pilot Knob, which rises 500 feet, is partly, and the Iron Mountain, 300 feet high and two miles in circumference, is entirely composed of this ore. Vast deposits of iron ore exist also in the northern parts of New York. Many valuable gems are found in these rocks.

Azoic Rocks of later Age.—As yet, the only rocks of the age of the Cambrian discovered in this country are the azoic rocks about Lake Huron. These have already been described. The latest researches render it probable that most, if not all of the azoic rocks of New England and the British Provinces, which are continued along the eastern coast of the continent to Alabama, are of palæozoic age.

There are two methods employed to prove that the Appalachian azoic rocks are palæozoic. 1. The northern extremity of the ranges gradually merge into the unaltered silurian and devonian members. For example, a range of mica schist in Connecticut becomes calciferous in Vermont, and just over the Vermont and Canada line, Favosites Gothlandica, and other silurian fossils, are found in it. 2. Crystalline schists both overlie and are interstratified with fossiliferous deposits. For example, there is a belt of Upper Helderberg limestone underlying the talcose schist of western New England, in the northern part of Vermont—and in such a way as precludes the idea of inversion. It was this phenomenon of the interstratification of these two kinds of rock that first unsettled the old ideas of geologists in regard to azoic rocks.

The Professors Rogers suppose that the eastern part of the azoic rocks of the Appalachian range are hypozoic, and that a part of the western border is Cambrian.

Taconic Rocks.—Along the western border of the azoic ranges just described, there is a succession of thick deposits, partially metamorphosed, which Professor Emmons has grouped under the name of *Taconic System.* They consist of quartz rock, limestones, dolomites, marbles, imperfect talcose and micaceous schists, and clay slate. They may be found along nearly all the Appalachian range, and, according to Professor Emmons, also upon its eastern side in Maine, Rhode Island, and North Carolina.

Professor H. D. Rogers has described these rocks in Pennsylvania as Lower Silurian. The quartz rock he calls Potsdam or Primal Sandstone, the limestones the Auroral or Lower Silurian Limestones of New York; the schists and slates as the Hudson River Group, or Matinal Shales.

The authors of this book have been examining these rocks as they are developed in Vermont, and take the following positions, the details of which are not yet published: 1. Some of the slates in a few localities, pronounced Taconic by Professor Emmons, belong to the Hudson River group of New York. 2. The remainder, including the typical localities, are of Upper Silurian and Devonian age. 3. The slates and schists are at least as high as Upper Silurian, overlying the Oneida conglomerate. As they so much resemble the Hudson River group, and are the rocks from which the name is derived for the Lower Silurian member, the name *Upper Hudson River Group* may be assigned to them. 4. Some of the limestones contain fossils, apparently identical with certain Devonian forms. Hence they are regarded as Devonian; and as the place in that series is yet uncertain, the name *Dorset Limestone* may be applied to the group, from Dorset Mountain in Vermont, where the whole series is beautifully developed. 5. The quartz rock, being associated with the Dorset limestone, must be newer than Lower Silurian.

SILURIAN SYSTEM.

During the hypozoic period, and at its close, the strata were disturbed by forces of elevation, so that the more elevated parts assumed a V form, as in the northern part of the continent, and there were several islands in the southern part. The Cambrian period seems to have been one of general inactivity; but strata were deposited unconformably upon the older rocks about Lake Huron.

Lower Silurian.—The Huronian rocks were also elevated before the deposition of the Silurian, as is seen at Lake Huron. The first positive evidences of the introduction of life in North America are found in the Potsdam sandstone. In New York the Lower Silurian is divided into the Potsdam sandstone and calciferous sandrock, which form a separate group by their structural and palæontological affinities, which may be called the *Potsdam Group;* the Chazy limestone, Birdseye limestone, Black River limestone, and Trenton limestone, or the *Trenton Group;* the Utica slate and Hudson River group, both of which may be termed the *Hudson Group.*

These members are distributed over most of the continent, and may generally be recognized by tolerably constant lithological characters. The limestones in the Western States are often magnesian, instead of the simple carbonate of lime. The Hudson River rocks are the most variable. Typically, they are slates, with a few strata of conglomerate and limestone, or dolomite. In Canada, besides the slates there are great accumulations of conglomerates and sandstones, often composed of pebbles of limestone, and called provincially the *Quebec Group*. In Ohio and other Western States, the whole series is changed into the *blue or Galena limestone.*

Along the eastern coast north of Cape Cod, and as far as Nova Scotia, various slates occasionally occur containing the *Paradoxides Harlani*, which is at the base of the silurian system in Europe. These slates are supposed to be of the same age as the Potsdam sandstone.

The Lower Silurian rocks are not distinguished upon the map, but their general position is at the edges of the palæozoic rocks, contiguous to the oldest azoic rocks. Often, as in northern New York, they encircle an insulated portion of the older groups.

Upper Silurian.—The Lower Silurian periods were closed by a revolution or disturbance of the strata, so that in some localities, as at Gaspé, in the Gulf of St. Lawrence, the Upper Silurian strata rest unconformably upon the Lower Silurian.

There are three periods in this system of rocks. The first embraces the Oneida conglomerate and the Medina sandstone, formations quite variable in composition and thickness. The local name of *Sillery sandstones* has been applied to the Oneida conglomerate in Canada, where it attains a thickness of 4,000 feet. As the range passes through Vermont, it becomes by turns silicious and calcareous, or dolomitic, and is exceedingly variable in thickness. A mere knife-edge thickness of limestone may suddenly expand into 100 feet within the distance of a mile. The range continues in this erratic way along the whole Alleghany range, and, in conjunction with the Medina sandstone, is found from western New York to Wisconsin, through Canada West.

While a somewhat turbulent agency was depositing this curious rock within the continent, at its border, near the mouth of the St. Lawrence there was a quiet accumulation of limestones, under conditions suitable to the development of life. As the lithological characters are so distinct, the group, which consists of six divisions, embracing the equivalents of the upper part of the Hudson River group, the Oneida conglomerate, Medina sandstone, and the Clinton group, has received a distinct name, the *Anticosti Group.*

The first period embraces besides the conglomerates, and having the same general geographical distribution, the Clinton and Niagara groups, consisting of alternate layers of shales and limestones. They are quite productive in interesting forms of life.

There is also a belt of Niagara limestone in Canada, running down to Memphremagog Lake, which seems to pass into calciferous mica schist, an azoic rock lying east of the Green Mountain range.

The second great period of these upper rocks embraces only the Onondaga salt group, a series of limestones and shales 1,000 feet thick. Its development is less extensive than the preceding groups.

The third great period is now called the *Lower Helderberg group*, embracing the following earlier divisions: Pentamerus limestone, Delthyris shaly limestone, Encrinal limestone, and the Upper Ponent series. These rocks are distributed in general conformity with the Niagara group west of Canada East.

Mineral Deposits.—There are some remarkable mineral deposits in the silurian rocks. In the Northwestern States there occurs one of the most remarkable deposits of lead in the world, in the Hudson River group. It covers 3,000 square miles, chiefly in Wisconsin, but found also in Iowa and Illinois. The greatest amount of lead produced from it in any one year was 315,700 tons.

Copper is abundant in the same vicinity; but especially about Lake Superior, in Potsdam sandstone. Masses of native copper have been uncovered there weighing fifty tons; and bowlders from the lodes are scattered over an area of several thousand square miles.

The salt springs of the United States issue invariably from the silurian rocks.

Probably all the gold in the United States, both along the eastern and western shores, is located in metamorphic rocks of Silurian or Devonian age. Along the Appalachian ranges it has been found all the way from Canada East to Alabama, being particularly abundant in North Carolina. The California deposits are the most extensive in the world.

DEVONIAN SYSTEM.

The formations of the Devonian system are ten in number in New York, which may be arranged, by structural resemblances, into five groups.

The *Oriskany Group* embraces the Oriskany sandstone and the Cocktail grit. The group extends from southern New York southwesterly to Tennessee, and westerly about 300 miles. The upper part is characterized by a fucoid resembling the tail of a cock, whence its name.

The Upper Helderberg group embraces the Scoharie grit and the Upper Helderberg limestone, or, as at first divided, the Corniferous and Onondaga limestones. The limestones of this group are widely developed through the Appalachian chain south of Hudson River, and westward, both in the United States and Canada, as far as the palæozoic rocks have been explored. It is also represented in the Dorset limestone of New England, a belt of limestone in Northern Vermont east of the Green Mountains, and probably at Bernardston, in the northern part of Massachusetts, upon Connecticut River. It is the lowest rock in North America which contains ichthyolites.

The three remaining groups are mostly sandstones and shales. The *Hamilton division* embraces the Marcellus shales, Hamilton group, and Genessee slate. This division is best developed in the Appalachian chain in Pennsylvania and Virginia, thinning out gradually between the Hudson and Mississippi Rivers. The Hamilton group in Iowa is mostly calcareous, containing many interesting fossils.

The fourth group embraces the Chemung and Portage rocks of New York. From New York they extend southwesterly to Tennessee. In Ohio they have received the provincial name of Waverly sandstone. Farther west and north their equivalents have not been ascertained.

The fifth group is the Catskill red sandstone, lithologically the *old red sandstone* of Europe. This is principally developed in New York and Pennsylvania, so far as its equivalency has been determined.

Devonian rocks are found in Eastern Massachusetts, Maine, Canada East, and in the more northern parts of the continent, but their equivalency with the deposits already specified has not been determined.

From the study of the Silurian and Devonian systems the following conclusions have been reached: 1. Until we reach the Genesee slate, all the organic remains found in these two great systems are marine. Here the first land plants are found. The *formations* appear to have been deposited successively near the shore of the palæozoic ocean, for besides the fossils ripple marks and shrinkage cracks occur, even in some of the limestones. 2. The rocks are thickest near the borders of the continents. This is *certain* respecting the eastern side, and *probable* respecting the western. 3. The rocks of the east coast are mostly silicious—shales, sandstones, or their altered forms: those of the interior are mostly calcareous. 4. The outline of the future continent is strongly marked at the close of the Devonian periods. The Appalachian, and perhaps the Rocky Mountain ranges, form long sand reefs, hemming in more or less perfectly an interior sea, covering the area now occupied by the Mississippi and its branches. At the same time the two other smaller northern basins received their outline.

CARBONIFEROUS SYSTEM.

The best classification of the Carboniferous system is that adopted by the Professors Rogers. They divide into the Vespertine, Umbral, and Seral series, or the Lower Carboniferous, Middle Carboniferous, and Coal Measures. The lowest division consists of sandstones and conglomerates, with dark-colored slates sometimes containing beds of coal. This division is most abundantly developed upon the eastern side of the Appalachian basin, sometimes being entirely wanting elsewhere. In Ohio it constitutes the upper portion of the Waverly sandstone, and in Tennessee it is a buhrstone and limestone.

The Middle Carboniferous strata are quite variable in composition. In Nova Scotia, etc., they are red shales, sandstones, and various marls. In Pennsylvania they are red shales, associated farther south with a bed of limestone, which continues to increase in thickness southward. These strata are nearly all limestone in the Western States, where they are thicker than in the eastern coal fields.

In the true Coal Measures the rocks are sandstones, conglomerates, shales, limestones, and beds of coal. Some have traced resemblances between some of the conglomerates and the millstone grit of England. Upon the map the area covered by this division is represented as perfectly black. Coal fields are represented in the islands of the Arctic Ocean, in Newfoundland, New Brunswick, Nova Scotia, New England, the Appalachian Basin, in Michigan, Illinois, Iowa, Missouri, and Texas. The amount of coal is almost inexhaustible.

In Iowa the Carboniferous system is divided as follows: Burlington limestone, Keokuk limestone, St. Louis limestone, Kakaskia limestone, and Coal Measures. These members are remarkably prolific with beautiful remains of radiate animals.

Permian Series.—Until recently the existence of Permian strata in the United States was unknown. Professor Emmons first announced that the fossils of the red sandstones of North Carolina corresponded with known Permian types. These fossils were plants and Thecodont saurians. If this position can be settled, then at least the lower part of the Mesozoic conglomerates east of the Appalachian range are Permian.

Some geologists doubt the correctness of this view. But every one admits the discoveries of Messrs. Hawn, Swallow, Meek, and Hayden in Kansas, etc., to be genuine. These gentlemen have established the Permian character of many deposits in Kansas, Nebraska, and Illinois, which were at first confounded with the Coal Measures. There is an excellent development of these rocks at Leavenworth, in Kansas. They are 861 feet thick, and have been divided into Upper and Lower Permian. About 100 species of fossils have been collected and described from these strata.

Subsequently to the deposition of the Coal Measures, and previously to the production of the Appalachian Mesozoic strata, the numerous plications in the Appalachian ranges were produced. The time of plication is known by the fact that there are folds in the Upper Carboniferous strata and none in the later rocks. Trunks of carboniferous trees, originally upright, are inclined at various angles, according to the amount of dislocation.

Inasmuch as fissures would be produced during these disturbances, through which heat would escape from the interior and penetrate through all portions of the strata, perhaps in connection with water and other essential agents of alteration, we may suppose that the azoic rocks along the Atlantic coast were metamorphosed during the time of these disturbances, or the Permian period. In the absence of any evidence, we may conjecture that a large part of the azoic strata forming the basis of the Rocky Mountain ranges were metamorphosed during the same period. But there is evidence to show that rocks are undergoing rapid alteration on the Pacific coast still later, even during the Alluvial period, as the action of heat is very great there at the present day. The metamorphism of these palæozoic rocks along the coasts must not be confounded with the alteration of the Laurentian series, for the latter were elevated and metamorphosed previous to the deposition of the Potsdam sandstone.

At the close of the Palæozoic periods the form of the continent resembled its present shape, but the amount of land above the ocean covered only about two thirds of its present surface. Yet the general continental features, the mountains and plains, were the same as now.

OLDER MESOZOIC SYSTEMS.

The older Mesozoic rocks are included upon the map with the Cretaceous groups, which are the most abundant, the former occupying comparatively little space. In the eastern British Provinces, New Brunswick and Nova Scotia, we first find the red sandstones, conglomerates, and shales of Mesozoic age. They are next admirably developed in the valley of Connecticut River in Massachusetts and Connecticut, where the most remarkable fossils, the ichnites, have been discovered. The dip of the rocks in this terrain is almost invariably to the east.

In going south of Hudson River the same rocks are found in several basins, all dipping to the west. The series appears first in New Jersey, where it forms a wide belt southeast of the Highlands. Thence it passes through Pennsylvania, from Bucks to York counties; thence into Frederick county, in Maryland; thence into Virginia. Throughout the whole extent of this deposit, from Nova Scotia to Virginia, ores of copper, bituminous shales and limestones, and protruding masses of greenstone, are associated with it. In Virginia the deposit appears to be eminently calcareous; and one of its lowest beds is the well-known brecciated Potomac marble. In North Carolina there is another basin, somewhat irregular in its shape, 150 miles long, from Tar River to Wateree River. Here the strata also dip west. This is the basin examined by Professor Emmons.

Several characteristic fossils of the Lias were found by Captain McClintock in the Arctic basin. The extent of the strata containing them was not ascertained, and hence they are not represented upon our map.

There has been, and still is a great variety of opinions expressed in regard to the age of these sandstones, it being generally assumed, perhaps incorrectly, that all the terrains are precisely contemporaneous. The older geologists (Maclure, Eaton, Silliman, and Cleveland) regarded those in the Connecticut valley as old red sandstone. In 1833 the senior author of this work, in his Report on the Geology of Massachustts, presented arguments to show that the upper beds were the equivalents of the new red sandstone; and that opinion was generally adopted in this country and in Europe after they had been examined by Sir Charles Lyell.

Professor W. B. Rogers subsequently maintained that the sandstone containing beds of workable coal near Richmond, Virginia, an isolated deposit, was of the age of the European Oolite. The fossils relied on to prove this are species of plants, viz., Zamites, Calamites, Equiseta, Tæniopteris, Pecopteris Whitbyensis, Posidonia—a species of mollusc—and several Fishes, as a Tetragonolepis. E. Hitchcock, Jr., has discovered a tree fern near the middle of the series in Massachusetts, the *Clathropteris rectiusculus*, some specimens of which can with difficulty be distinguished from the European species C. meniscoides—a characteristic fossil of the beds of passage between the Trias and Lias. Hence he argues that the ichniferous or upper beds of the series are Jurassic or Oolitic. Professor Emmons has discovered, in North Carolina, species of plants and Thecodont Saurians, which, with several European authorities, he regards as distinctly Permian. Professor Agassiz considers the ichthyolites of New England and New Jersey, occurring in these rocks in connection with the ichnites, as corresponding best, by their structure, with European specimens from the Upper Trias. The Messrs. Redfield find some traces of Oolitic structure among the fishes. The heteroclitic forms of the Lithichnozoa correspond best with the bizarre forms of the Oolite.

From these discoveries and opinions we regard one point as settled, and a second as rendered probable. 1. A belt of rock, occupying the middle portions of the Connecticut River sandstone, below which no tracks are found, is of Upper Triassic age. The ichniferous strata above are either Liassic or Oolitic. 2. Probably the whole series of rocks, from the Permian to the Oolite inclusive, are represented in these strata. The strata are at least 5,000 feet thick in Massachusetts, and this is adequate to embrace the whole, so far as they have been measured in other countries.

In connection with palæozoic and cretaceous rocks in Kansas and Nebraska certain rocks have been described, which, upon careful examination, may prove to be Triassic or Jurassic.

These different basins of older Mesozoic rock were probably formed in estuaries; or, as the Professors Rogers maintain, in some of the basins there may have been large rivers, depositing the materials in their beds, without any marine deposits. The physical features of the continent were being perfected while these deposits were forming. The lower layers have a higher inclination than the upper, amounting to absolute unconformability in some parts of the basin along the Connecticut River valley. If the lower be Permian or Palæozoic, and the upper Triassic or Oolitic, we should expect such a difference of dip.

CRETACEOUS SYSTEM.

The varieties of rocks composing this system, and the comparison of the different members in the different parts of the continent, are treated of in Section III. of Part I.

The Cretaceous system occupies more territory, perhaps, than any other system in North America. It probably commences as far east as Nantucket and Martha's Vineyard, and extends continuously from New Jersey along the Atlantic coast to Mexico, and then covers nearly one third of the width of the continent, from near the Gulf of Mexico to British America, with occasional interruptions of older or newer strata. Along the Atlantic seaboard it is not represented upon the map, because the greater part is covered by t rtiary deposits, but may be occasionally observed in deep excavations. There is some uncertainty respecting the northwestern limit of this system, but it is no doubt of as much extent as is indicated upon the map. Other Cretac ous beds are marked upon the map in Yucatan, Mexico, and the north part of South America. Dr. J. S. Newbury, United States Geologist, who has just returned (1860) from exploring the San Juan and Upper Colorado Rivers, in Utah and New Mexico, found the Cretaceous system there 4.000 feet thick, and "occupying an immense area west of the main divide of the Rocky Mountains."

About 100 species of shells have been discovered in this system, of which twenty-five per cent. are identical with European forms. Several interesting forms of vertebrate life have been discovered, as the *Hadrosaurus* in New Jersey—an animal resembling the Iguanodon of England.

The area occupied by the Mesozoic upon the map, shows what were the outlines of North America in the Cretaceous period. The Atlantic coast was at the western margin of this group, and the Gulf of Mexico extended even into British America, covering the cretaceous rocks. A part of the Rocky Mountains was also beneath the water, as some of their summits contain marine shells of Cretaceous age. Yet the interior ocean may have been shallow, and thus the continental area have been substantially the same as at present.

Gypsum is found in small quantities in Mesozoic rocks in North America. But the most extensive deposit is probably of Cretaceous age. Captain Marcy, in exploring the sources of the Red River in 1852, traced out a thick deposit of this substance extending from the Canadian River, in 99° W. longitude, nearly to the Rio Grande, at least 350 miles long, and from 50 to 100 miles broad.

TERTIARY SYSTEM.

The surface represented as Cainozoic upon the map is mostly underlaid by the Tertiary system. There are three great deposits. 1. Along the Atlantic coast, outside of or covering the cretaceous rocks, from Boston to Southern Mexico, including the whole of Florida and large parts of Louisiana and Mississippi. 2. Along the Pacific coast, from Lower California to Russian America. 3. Occupying the great table lands of the Rocky Mountains, covering more square miles than the Cretaceous system, though not as wide, though we suspect, since Dr. Newbury's researches, that here is some mistake. Other deposits of small extent are found along the Alleghany ranges, upon the shores of the Arctic ocean, in Yucatan, and in South America. There are also several detached tertiary tracts upon the great interior cretaceous deposit which are not represented upon the map.

The latest researches show that the European divisions of Eocene, Miocene, and Pliocene, can be traced upon this continent. The deposits along the Atlantic seaboard (including the Mexican Gulf) have received local names. The *Clairborne group* corresponds to the older Eocene, having the following characteristic fossils: Cardita planicosta, C. Blandingii, Crassitella alta and Ostrea sellieformis. The *Vicksburg group* corresponds to the newer Eocene, containing the following characteristic fossils: Dentalium thaloides, Sigaretus arctatus, and Terebra costata. The *Yorktown group* embraces both the Miocene and Pliocene.

The Eocene strata are in too many localities to be here specified. The great Zeuglodon cetoides is found in them in the Southern States.

A most extraordinary Miocene deposit has been brought to light in one of the insulated basins in Nebraska. It is in the *Mauvaises Terres*, or Bad Lands, on White River. There is a basin 300 feet below the general level, in which there are thousands of abrupt, irregular, prismatic, and columnar masses, frequently capped with irregular pyramids, and stretching up to a height of from 100 to 200 feet, or more. So thickly are these natural towers studded over the surface of this extraordinary region, that the traveler threads his way through deep, confined, labyrinthine passages, not unlike the narrow, irregular streets and lanes of some quaint old town of the European continent.

But the most interesting facts there brought to light are the bones of numerous extinct quadrupeds, some of them of enormous size, and differing from all fossil animals hitherto described, though of the same families. Species of rhinoceros and tiger, large tortoises, a palæotherium eighteen feet long and nine feet high, the archæotherium, the oreodon, machairodus, etc., are described by Dr. Leidy; and doubtless many more will soon be brought to light from this singular fresh-water deposit, where some of the same genera of animals occur as in the Paris basin.

Miocene.—The other Miocene deposits of the continent, as at present known, are confined to the two oceanic tertiary belts—upon the Atlantic and Pacific coasts. The strata consist largely of sandstones, conglomerates, and shales, with scarcely any limestone.

Too little is known respecting the immense area of tertiary desposits in the Rocky Mountain district to pronounce with certainty their age. The presumption is in favor of the older tertiary.

Pliocene.—The Pliocene strata have not yet been much studied in this country, except at certain localities. In distribution they correspond to the Miocene; viz., along the coasts.

But there are, along the Appalachian chain, occasional beds of clay and iron ores with which no fossils are associated. In 1852, however, the senior author of this book explored a bed of lignite associated with these beds of iron, in Brandon, Vermont, and found in them a large number of fruits (figured in Part II.) which are evidently of the age of the Pliocene. From these data we have supposed all the deposits in similar positions throughout the range to belong to the Pliocene, although no lignite has been discovered in connection with them.

Alluvium.—The drift and alluvial deposits of North America have been so largely treated in Section IV. of Part I., that we simply refer to them in this connection, with no additional remarks.

IGNEOUS ROCKS.

That part of the map which is covered with small dots represents the distribution of the more recent igneous rocks, the traps and basalts. They are most abundantly developed along the Pacific coast. The largest area occupied by them is along the Columbia River in Oregon and Washington Territory. There are multitudes of smaller deposits along the Rocky Mountain ranges in Mexico and the Territories, which are not shown, three or four dotted areas standing as representatives of a large number. Farther north a similar series of trappean rocks is represented as a continuous belt. Igneous rocks are abundant, also, about the Arctic Ocean, especially in Greenland.

Trappean dykes are abundant along the Appalachian ranges and about Lake Superior; but they do not overflow the surface like those just described, and occupy too little space to be indicated upon the map. The granitic and the oldest igneous rocks are included under azoic rocks.

NOTE.—Captain J. H. Simpson, who in 1858 and '59 explored the *Great*

Basin lying between the Wahsatch and Sierra Nevada Mountains, "never before traversed by a white man," has just published (April, 1860) a note from Messrs. F. B. Meek and H. Englemann, respecting the new geological discoveries made by them in those *terræ incognitæ*. In west longitude 116°, they found, near the Humboldt Mountains, extensive deposits of Devonian rocks, 1,200 miles farther west than ever before known. Nearly as far west they found extensive Carboniferous formations, though not much coal. In several places east of Lake Utah they found Triassic red sandstone, with numerous beds of gypsum and rock salt, as in Europe; which, according to Marcou, Blake, and Newbury, is developed on a grand scale in New Mexico. Jurassic rocks occur, also, near the same place in Utah; also on Weber River Cretaceous strata; also both Eocene and Miocene tertiary near Fort Bridger and elsewhere; all of which, like the tertiary of Nebraska and elsewhere at the West, seem to have been deposited in brackish waters. What an interesting field for American geologists opens in these vast western regions!

INDEX.

A.

B.

H.

I.

P.

Q.

R.

T.

U.

ROBINSON'S
COMPLETE MATHEMATICAL SERIES.

By HORATIO N. ROBINSON, A.M., *late Prof. of Mathematics in the U. S. Navy.*

Most of the books of the series have been thoroughly revised and corrected, and published in superior style. For extent of research, and facility and aptness of illustration, the author is unsurpassed, and his long experience as a practical mathematician and teacher guarantees the utility and sound science of his works. They have received the unqualified approval of most eminent educators, including several STATE SUPERINTENDENTS OF PUBLIC INSTRUCTION, and many Presidents and Professors in Colleges. They are already used and preferred to any others in thousands of the best Colleges, Academies, and High Schools throughout the Union.

ROBINSON'S PROGRESSIVE PRIMARY ARITHMETIC	$ 15
ROBINSON'S PROGRESSIVE INTELLECTUAL ARITHMETIC	25
ROBINSON'S THEORETICAL AND PRACTICAL ARITHMETIC	56¼
ROBINSON'S KEY TO PRACTICAL ARITHMETIC	50
ROBINSON'S HIGHER ARITHMETIC. (*In press*)	75
ROBINSON'S KEY TO HIGHER ARITHMETIC. (*In press.*)	75
ROBINSON'S ELEMENTARY ALGEBRA	75
ROBINSON'S NEW ELEMENTARY ALGEBRA	75
ROBINSON'S KEY TO NEW ELEMENTARY ALGEBRA	75
ROBINSON'S UNIVERSITY ALGEBRA	1 25
ROBINSON'S KEY TO UNIVERSITY ALGEBRA	1 00
ROBINSON'S GEOMETRY AND TRIGONOMETRY	1 50
ROBINSON'S SURVEYING AND NAVIGATION	1 50
ROBINSON'S ANALYTICAL GEOMETRY AND CONIC SECTIONS	1 50
ROBINSON'S MATHEMATICAL OPERATIONS	2 25
ROBINSON'S ELEMENTARY ASTRONOMY	75
ROBINSON'S UNIVERSITY ASTRONOMY	1 75
ROBINSON'S DIFFERENTIAL AND INTEGRAL CALCULUS. (*In press.*)	
ROBINSON'S KEY TO ALGEBRA, GEOMETRY, SURVEYING, AND CALCULUS.	

SPENCERIAN PENMANSHIP.

The Spencerian System of Penmanship is a scientific, methodical, beautiful and complete series of writing books, eminently practical, and admirably adapted to impart a correct knowledge of the art. They have been adopted as the text-book on penmanship by many of the Commercial Colleges in the United States, among which are Bryant & Stratton's *Chain of Commercial Colleges* at New York, Philadelphia, Albany, Buffalo, Cleveland, Detroit, Chicago, St. Louis, and Pittsburgh (the most extensive Chain of Colleges in the world), and have recently been adopted in the Public Schools of the cities of Buffalo, Cleveland, Detroit, St. Louis, Rochester, Oswego, and other large cities and towns. The School Series is complete in the first five Nos.

SPENCERIAN PENMANSHIP. Newly engraved. In 9 Nos.

Nos. 1, 2, 3—BEGINNERS' COURSE	$ 12½
No. 4—CAPITALS	12½
No. 5—SENTENCES AND REVIEW OF CAPITALS	12½
No. 6—ANALYTICAL AND SYNTHETICAL REVIEW OF THE SYSTEM, with Short Business Forms	12½
No. 7—GENERAL BUSINESS FORMS (containing nearly 150 lines)	12½
No. 8—ANALYSIS OF THE SYSTEM IN ITS APPLICATION TO LADIES' PENMANSHIP	12½
No. 9—LADIES' BOOK OF FORMS FOR PRACTICE	12½

(The first five Nos. constitute the "'Common School Series;" the last four Nos. the "Commercial and High School Series.")

COMPENDIUM OF THE SPENCERIAN SYSTEM. 60 pages (over 400 lines), illustrating Chirography in its analytical, practical, and ornamental forms.

Paper	1 00
Cloth	1 50
Cloth, full gilt	2 00

SPENCERIAN CHART OF LETTERS. Size, 38 by 44 inches:

Mounted on rollers and varnished	2 00
Canvas back (not mounted)	1 50

This Chart is designed to suspend in the School-room, from which to give general explanations of the *Elements* and *Rules* of the System. The *Elements* are all given and numbered,—their slope, shading, and proportions, together with their combinations, giving the perfectly formed letters, are fully elucidated; so that when the pupil has fixed in his mind a perfect *ideal* of the elements and letters as here exhibited, *he becomes a teacher unto himself.*

www.ingramcontent.com/pod-product-compliance
Lightning Source LLC
LaVergne TN
LVHW021148110826
845150LV00005B/1154

* 9 7 8 1 4 2 5 5 4 7 3 2 5 *